D1087656

The New Science of Cities

The New Science of Cities

Michael Batty

The MIT Press
Cambridge, Massachusetts
London, England

MIT Press books may be purchased at special quantity discounts for business or sales promotional use. For information, please email special_sales@mitpress.mit.edu.

This book was set in Sabon by Toppan Best-set Premedia Limited, Hong Kong. Printed and bound in the United States of America.

Library of Congress Cataloging-in-Publication Data

Batty, Michael.
The new science of cities / Michael Batty.
    p.   cm
Includes bibliographical references and index.
ISBN 978-0-262-01952-1 (hardcover : alk. paper)
1. City planning. 2. Cities and towns—Growth. I. Title.
HT166.B38667   2013
307.1'216—dc23
2013004383

10 9 8 7 6 5 4 3 2 1

Science may be described as the art of systematic over-simplification—the art of discerning what we may with advantage omit.

—Karl Popper, *The Open Universe: An Argument for Indeterminism* (1992)

Contents

List of Figures

List of Tables

Preface

We live in an age of cities. By the end of this century, it is likely that most of the world's population will be living in one type of city or another, as urbanization and globalization become the norm and our past rural pursuits slowly disappear. When all the world's a city, it is questionable whether the term "city" will continue to have the same resonance it has had for the last 5,000 years. Our focus on individual cities will probably lessen as we explore the way a system of cities composed of most of the world's population actually functions. To understand what is happening, we need to take seriously the idea that cities are places where people come together to "interact" with one another. As our technologies enable people to connect ever more easily and in many new ways, our understanding must be enriched by studies of networks, interactions, connections, transactions, and every other possible way in which we are able to communicate with one another.

This book takes literally the idea that cities are devices that enable us to communicate. In doing so, we are able to use them to increase our prosperity by providing environments in which we can work together, innovate together, and generally share the fruits of our labors. Until quite recently, most of this sharing took place in individual cities, but increasingly our technologies enable us to share and communicate at a distance. At one level, this is trade, and although fundamental to ancient and modern societies, the global world that has emerged is dissolving our reliance on material movements in favor of the ethereal and the social. Information is replacing as well as complementing energy. Thus, to understand the contemporary city—to fashion a science that is able to explain city growth, sprawl, decline, and so on—we must underpin our theories with ideas about how we relate to one another.

These ideas have been anticipated before. Fifty years ago, Jane Jacobs, in her seminal book *The Death and Life of Great American Cities* (1961), argued that it is not enough to simply study the location of things in cities as we have done in the past; instead, we need to consider location as lying at the heart of how evolving networks of relationships provide the cement that holds people together in cities.

Richard Meier, in the preface to his book *A Communications Theory of Urban Growth* (1962), talked of urban environments "continually bombarded by messages," and he produced an outline for a theory of cities as kaleidoscopes of information. Peter Haggett and Richard Chorley in their book *Network Analysis in Geography* (1969) produced a wonderful and challenging review of the way networks were being used in geography that focused very much on parallels between cities as human systems and geomorphologies as physical systems. Around the same time, Waldo Tobler produced a long line of influential papers on geometry and flows that served to keep the morphological message alive in geography until recent times, when these ideas began at last to connect up with the new science introduced here.

Yet despite these occasional and prescient contributions, we have continued to think of cities as spaces and places. We recognized that these were stitched together by transportation, but we have never attempted to see the city as a set of networks from which locations naturally emerge. We have always tried to unpack actions and activities at locations into interactions rather than the other way around, by attempting to see locations as the product of interactions. In this book, I switch our traditional focus from locations to interactions and, in so doing, invoke many ideas about networks and flows.

There is another traditional dimension to cities relating to their physical or spatial form, and we will not downplay this here. It is quite possible, of course, to study cities as networks and interactions without focusing on their form. Indeed, much of the literature on urban studies tends to treat urban phenomena and cities as being about aspatial and nonspatial issues, about processes, rather than about physical or spatial form. But urban planning and design, which is the most obvious and perhaps least intrusive way of intervening in the evolution of cities, is based on physical rather than social or economic instruments of control and management. I will keep this perspective here, while acknowledging that we could treat the material of this book in a very different manner, notwithstanding that the focus of our discussion might be the same. My perspective is thus unashamedly about the physical and spatial artifacts that define our cities. The tools I will introduce that underpin the new science I argue is needed to grasp the challenges of the near future are thus manifestly physical and spatial in their treatment of systems of cities and cities as systems.

This is not a book that will tell you what the future city will be like or even how one might create better cities, but it will provide what I consider tools for and perspectives on a science that can be used to explore these futures. Yet this world of science will inevitably be incomplete, as science always is. Karl Popper's wonderful insight that science is "the art of discerning what we may with advantage omit," which is the quote I use to introduce this book, provides one of the leitmotifs for this work. Readers need to be aware that this is a partial set of insights and that there are many directions in this science that are inevitably inconsistent with one another, as I am at pains to emphasize in the concluding chapter. This science, however, has not sprung Phoenix-like from the ashes of an earlier world, but builds on strong traditions in social physics, urban economics, and transportation theory, on regional science, urban geography, and of course on the systems approach to physical planning, which has now morphed into the complexity sciences. I have a deep respect for these traditions, and they are echoed everywhere in the text that follows. More particularly, many of the ideas here build on my previous forays into this way of thinking; as first represented in my book (with Paul Longley) *Fractal Cities: A Geometry of Form and Function* (Academic Press, 1994; available for download at http://www.fractalcities.org) and in my book *Cities and Complexity: Understanding Cities Through Cellular Automata, Agent-Based Models, and Fractals* (MIT Press, 2005). A summary of these ideas can be gleaned from http://www.complexcity.info. Far be it for me to instruct you to read these, but the material therein does underpin many of the approaches and methods introduced here.

There are many developments in this new understanding of cities that are currently emerging. I hint at these here, but they are implicit in the development of this science and its application. The digital revolution has now penetrated our culture so deeply that the many new forms of communication transforming our cities are now yielding up their secrets in the form of very large databases, providing us with opportunities for analysis and modeling that are quite different from those available in an earlier age. Cities are becoming "smart," and in many ways the kind of science reported here can inform how they might become "smarter." But more particularly, the dissemination of this science using new forms of visualization offers new ways of thinking about the design of future cities. I will not venture here into planning support systems, participation in design using new online technologies, and the like, but I am well aware of this movement, which forces us to embed these ideas into the wider policy context. This, in fact, is the focus of my research group in the Centre for Advanced Spatial Analysis at University College London, where various research projects have provided me with some of the illustrations of "big data" pertaining to networks and flows in cities used here. I acknowledge particular contributions below.

Many others have helped me get this far. Paul Longley, who worked with me on fractals in the 1980s, has been my colleague for nearly thirty years, and his dry humor continues to remind me that science is contingent and always temporary but, nevertheless, that we should strive for the best. Peter Hall, my longtime mentor from my first permanent appointment at Reading University in the 1970s, continues to support this way of looking at cities while supporting many others who follow his pluralistic approach, which is entirely consistent with the sentiments I express here. There is a good group of urban morphologists at University College London whose work is complementary to these ideas. Stephen Marshall's views on evolution and cities, Bill Hillier's ideas about how the physical and spatial syntax of cities molds their function and vice versa, and Phil Steadman's work on configurational forms all find implicit expression in this work, and I thank them for their insights and influence. I must also thank my original teachers George Chadwick and Brian McLoughlin, who in the late 1960s introduced me to the systems approach in planning and set me on my way to where I stand on these matters today. They are sadly long gone, but I hope they would recognize a little of their contributions in what follows in this book. I fear they might not, for the terrain of how we understand cities has changed radically and is still changing fast. However, in the methods and models we use, there is still synergy and complementarity with their earlier era.

I want to thank the editorial team at MIT Press, particularly Susan Buckley, Virginia Crossman, and Kristie Reilly for their splendid work on bringing the book to publication.

My wife, Sue, helped me with some of the figures and proofs of the book, which would not have been possible without her, and it is as much hers as mine. My son, Daniel, remains a clever skeptic on all matters of cities and continues to test me on the plausibility and wider import of the ideas therein. The book is for them.

Michael Batty

Welsh Saint Donats
Cowbridge, Vale of Glamorgan
October 2012

Acknowledgments

I wish to thank James Cheshire and Oliver O'Brien for figure 2.7(a); Bahaa Al Haddad for figure 1.3(c); Erez Hatna for figure 5.15(a); Robin Morphet for figures 3.7(a) and (b); Jon Reades for figure 2.7(c); and Joan Serras for figures 1.3(d), 1.4, 3.8, and 3.10(a), (b), (c), and (d). Parts of this book have appeared before. In chapter 1, some material is taken from a paper published in the journal *Cities* (29, S9–S16, 2012). In part II, chapters 4, 5, and 8, some sections were published in *Complexity* (16, 51–63, 2010), and in the edited books *Hierarchy in the Natural and Social Sciences* (edited by D. Pumain, Springer, 143–168, 2006) and *Embracing Complexity in Design* (edited by K. Alexiou, J. Johnson, and T. Zamenopoulos, Routledge, 1–18, 2007). Some of chapter 9 was published online in UCL CASA Working Paper no. 164. In part III, elements of chapters 10, 11, and 14 have appeared in *Architectural Design* (41, 436–439 and 498–501), and *Environment and Planning B* (1, 125–146, 1974; 2, 151–176, 1975; 11, 279–295, 1984), while parts of chapter 13 appeared in *Regional and Industrial Development Theories, Models and Empirical Evidence* (edited by A. Andersson, W. Isard, and T. Puu, North Holland, 309–329, 1984).

Some of the figures in this book were originally produced in color but are printed here in gray tones. Where possible, notes on the figure captions indicate the range of colors used and their translation into gray tones. Readers who wish to examine the figures in color should access the website http://www.complexcity.info/ newscience, where they will find the original color versions.

Preamble

A city is more than a place in space, it is a drama in time.
—Patrick Geddes, *Civics: as Applied Sociology* (1905, p. 6)

During the last one hundred years, if not before, there have been many attempts at fashioning a science of cities. All have emphasized different aspects of the city, but none have succeeded in being universally recognized as providing the kind of comprehension that other areas such as the physical sciences appear to display. There is no single paradigm that has come to dominate our understanding of cities, and as we will elaborate here, there is unlikely to be one. So what is our rationale for a "new science"? We believe it axiomatic that to understand cities, we must simplify, we must abstract. We must dig below the surface of what we see and reveal the foundations of how cities function. Most serious commentators who take the idea of a science of cities seriously would not disagree with this strategy. We quite literally need to think of cities not just as artifacts—buildings that endorse the long-standing idea of the "city beautiful," although this is a perfectly respectable agenda—but as systems built more like organisms than machines. A new science is timely at this point, for in the last twenty years, there has been a sea change in the way we have come to understand how cities function. As Patrick Geddes (1905) so presciently remarked in the above quote, the idea that a city is simply "a place in space" completely ignores the fact that cities change in time. Such changes involve movement through space, and this means that time is composed of flows between places. In fact, Manuel Castells (1989) has defined contemporary cities as being a "space of flows," thus nicely augmenting Geddes's insight and linking our idea of place and space to ways in which places and spaces are related to one another and to the activities that compose them.

In short, our argument for a new science is based on the notion that to understand place, we must understand flows, and to understand flows we must understand networks. In turn, networks suggest relations between people and places, and thus the central principles of our new science depend on defining relations between the objects that comprise our system of interest. This may seem obvious, in that all sciences rest on such relational foundations, but in the case of cities, this abstraction is more remote, for it depends on us suspending our belief that we can observe how cities work by simply using our immediate senses to underpin our understanding. What you see is not what you get when it comes to cities. The idea that it is relations—or, rather, networks—between places and spaces, not the intrinsic attributes of place and space, that condition our understanding, is the first principle that we will use in our new science. It lies at the basis of all the theories, methods, models, and tools that we will introduce here.

Our second principle reflects the properties of flows and networks. We will see that there is an intrinsic order to the number, size, and shape of the various attributes of networks and thus, in turn, of spaces and places that depend on them, and this enables us to classify and order them in ways that help us interpret their meaning. In essence, the distribution of elements that compose the city—the hubs or nodes of the networks that sustain them—present us with highly skewed distributions , reflecting the essential economic processes of competition that drive a city's functions and determine its form and structure. These distributions usually describe large numbers of small objects and small numbers of large, following what are called scaling laws that, in turn, are usually configured as power laws. Power laws reflect processes that scale, that in some senses are self-similar, and this signature of a system's function implies that the system's subsystems, components, elements, and so on are ordered hierarchically. These processes generate urban growth and underpin the city's evolutionary architecture, opening up our theories and models to the world of complexity theory, to highly ordered forms such as fractals (Batty and Longley, 1994), and to models that are defined on the most elemental units comprising the city, such as cells, agents, and actors, which are the building blocks of scale (Batty, 2005).

Our focus here on flows and networks is by no means the full story of how cities grow and evolve into different forms and functions. Networks are of two types: first, those that are interactions or relations between all the elements that compose a set of objects—for example, networks between all locations, say, that are interactions such as journeys from home to work places, or migration from one place to another, etc.—or, second, those that are relations between two different sets of objects, such as between people and places. If we have such bipartite relations, we can construct other sets of relations or interactions—between, say, people and people through the places they have in common, or places through the people they have in common. In short, we have a means, albeit hypothetical, of predicting sets of interactions from more elemental relations. This leads to our third principle, that an understanding of cities must extend to predicting flows and networks rather than merely observing them.

There are many more constructs to our science than the basic notions we enshrine in these three principles. But we will begin by elaborating these as our foundations and prerequisites, which form the core of part I. We define flows and their networks as synergies of one another and introduce tools that enable us to adapt and apply them to many different aspects of city structure in parts II and III. There are many other theories that are complementary and consistent with our approach here. Complexity theory and the models that are central to its application are intrinsic to this science, and readers are referred elsewhere to this literature (Batty, 2005). Urban economic theory is important to understanding the way competition orders space, and the work in trade theory of Fujita, Krugman, and Venables (1999) must be noted. In intra-urban economics, the monocentric city model is key (O'Sullivan, 2011), while transportation theory and modeling also need to be considered (Ortuzar and Willumsen, 2011). Generally we will assume that our readers are able to grasp the rudiments of this literature without necessarily understanding the complete edifice, and, in general, readers should be able to absorb the material here without recourse to previous study of urban economics, regional science, urban geography, and transportation. More-over, readers should be able to absorb most chapters independently of any other, with the possible exceptions of chapters 2 and 3 on flows and networks, which build the tools to be used in subsequent chapters; chapters 6 and 7, which need to be read in sequence; and possibly the chapters in part III, which tend to build on each other in sequence, notwithstanding that the arguments are repeated in each chapter.

Parts II and III develop applications of the basic tools to different perspectives on how cities function and how they are designed. Just as flows and networks are opposite sides of the same coin, so too are the sciences for understanding cities that compose part II and those for their design that compose part III. This is a switch that some readers might find unexpected, for often those interested in cities are not so interested in their planning, and ironically the same can hold true the other way around. In part II, we examine the size of cities, their internal order in terms of hierarchies, the transport routes that define them, and the locations that fix these networks, as well as the form and then structure of such networks. These define the six chapters that also introduce methods of simulation, from simple stochastic models to bottom-up evolutionary cellular and agent-based models to, ultimately, more aggregate land-use transportation models. Throughout we emphasize the three principles that we have defined—networks, scale, and the prediction of interactions or flows. This part of the book is far from a comprehensive theory of how cities evolve, are structured, and function, but links to this larger body of knowledge are clear. The fact that the chapters can be more or less read independently tends to suggest a degree of immaturity in this science. In time, the edifice will be filled in and then the arguments will be more elaborate, thus building more comprehensively on one another.

The switch we make in part III from cities to their design, or rather to design and decision-making models that can be used to fashion future cities, although a little abrupt, is consistent with our treatment to date. The tools that we employ to build our design and decision networks are largely the same as those we use in earlier chapters. But this time, the networks function as communications media between actors or agents who represent the stakeholders attempting to design policies that will generate more efficient and equitable cities. The focus, of course, is still physicalist, and in this sense perhaps a little narrow. To an extent, the models are less real than those in part II, but they do impart the notion that the design and planning of cities is collective action: that consensus about the future must be reached by pooling opinions. In this sense, the models are both dynamic and equilibrating. We also exploit rather heavily the idea of bipartite relationships in these models and suggest how we might extend our explanations as well as our models to exploring the deep structure of city systems and ways we might best engage in their design. Although our models of design tend to be hypothetical, we do attempt to square the circle in the last two chapters, in which we push these ideas to the point where our models reflect the ways in which actors make decisions about future cities, focusing on the location of activities. This provides us with a direction forward.

Before we begin our exposition, some explanation is needed of the way this book might be read. Although it is possible to dip into the book at various points and take meaning from its sections, there is a loose chronology of ideas, tools, and methods that suggest the best way is to read the various chapters is in sequence. In this way some of the messages that we have noted in this preamble are best understood. The treatment is formal, at times, but the level of mathematics ranges from quite elementary to intermediate, and a basic understanding of simple calculus and matrix algebra will suffice. In terms of notation, we cannot use a completely consistent set of symbols across all chapters, for there will be some repetition of the same symbols for different equations, but whenever this occurs, the reader will be alerted. In fact, we define standard variables such as population, P; employment, E; land, A or L; and trips or flows as T and S. Probabilities are p and q and frequencies usually f and F, while matrices are given in appropriate terms but are in **bold** type. Arrays that are often vectors or matrices are given by variables in brackets, such as $\{\cdots\}$ or $[\cdots]$. Individuals, locations, problems, policies, and so on, which are defined over appropriate ranges, are defined by subscripting these variables from the range i, j, k, ℓ, with occasional use of $k, m, n,$ and z. Parameters and some constants are defined using the Greek alphabet. As far as possible, our notation is consistent with that used in the related literature, particularly in spatial interaction modeling and demographic analysis. As the treatment is generally informal, readers will find that the logic of developing these ideas is quite straightforward. One final note: many of the figures in this book were originally produced in color but are printed here in gray tones. Readers who wish to examine the figures in color should access the website http://www.complexcity.info/newscience, where they will find the original color versions.

I

Foundations and Prerequisites

Jane Jacobs once said, "Cities have the capability of providing something for everybody, only because, and only when, they are created by everybody" (1961, p. 238). Her insight is one that we feel deeply because all of us articulate our understanding of the city in different ways, thus implying that cities are kaleidoscopes of plurality, a multiplicity of ideas, perceptions, theories, models—indeed of every and any abstraction that we might imagine. It is thus rather difficult to convince ourselves that any one approach to understanding the city takes precedence over any other, and this makes the development of a science of cities built in the traditions of classical science from the time of Newton simply impossible. When we step back and consider how cities function and how they might be made more livable through their planning, we inevitably abstract by peeling back layer after layer of complexity until we alight upon what we might consider fundamental ideas and techniques that compose the foundations of our understanding. In doing this, we expose the building blocks we can use to construct a new science, one that is built from the bottom up, and is robust and consistent with the way we consider cities to function, change, and evolve.

To do this, we adopt the contemporary approach of complexity theory, which treats systems as being constructed from the bottom up, in a hierarchical fashion in which their basic components—functions that relate to how populations interact with one another—determine the networks on which individuals and groups engage with each other through social and economic exchange (Batty, 2009a). Systems such as cities evolve in time. They usually grow larger, since the pressure in modern society is to agglomerate so that scale economies might be realized, but cities are also the product of many top-down processes that operate at every level, thus complicating this nexus of interactions. Complexity theory is essential in this interpretation, for cities are not only growing larger, with the prospect that by the end of this century most of us will be living in one kind of city or another, but they are in the process becoming more complex. More and more layers of interaction between their populations are being fashioned as society shifts from flows of energy to information, and this is forcing us to consider our understanding of cities as being a temporary phenomenon, always changing and evolving like the very city systems we are seeking to understand. This, of course, is not new. Popper (1959) implied as much when he laid out his notion that since the future is intrinsically unpredictable, all our theories are contingent and all must be expected to be falsified at some point, because our own actions in the future are unknowable.

In establishing foundations and introducing prerequisites that serve to build this new science, we will begin in chapter 1 by introducing the idea that cities can be constructed in modular form, as hierarchies that reflect subsystems of interactions on which processes of change take place and evolve. In this conception, the idea that locations, places, and spaces are the dominant way we articulate these systems is somewhat passé. Locations are important, but only as places that anchor interactions. In our first chapter, we make the case for peeling back the city to expose locations as patterns of interactions acting as the glue that holds populations together through flows of material, people, and information. Thus, networks based on bringing people together to work and play—economic networks and social networks—will be critical to this new science. Chapter 1 establishes this perspective. From the ideas we outline comes the notion that the modules that compose the city are hierarchically similar across different spatial (and indeed) temporal scales. The best example reflects the way clusters of economic activities are organized, with market centers getting larger and less frequent but more widely spaced as the spatial scale expands. There is an intrinsic self-similarity to such aggregations that we can easily exploit in terms of the morphology of fractal patterns (Batty and Longley, 1994). But this kind of scaling is a basic signature of complexity in that it reveals how simple processes—in this example of trade and interaction—are intrinsic to activities at all scales and simply transform themselves as the spatial scale changes. Here we will introduce the key notion of scaling, which is central to the way we portray relations across space and time, in locations and networks, as power laws. In chapter 1, we show there are many such scaling relationships that reflect the order that is a feature of the way populations compete for land and location in cities. For example, as cities get larger, they "tend" to get wealthier, greener and denser, thus implying some qualitative change in their metabolism as they scale. This is urban allometry, which we will employ in various parts of this book.

The key to cities, then, is the way we can unpick their physical form to reveal the networks that enable them to function at different scales. In this sense, cities are agglomerations or clusters of individuals who are interacting to some purpose, and thus one of our central ideas in this new science is that locations are really the nodes that define the points where processes of interaction begin and end. In one sense, instead of thinking of cities as sets of spaces, places, locations, we need to think of them as sets of *actions, interactions,* and *transactions* that define their rationale and relate to the way scale economies generate wealth in social and economic terms. In this sense, we begin to think of cities as patterns of flows, of networks of relations, pertaining both to physical-material as well as ethereal movements. In chapter 1, we establish this perspective, while in chapters 2 and 3, we introduce cities first as flow systems and then as network systems. To an extent, these involve looking at the city using slightly different but complementary perspectives. Flows and networks are opposite sides of the same coin, so to speak.

In chapter 2, we introduce the idea of flows. Flows between locations are not strictly embedded in the physical fabric of urban space itself, but float somewhat freely across the space. Models of such flows do not usually relate directly to the physical networks on which they actually take place, or if they do—as in transportation planning—they focus on physical issues involving networks, such as capacity, which are entirely separable from the way flows themselves might be simulated. We begin our treatment of flows with typical examples, such as the journey to work. This illustrates immediately the difficulties even with free-floating flows in presenting them as flow systems on the Euclidean space of the map, which, as we will see, becomes even more intractable once we deal with networks. In fact, visualizing cities as flow systems goes back many years, and we provide some examples of how flows have been presented by transforming them in geometric terms, assigning them to network links, and converting them into vector spaces or fields. In fact, the main purpose of chapter 2 is provide a set of flow models, and we thus begin by outlining spatial interaction theory. We show how different kinds of flow models can be constructed, and we particularly focus on how flows are represented as functions of locations in terms of accessibilities and potential. One key issue is that we construct a baseline model of flows that treats them as symmetric. Of course, real systems depart from this baseline, and this introduces the notion that whenever we look at flows in cities, we need to consider how they distort the hypothetical isotropic plane that was often taken as the starting point for urban economic theories of the city fifty years or more ago, and still has an important role in the construction of such models.

In chapter 3, we flip the coin from flows to networks and develop the machinery of network science insofar as we need to employ it in this book. The key idea is that networks, which are more closely embedded in the two- (and perhaps three-) dimensional structure of the physical city than flows, represent the structures on which flows take place. These in turn reflect the processes that enable the city to function socially and economically. We use graphs to represent networks, and although graphs are usually developed to simulate networks where one set of objects interacts with itself—for example, as in social networks, in which members of a group interact with each other—we introduce the key idea of the *bipartite graph*. This is a set of relations between one set of objects and a different set—between, say, individuals and their activities, or between activities and the places they populate, and so on. This is an extremely powerful idea, because it enables us to construct graphs between one set of objects and itself—the usual *unipartite graphs*—through the relations specified by the bipartite graph: for example, networks between individuals through the medium of their activities, or the places they populate, and so on. Throughout this book, this idea will appear again and again. In fact, it is no more or less than the notion used in all multivariate analysis where relations are sought between one set of objects and their attributes, but in this context, we will exploit this concept in network terms.

We will use these ideas to define processes in networks. These processes all pertain to diffusion from nodes through links and treat the idea of the network as a communications media, which indeed is exactly what it is in terms of cities. Typical processes of routine diffusion on networks are those such as innovations, epidemics, and growing cities (Batty, 2005), and there is substantial work beginning on disruptions to such networks as well as cascading effects from pulses introduced at specific nodes. We will not develop any of this here, since it is an emerging field, but in time it will certainly become part of this new science. We do, however, define processes of communication in such networks, which lead to changes in the messages that are communicated. In particular, averaging processes that bring initial differences at nodes to some form or steady state, equilibrium, or consensus are introduced in chapter 3 and are taken forward quite explicitly in part III, where we examine models of reconciling conflicting planning and design issues held by different stakeholders. In chapter 3, however, we suggest that this is a generic process, and bringing it together with the idea of the bipartite graph defines many processes on networks that we discuss in later chapters. The other thread we need to develop is ideas about spatial networks. These are essentially transport networks in cities, which are embedded in the two-dimensional Euclidean space of the map but are not always planar, because they sometimes extend into the third dimension as routes cross. We illustrate key notions in terms of road and rail networks, and we conclude by showing how scaling relates to these forms.

These are foundations. The tools we introduce are supported by other theories and models of how cities organize themselves that we will take for granted in this book. We do not say much about dynamics here, even though the temporal dimension is crucial to complexity theory as it is applied to cities. We have explored dynamics elsewhere (Batty, 2005) and there are enough references to these ideas that the reader can explore these further. Spatial economics and regional science also provide many theories and tools that are important, particularly in the area of urban economics, while new forms of modeling, particularly cellular automata and agent-based models, enter as examples, but we do not elaborate them in any detail here. There are treatments elsewhere, just as there are treatments of land use transportation theories and models (Heppenstall, Crooks, See, and Batty, 2012). We will develop some of these related strands in part II where we focus on simulation, but for many of the ideas in this book, the theories, methods, and tools introduced in the next three chapters will suffice.

1

Building a Science of Cities

The building of cities, if it is not so already, will soon become a genuine science calling for major and profound search in every branch of human knowledge and, most especially, into social science.
—Ildefonso Cerda, *The Five Bases of the General Theory of Urbanization* (1859, p. 66)

A science of cities has taken a long time coming. More than 150 years ago, Ildefonso Cerda anticipated a world of cities based on a science of geometry and form as the substructure for social behavior, while some fifty years later, Patrick Geddes (1915/1949) switched the argument away from notions of mechanism to flows and fluxes that had begun to dominate the life sciences. In his *Cities in Evolution*, Geddes focused on the theory and practice of such science when he said: "Thus, in fact, appear the methods of a Science of Cities—that our cities should be individually surveyed, scientifically compared; as their architecture long has been—cathedral with cathedral, style with style" (1915/1949, p. 269). In one sense, these prescient statements heralded a false dawn in the development of this new science, for only now, in the early twenty-first century, do we have the elements to begin to fashion such an effort. There is now considerable momentum in developing formal ideas about how cities are ordered and structured, which are part of the rapidly expanding sciences of complexity (Batty, 2005, 2009a). In short, a new paradigm is emerging.

In this introductory chapter, I will sketch the context for the science that I will slowly develop during these pages. To lay the foundations, I will first introduce the key notion that cities must now be looked at as constellations of interactions, communications, relations, flows, and networks, rather as than locations, and argue that location is, in effect, a synthesis of interactions: indeed, this concept lies at the basis of our new science. We will then explore ideas about the size and scale of cities, introducing rather general notions of agglomeration, globalization, and decentralization. This leads us to the idea that we approach cities as physical systems in terms of their geometry and geography, which we use as the lens to describe, explore, and analyze the focus of our interest. The history of how we arrived at this point is then outlined through the systems approach, which leads us quite easily into complexity theory. Aside from our concern with interactions, there are three related themes that play out again and again in this book. These are questions of *temporal dynamics*, or how things change; *flows*, or how strongly locations interact with one another; and *size and scale*, or how things change as they get bigger or smaller. We elaborate our ideas about complexity first in terms of dynamics, emergence, patterns, processes, flows, and interactions, and then introduce various laws of scaling, which can be used to describe how cities change qualitatively as they change at different scales. In this chapter, I seek to simply focus the reader on the much larger literature of cities and complexity that has already been developed, without anticipating the way we will link all this theory together in this book. In so doing, we move to the point where we can begin to develop the ideas about flows and networks that dominate chapters 2 and 3, before we then launch into applications to cities in part II.

A New Paradigm

Half a century or more ago, cities were first formally considered as "systems," defined as distinct collections of interacting entities, usually in equilibrium, but with explicit functions that could enable their control, often in analogy to processes of planning and management. These conceptions treated cities as being organized from the top down and distinct from their wider environment, which was assumed largely benign, with functioning that was dependent on restoring equilibrium through various negative feedbacks. As soon as this model was articulated, however, it was found wanting. Cities do not exist in benign environments and cannot be easily closed off from the wider world. In addition, they do not automatically return to equilibrium, for they are forever changing; indeed, they are always far from equilibrium. Nor are they centrally ordered, but evolve mainly from the bottom up as the products of millions of individual and group decisions, with only occasional top-down centralized action. In short, cities are more like biological than mechanical systems. The rise of the sciences of complexity, which have changed the direction of systems theory from top down to bottom up, is one that treats such systems as open, based more on the product of evolutionary processes than of grand design (Portugali, 2000). During this time, the image of a city as a "machine" has been replaced by that of an "organism," while the origins of these ideas remain firmly embedded in past developments (Berry, 1964).

This developing science has not abandoned more formal approaches to understanding cities, for the new science builds on this edifice while changing its emphasis. In essence, what is being forged is a much more comprehensive set of structures that allow us to understand the many perspectives on the city that reflect its diversity and plurality. However, there is one key theme that breaks with the dominant approach to cities that has been adopted hitherto. Throughout recorded history, cities have been studied, planned, and imagined as places whose form and structure can be represented as models, maps, and pictures of locations. This obvious physical representation provides our most immediate and intuitive understanding of the spatial implications of equity and efficiency that dominate our quest to design better cities, and our understanding of cities has been framed in terms of the way we might manipulate urban activities physically and geographically. But plans based on moving activities and their land uses into ideal configurations, or on imposing constraints on what activities can locate where, rarely grapple with the essence of what cities are all about and how they evolve. Location encapsulates the workings of such urban activities, but does not reveal the relations and interactions between populations that represent the rationale for living and working in cities. In a world now dominated by communications and in a world where most people will be living in cities by the end of this century, it is high time we changed our focus from locations to interactions, from thinking of cities simply as idealized morphologies to thinking of them as patterns of communication, interaction, trade, and exchange; in short, to thinking of them as networks.

This book outlines a science no longer exclusively based on theories of location, ranging from the economic to the aesthetic, which have hitherto formed the *modus operandi* of our understanding and design. It suggests, albeit in a preliminary way, that if we approach cities by manipulating patterns of location, we are in danger of devising solutions to problems of urban equity and efficiency that simply miss the point of why cities exist in the first place. We need a science built on concepts of why people come together to trade and exchange commodities and ideas, to realize social contacts, and to procreate; in short, to relate. Interactions, hence networks, are therefore considerably more important to our understanding and planning of cities than are locations. This is not a book, however, that throws away all we have learned about location in favor of networks. Locations are built on interactions. What happens in locations is a synthesis of what happens through networks and of how activities interact with one another. To build a science of cities that meets the challenge of seeing locations as patterns of interactions, this book will hopefully provide a set of elementary but intuitive tools for representing, analyzing, simulating, predicting, and creating urban structures. These tools will be constructed around how we interact, primarily expressing such interactions through physical realizations and metaphors that are built on analogies with networks, flows, and, indeed, relations of many kinds.

There is another reason we must move from an approach based on locations to one based on networks. In the medieval city, interactions were for the most part local, immediate in some sense, and highly clustered. Physical interactions were limited, and it was not until the Industrial Revolution and the invention of the internal combustion engine that cities were able to grow, thus realizing much greater interactions among their populations. Technology clearly imposes enormous limits on the extent to which we can communicate. With the move from physical interactions to digital, from energy to information, from "atoms" to "bits," as Negroponte (1995) has so persuasively characterized this transition, and from the industrial to the postindustrial world, the technological limits on interaction are being massively relaxed. Recently, with the convergence of computers and communications through the Web and wireless, transaction costs have fallen dramatically, making possible all kinds of interactions that hitherto were too expensive, impossible, or often simply never imagined. The need to focus on costs of interaction in a world where these are being dramatically changed through new information technologies means that by the end of this century, the physical form of cities in terms of their locations is likely to be massively different from the urban patterns we see today. In fact, this change is likely to be so great that we have no idea as to what the city of the medium-term future will look like, for we are only now acquiring the concepts and tools to think of the world in terms of interactions and networks rather than locations. Laying out these tools and demonstrating how we can use them effectively is one of the central missions of this book.

In arguing for what is, in effect, a new paradigm—that we think of cities as sets of interactions rather than locations—we face the inevitable charge that we abandon the familiar in favor of the new. In fact, we will show that locations are intimately related to interactions and networks, and that the patterns that characterize urban growth and form emerge as intrinsic consequences of interactions, flows of energy, and information. Ideas about the shape of cities, about urban sprawl, about public versus private transit, about the integrity of neighborhood and so on, which have been used in the past in illustrating density and accessibility, follow naturally from the approach advocated here. Although this book will not dwell on conventional pictures of urban form and will not provide a detailed social articulation of the way cities function spatially, it will provide tools for exploring such understanding. If this approach is to be effective, then readers should be able to exercise their own ingenuity in extending these ideas to many aspects of the contemporary city in terms of building an understanding for better design. This book does not seek to replace what has gone before, but to argue that there are deeper and more fundamental patterns in cities that must be unraveled if we are to build an effective science of cities and city planning.

Agglomeration, Globalization, and Decentralization

Cities exist primarily to bring individuals together to trade the products of their labor. Historically, clusters of individuals formed at points where their labor could best exploit natural resources while providing sustainable locations where production could be achieved, generating a surplus that could be distributed over a wider area. Our world is littered with examples of cities that have formed at locations accessible to wider populations, at the confluences of rivers, in natural protected harbors, adjacent to places where fossil fuels could be economically exploited, and so on. In this way, the idea of a central city and its hinterland was formed. As soon as cities appeared, the spatial system that emerged was explicable in terms of how the economic advantages of local production and exchange could be traded off against their distribution to wider populations. The essence of the city is thus its economies of scale, which conventional wisdom suggests increase more than proportionately as cities grow in size (Glaeser, 2008). We will exploit these ideas extensively in this chapter and elsewhere in this book when we formally introduce notions of scaling.

This is *agglomeration*, as Marshall (1890) defined it more than one hundred years ago, and it implies a network effect. As individuals come together to trade, traditionally in the agora or medieval market place, the fact that they are in close proximity increases the number of potential contacts. The number of potential interactions between people in close proximity—the population that we will call P—increases as the square of the number P^2, and this positive feedback effect is the essence of the economies that are gained. But it is easy to see that the number of people T who can communicate face to face is much less than this; therefore $T \ll P^2$, even if they are clustered together. Thus to achieve these economies of agglomeration, new forms of communication are required. In essence, the larger the city, the more likely it is for the city to have mechanisms in place that enable more and more people to communicate with one another efficiently. Moreover, these agglomeration economies that are achieved as cities grow bigger imply *increasing returns to scale,* which in turn imply falling transaction costs. There are important spatial and physical consequences of this kind of economy, which we will describe in more detail a little later and explore in depth in the following chapters.

There is an important corollary to this tendency toward agglomeration that is a countervailing force. Resources are always limited, and thus cities, like any other type of population, compete with one another. Cities cannot grow indefinitely, since technology puts a limit on the extent to which transaction costs can fall, while all cities cannot be of the same size, as this would imply that there were no increasing returns to scale to be gained. What we observe, in fact, is that city sizes vary from a few larger cities to many smaller ones following a progression that is highly regular, reflecting a competitive process in which cities sort themselves out into a hierarchy of sizes. Much location theory deals with the way the spatial economy organizes itself according to a theory of sizes, which is reflected in the spatial clustering of cities and their hinterlands. An unusual and somewhat counterintuitive

L.A. (#2) = ½ of NYC (#1)
Chicago (#3) = ⅓ of NYC

observation for some parts of the world, such as the United States, is that cities follow a pure *rank-size rule*. This says that if we rank cities from the largest to the smallest, the city at a particular rank in the size hierarchy is the size of the largest city divided by its particular rank. We will elaborate these ideas later in this and the following chapters, as they are key determinants of how cities of different sizes imply different regimes of interaction and different network structures.

2

Transaction costs and the ability to communicate over ever-longer distances lie at the heart of our second theme: *globalization.* Cities are increasingly specialized through interactions on a global scale in which production and consumption can be optimized in terms of added value by spreading activities well beyond their traditional hinterlands. In a sense, this has been true since classical times with the largest cities, beginning perhaps with Rome or Nanjing, which served empires and territories well beyond their local or regional communities. Most industrial cities of the nineteenth and much of the twentieth century were not truly global, for globalization has only become of central consequence in the modern age with the invention of new information technologies that have changed the function of space and distance as we have traditionally perceived them. Globalization, in fact, has gone hand in hand with falling transaction costs and the invention of new information and communications technologies. Although the somewhat impressionistic "death of distance" only captures one element of this transition (Cairncross, 1997), the development of the Internet during the last quarter-century has brought the role of networks, communications, and interactions to the fore in thinking about the contemporary and future city. Any activity which seeks to manage or control activities in cities must thus depend on how interactions and networks evolve and are managed and controlled, and any kind of intervention is doomed to failure if it does not account for networks that span the globe.

3

Decentralization is our third theme, and this is resonant with both agglomeration and globalization. As transaction costs have fallen for a variety of communications media, it has become ever easier to spread urban activities over larger and larger areas. The first evidence of this appeared a century or more ago, when industry that traditionally was arrayed around the central core of the industrial city began to move to the periphery. This was as much occasioned by falling costs of transportation and the rise of the automobile as by the need for space, but such decentralization has continued apace, and by the late twentieth century, entire commercial areas had appeared as "edge cities," heralding a polycentric pattern of urban development composed of many clusters and cores. In fact, the dominant mode of urban development is no longer one that can be simply pictured as a system of highly centralized cities. Local economies of agglomeration associated with putting activities in one specific location have been tempered by wide-area economies, in which local transport binds activities together as efficiently, or even more so, than the much higher-density cores that were traditionally sustained through face-to-face contacts.

One might argue that the two key forces holding a city together are the opposing forces of decentralization and centralization. Physically decentralized activities are clearly made possible by falling transportation costs, but this is also determined by the need for more space as cities grow. The need for space is clearly one of feasibility in locating activities at appropriate densities, but there is also an intrinsic need or preference for more space, per se. This is certainly the case for residential populations, and thus we might picture the form of a city as an essential tension between the desire to be as close as possible to everyone else, which is the very idea of a city, and the desire to be as accessible as possible to as much space as possible. Indeed, a simple model of constrained diffusion, which we will introduce in chapter 8, embodies these ideas perfectly. As transaction costs, particularly transportation costs, change, this tension is resolved in different ways. When it becomes possible to attain the economic and social advantages of exchanging activity with different cities, then urban patterns become global and decentralization merges into globalization. Some of the ideas I summarize in our foray into complexity theory later in this chapter build deeply on this notion of decentralization, but before we enter that territory, it is important to articulate how we are able to express these ideas in physical terms. In fact, my premise in this book is that our concern is still with the physical structure of cities in terms of their geometry and morphology, but that the traditional focus on manipulating locational configurations in cities to achieve our goals is passé: the new paradigm emphasizing interactions still requires us to express our understanding and designs in physical terms, often as networks, but networks built on strong and significant socioeconomic relationships.

Physicalism: Geometry, Morphology, and Urban Form

Public responses to the condition of cities before the Industrial Revolution were particularly muted with respect to social concerns. Grand plans in which parts of the city were laid out geometrically in monumental style and implemented by diktat have existed since prehistory, but it was not until the mass migration from the countryside to fuel the growth of the industrial city, where modern technological innovations were being developed and applied, that public responses to the obvious evils of overcrowding became vocal and significant. This was a response to high densities and all the attendant consequences in terms of health and overcrowding. In Western countries, it led to the kind of institutionalized physical planning that underpins the framework of this style of public intervention even today. Initially, physical planning addressed the related foci of *density* and *segregation*: plans were fashioned to reduce densities by simply providing more space at much lower densities in the form of "garden cities" and "new towns" that mixed countryside and town in a much more even proportion. Plans also proposed that distinct areas be segregated from one another, residential from industrial for example, using instruments that came to be called "zoning." Much of this in fact has now been turned on its head, for far from being associated with pollution and overcrowding, the contemporary high-density city tends to be greener and more sustainable than the urban sprawl that is the dominant characteristic of the late-twentieth-century city. In the nineteenth and early twentieth centuries, *accessibility* or nearness, which is a key feature of agglomeration, was rarely thought about; towns were much more limited in aerial extent and smaller in population size. Insofar as this entered our thinking, the advantages of agglomeration were not well understood and plans were often devised to explicitly increase the distances between populations rather than decrease them. It was only with the rise of the automobile in the early and mid-twentieth century that accessibility became a more significant feature in our understanding of city systems.

These elements—density, segregation, and accessibility—remain the key elements in the vocabulary of what we refer to here as "physicalism." In its strictest form, physicalism assumes that the only things of significance have a physical nature, but clearly this is an extreme position, and what we mean by the term here is that physical form is the appropriate way to represent cities, in terms of the city's geography, geometry, and associated attributes. It does not exclude the drivers of cities such as agglomeration, decentralization, and globalization, which clearly originate in human behaviors and social structures, but it does focus on what can be immediately observed and hence manipulated through city planning. Nor does it exclude other forms of planning, although we would argue that the manipulation of the city's physical form is still the most obvious, appropriate, and least controversial approach.

Area and distance are basic elements in defining density and accessibility, and in geometric terms, this representation is composed of points, lines, areas, volumes, and various combinations thereof. There are various theories of these expressions of physicality, which determine the costs and benefits of the human functions and behaviors that determine the way cities appear. Many of these lie in location theory, urban economics, and regional science, which are built on foundations in microeconomics and in turn relate to more generic concepts such as utility. I will do no more than refer to this great edifice of theory and knowledge in developing our tools to explore the physical form of cities and their networks, but I will assume that in grounding this all in practice, these ideas are relevant and useful. The most significant syntheses of this body of work involve applying trade theory—the theory of exchange—to cities and regions (Fujita, Krugman, and Venables, 1999) and growth theory to urban economies (Glaeser, 2008).

Turning to interactions, we find their infrastructure is established in terms of physical networks, which most clearly involve different kinds of transportation systems. Flows existing on these networks measure the volume and type of activities, which in turn impact locations. These all exist in two- and three-dimensional space and have physical presence where flows are measurable. This is in contrast to information, which flows on physical networks but whose volume and, often, network infrastructure remain hidden or invisible to direct measurement. With respect to social networks that overlay the city in diverse ways, these are usually studied outside their embodiment in space, and the focus is mainly on their topological or relational structure. In the next two chapters, we will discuss all these networks in terms of how we locate them in either geometric or relational space using the vocabulary of flow theory and then graph theory to represent them as nodes and arcs, hubs and links, vertices and edges, with these descriptors being interchangeable. We will show how nodes relate to locations, and the links or arcs to flows between locations, which define bundles of interactions between activities.

Agglomeration implies location, but in this context, we will think of such clustering as densities of links around nodes or hubs, while physical decentralization can be pictured as the spreading of network links across geographic space. Globalization too can be seen in the same way but on a greater scale. Fashioning our understanding of cities in terms of these kinds of interactions is absolutely essential in making sense of the contemporary city, which has links on all scales from the most local to the global. In fact, the network-interaction paradigm is the only one that can make sense of the transition from the medieval to the industrial city and thence to postindustrial urban development, where the concept of the city as a distinct node in a sea of rurality no longer applies. Increasing density and scale of networks, which imply increasing urbanization and wealth, also says something about increasing complexity, and it is this theme that we will use to tie together these ideas and provide directions to this book in the rest of this chapter.

The Systems Approach

There are almost as many approaches to understanding cities as there are commentators trying to make sense of this complexity. This comes as no surprise, for every individual has a different perception of what it means to live in a city, notwithstanding common agreement about certain salient features. Given such dimensionality, one clear strategy for progressing our understanding is to use more generic theory that is independent of the domain in question to provide a framework. Some would argue that theory that is domain-independent often misses the essence of the system, but I argue here that a generic approach has the initial advantage of posing some degree of neutrality on the structure and behavior of the system in question. Fifty years ago, the emerging theory of general systems that had originated in biology and engineering was adopted by the weaker of the social and professional sciences such as sociology, management, and city planning. The approach found favor in making sense of situations for which there was no strong and established theoretical edifice of knowledge and practice. When ideas about cities and city planning moved from that strange mix of architectural determinism and social administration that had dominated their early history in the twentieth century to more systematic social science approaches, the systems approach was briefly fashionable (Chadwick, 1971). Since then, it has remained implicit in our thinking about how cities function.

Systems are generally defined as organized entities that are composed of *elements* or objects and their *interactions*. The idea of a network is deeply embedded in this way of articulating systems, with their organization being composed of aggregates of entities at different scales usually forming distinct structures, clusters, or *subsystems* that can be arranged in a *hierarchy*. In nature and in life, many systems display such hierarchical organization. In some contexts, the notion of strict hierarchy is assumed to imply an optimal order, something to be achieved or aspired for. This order is the usual characterization of system *structure*, as distinct from system *behavior,* which in early approaches tended to be focused on the way the system's structure was maintained through routine actions that defined the way the elements of the system interacted with one another. A good example is the notion that cities are composed of locations such as places of work and residence. The interactions that link these locations are the traffic flows that define the journey to work, which maintain the system's structure as it occurs routinely day by day. Changes in these patterns over the short term tend to retain integrity of the structure in such a way that these patterns slowly evolve, continually readjusting to jobs and houses that are being destroyed and created through time. The feature of this kind of behavior is feedback to restore the equilibrium; it is often called *negative feedback,* since it tends to dampen any volatile behavior that threatens to push the system away from its long-term steady state.

This conception of hierarchy, which still resonates strongly with more modern approaches, as for example in complexity theory, was beautifully captured by Simon (1962) in his example of how hierarchy must be a fundamental property of how systems hold themselves together, persist, and are resilient to outside influences. His example is worth repeating. Imagine, he said, two Swiss watchmakers, Hora and Tempus, who both constructed identical but intricate watches, each made up of 1,000 pieces. The only difference between the two watchmakers was in the way they constructed the watches by adding the pieces together. Tempus simply added a new piece each time to the entire assembly, whereas Hora produced subassemblies of ten pieces each, which he then added together into one hundred pieces once he had ten subassemblies available. His watch was thus created by adding the ten intermediate assemblies of one hundred pieces together in a final stage. All worked well in an environment that was peaceful, but as the watchmakers became more famous for their wonderful watches, they received more and more orders. As the orders only came in by telephone, each time the telephone rang, Tempus had to put down his assembly and it dropped to pieces, whereas Hora only lost the subassembly he was working on. What ultimately happened is that Tempus found it harder and harder to complete a watch, while Hora could handle the increasing volume of orders much more effectively. The long-term outcome is obvious: Hora prospered and Tempus went out of business. The moral of the story, of course, is that hierarchical organization from the bottom up is essential for evolving systems and that hierarchical structure is the way nature and society develop robust and resilient structures. Networks of dense clusters or subsystems at successive levels of hierarchy must indeed be one of the signatures of complex systems, a notion that resonates in every chapter of this book.

The picture of a system implied by these early theories, reflected in the design of Hora's Swiss watch, is one of subsystems based on local interactions or networks whose sum of parts can be arranged as a "strict" hierarchy or treelike form, or "dendrite," as we will refer to it here. Figure 1.1(a) shows the subsystems and their interactions, and figure 1.1(b) the hierarchy. In fact, as soon as this notion of hierarchy became fashionable in thinking about cities and their neighborhoods, it was relaxed to reflect the notion that such strict subdivision could only be a simplification and that a better picture was of overlapping subsystems. Indeed, Alexander (1965) wrote a famous paper entitled "A City is Not a Tree" in which he argued that the kind of variety and diversity that was the essence of cities, as articulated by Jacobs (1961), was being destroyed by the implementation of city plans that imposed such a rigid hierarchy through, for example, zoning. He used the term "semi-lattice" to convey his view that although the notion of hierarchy is inevitably useful as a way of thinking about how systems might self-organize, in fact the world is considerably more complicated than any strict hierarchy implies. Figure 1.1(c) illustrates this conception in a simple diagram of system structure.

What was not taken forward within this body of ideas was any sense in which systems might change. Hierarchy and related treelike structures were associated with structures that had evolved to some equilibrium, as Simon's (1962) parable implies. Thus when systems theory was first applied to cities, it was widely assumed that articulating the city system in terms of some long-term equilibrium was the appropriate response. In fact, as we now know, this notion of observing the city at a cross-section in time as though it is in equilibrium is far from the mark. Cities might appear as though they are in equilibrium because of the enormous lag between the speed at which the physical built environment changes and the way

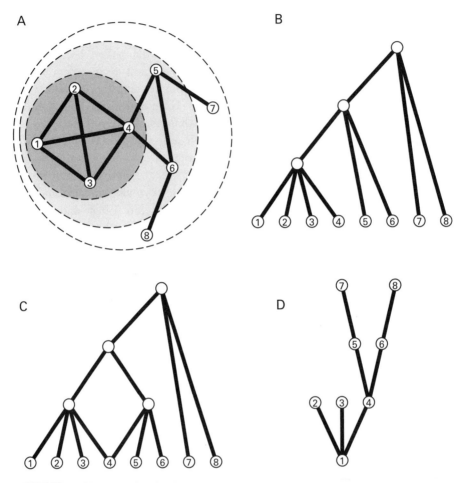

Figure 1.1
System structure and hierarchy.

human behaviors change, but our thinking has moved to the point where cities are now regarded as being in a perpetual state of disequilibrium. The focus is on ways in which growth or decline imply a positive feedback or cumulative causation, thus moving the system continually into a new equilibrium that, if fast enough, implies a continual state of disequilibrium. In fact, as cities are highly ordered entities, far from the kind of random system that is the default model of the physical world, the term *far from equilibrium* has come to characterize cities, with growth and change representing continual readjustments to the far-from-equilibrium state (Batty, 2005). The problem with a systems theory of cities is that it tends to view systems as being well-behaved, in the sense that external shocks

Buddhism: no Closed Systems

to the system tend to work themselves out, restoring the previous equilibrium or at least evolving to something that is close to the pre-existing state. What has been realized in the last fifty years, however, is that this notion of systems freely adjusting to changed conditions is no longer valid, and in fact never was. Cities admit innovation—indeed they are the crucibles of innovation; they generate surprise; they display catastrophes.

Early systems theory did not focus much on the question of system openness either. The idea that one could make effective progress by considering cities as closed in equilibrium—that is, closed in time—was paralleled by the general notion that although systems were inevitably open to their *environment*, to the outside world so to speak, systems could be defined as if they were closed, or at least could be defined so that interactions with their environment were minimal. This assumption, like that of systems in equilibrium, has been widely discredited for many human systems, and for cities in particular. Decentralization and globalization now make it almost impossible to find the degree of closure necessary to assume that such external interactions might be internalized in some way. This inability to find systems that are closed is one of the most pernicious problems in modern science, for it throws into doubt the idea that has been widely accepted since the Enlightenment that science can provide a technology to "solve problems." This view gained considerable momentum in the social sciences during the early to mid-twentieth century, but as it developed, doubt was being cast on our ability to provide closure and thus consensus about the scope and extent, indeed the very definition, of the systems in question. This, among other sources of uncertainty with respect to the nature of open systems composed of countless numbers of distinct and purposive elements and interactions, has led to a sea change in systems theory itself. The complexity sciences, which change the focus from highly organized, top-down conceptions to much more organically structured, bottom-up types of systems, grew from these concerns and now provide a framework that is considerably richer and more appropriate for the city systems that are dealt with in this book.

Complexity Theory and Cities

The idea that a system can be effectively understood through devising theories about its structure and behavior when it is closed implies that its environment acts benignly on its form. It also suggests that its structure is organized coherently from the top down. This is entirely consistent with the notion that a system has an ordered hierarchy of subsystems, all the way down to its fundamental elements, whose interactions function in such a way that the system is composed of aggregates or clusters at different scales. However, if the system cannot be closed, which is tantamount to a system existing in an environment of considerable volatility, then this notion of top-down order is no longer appropriate. In fact, top-down order implies that systems can be understood by their disaggregation and is entirely consistent with the reductionist program that has dominated science for more than three centuries.

Nevertheless, this idea of an order that is comprehensible from the top down has been questioned in general system theory from its inception. The hallowed cliché that "the whole is greater than the sum of the parts," which implicitly originated from the Gestalt movement in the 1920s, cast doubt on such order, although in the early days of systems theory, there was little sense in which such emergent order could be understood. As the concept of openness became significant and the idea that systems in equilibrium were by far the exception rather than the rule, the idea that systems grew from the bottom-up gained ground. In fact, most systems are not "manufactured," but are "grown." They evolve through countless decisions that have an impact on many scales, as Simon's (1962) parable implies. Biological systems grow from the cell and are relatively self-contained, although in the context of their environment, they field many influences that are hard to assess with respect to their development. In human systems, decisions about organization and behavior are made at all scales, from the local and routine to the global and strategic, but they tend to be made by individuals and in this sense are highly decentralized. In fact, cities have become ever more decentralized in their morphology, not because decisions have become more decentralized per se, but because individual decisions have broadened in scale in response to falling transaction costs. We must not confuse decentralization in terms of urban form with decentralization as a feature of the way systems grow from the bottom up, although there is a relationship.

The move to articulating systems as structures that grow from the bottom up immediately puts dynamics onto the agenda, and in turn leads to the notion that patterns and structure at scales above the most local, where individuals operate, "emerge" from these actions. Emergence, which embraces novelty, surprise, innovation—all the unexpected features that we see in systems that grow in this way—is the watchword of this new view of systems and one that will appear in many chapters in this book. It encapsulates rather nicely the notion that a system such as a city is not something that is planned from the top down but emerges organically. The networks and locational patterns that we will explore throughout this book are not planned by any supra-agency or one individual. Their structure, in so far as we can understand and articulate it, emerges through countless decisions in the context of physical constraints that limit the feasibility of certain patterns over others. In a sense, economic theory has always assumed that economies develop in this way. Adam Smith's "invisible hand," the life force of capitalism that keeps markets working, is the glue that holds the world together. We find it almost impossible to define this in formal terms, but the result is sets of ordered patterns that emerge from the actions of countless individuals who, insofar as they are coordinated, are only so at the most basic level.

A useful analogy for complexity is thus based on systems that grow organically. In reference to cities, this concept has been vigorously exploited since the late twentieth century, as much as a critique of the processes that city planning ignores as a means of understanding cities better. In essence, systems that grow from the bottom up expand and diffuse as subsystems and clusters that take on different morphologies at different scales. Many such systems in fact repeat their essential structure at different scales, as figure 1.1 implies, and this often can be seen as some kind of regularity or modularity, again detectable at different scales. This is referred to as self-similarity, the hallmark of fractal structure (Mandelbrot, 1982), which in cities can be seen in the organization of land uses and their economic activities, particularly in centers that exhibit ever more specialized functions at ever-larger scales. Central place theory and the distribution of city sizes that pertain to such organization are key characteristics of this fractal structure (Batty and Longley, 1994). In one sense, the hierarchy that emerges from the processes that define growth provides a pattern that can be read from bottom to top or vice versa. The top-down structure of a hierarchy implies a rather centralized order, but the processes of growth from the bottom up that generate such a hierarchy might be construed from the viewpoint of each element at the hierarchy's base. We can plot the links of base elements to all the others as a hierarchical tree, as we show for one of the elements in figure 1.1(b) in figure 1.1(d). This implies a divergence through the system rather than a convergence, but all it says is that hierarchies, which define structured patterns of interaction, are always relative to the organizing principle that is in focus. Cascades occur both up and down hierarchies and, as we shall see in later chapters, these constraints and conditions are always relative to the question at hand.

Complex systems that self-organize into clusters from the bottom up do show emergence, in that patterns established through rules at the lowest level repeat themselves at larger or higher scales. However, the simplest such systems do not display the kind of variety and heterogeneity that characterizes real systems like cities, and it might be argued that there are subtle changes in these patterns that emerge as the scale is changed. In a sense, fractal structures do not quite embrace this kaleidoscope of change, although the criterion of invariance at different scales can be relaxed to admit a weak kind of self-similarity. Our understanding of any kind of system depends on simplifications and principles such as those that pertain to fractal geometry, but one of the characteristics of complex systems is that they are intrinsically impossible to simplify in a way that makes their definition closed. To progress this argument, we will now examine in a little more detail elements that comprise complexity theory, quickly and briefly summarizing examples that pertain to the mainstream and support many of the ideas that will surface in the rest of this book. In later chapters, we will review networks, flows, and morphology extensively, although our focus will not be specifically on the types of dynamics that constitute an important strand in contemporary complexity theory.

Cities as Complex Systems: The Rudiments

Equilibrium and Dynamics

We could be forgiven for thinking that cities are in equilibrium, for the built environment that conditions our immediate responses changes only slowly in comparison to the functions that take place in such environments. Cities can be pictured as if they are in equilibrium, the focus being initially on representing and simulating static urban structures, such as population density profiles, cross-sectional patterns of movement, and the configuration and location of different land use types, often as concentric rings of use around the origin of settlement, invariably the central business district (Alonso, 1964). Insofar as change to these structures has been articulated, this is smooth change, but as soon as scholars first became aware of the nature of actual change, which was often discontinuous and lumpy, the need for widening the framework to embrace all kinds of non-smooth dynamics became obvious. In the 1970s, ideas about how cities developed in discrete jumps—via catastrophes in development reminiscent of housing price (and building) booms and crashes, and via spatial behaviors that were revealed to be chaotic, that is, qualitatively different from near-identical initial conditions—came to dominate our thinking about the dynamics of change (Wilson, 1981).

Urban simulation models, initially predicated on forecasting development at a cross-section in time, were the first to deal with such dynamics, but these developments were slow, for at the time, the wider theory of how change takes place in cities was barely developed. Forrester's (1969) exposition in his book *Urban Dynamics* was an early statement that urban change had an often counterintuitive quality, built as it has been around notions of positive feedback giving rise to exponential growth and decline, capacitated logistic growth, and oscillations as systems overshoot and undershoot their assumed equilibria (Batty, 1971). It was widely regarded that an acceptable science of cities must embrace a dynamics that directly enabled simulations to be made of widely differing growth scenarios. New developments in the mathematics of dynamics based on catastrophe, bifurcation, and deterministic chaos were quickly embraced by rudimentary complexity theory and have since become integral to the definition of every type of complex system (May 1976). To quickly illustrate the range of dynamics and how they relate to one another, we graph these kinds of dynamics for population change in figure 1.2.

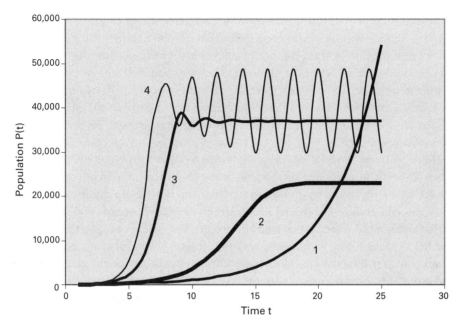

Figure 1.2
Varieties of population dynamics. Simulations from a generic model $P(t + 1) = \varphi P(t) z(\bar{P} - P(t))^{\gamma}$ of population growth, where P is population and t, $t + 1$, the timescript. The parameters φ, z, and γ control the positive and negative feedbacks that generate damped exponential growth, and $\bar{P}$ is the capacity of the population. Curve 1 is the classic exponential growth model with $\gamma = 0$, while curves 2–4 are purely logistic, then oscillate with damped exponential growth for different combinations of the parameter values.

Patterns and Processes

The functioning of cities in space and time is based on multiple processes of spatial choice in which individuals and groups in the population locate with respect to one another and their wider activities in the form of land use types. These activities tend to be dominated by trade-offs between agglomeration economies and diseconomies, which are often represented in terms of relative accessibilities between different locations. These trade-offs give rise to patterns of activity that reflect different levels of clustering, and in turn these imply different density levels associated with different locations. For a long time, locational patterns were represented in rather coarse, abstract terms as density profiles around key hubs, as nested hierarchies of central places, and as patterns of accessibility reflecting more polycentric forms, but with little sense of the morphological structures they actually represented. This meant urban researchers missed some significant signals that urban patterns manifest, such as their self-similarity or spatial invariance across different scales, which in turn imply that similar sorts of processes are operating across scales. Moreover, patterns that repeat in module-like form, such as those characteristic of central place theory (Christaller, 1933/1966), generate hierarchies that can be handled using new types of geometry. In the 1980s, onto this canvas, new ideas emerged about how modular patterns in cities were structured and, using concepts about self-similar processes of development, cities began to be interpreted as fractal structures (Batty and Longley, 1994). Figure 1.3 presents a potpourri of aggregate urban patterns that display such fractal structure with distinct hierarchical ordering that accords to the rank-size scaling we introduce below.

This idea of using morphology as a signature to detect the different underlying processes at work in cities relates very strongly to notions about how individual spatial decisions determine how cities grow from the bottom up and how patterns repeat themselves at different spatial scales. Urban development rarely fills the entire space that defines the wider hinterland of a settlement or city and, in this sense, it is regarded as space-filling in the same way that fractal forms fill space between the Euclidian (integer) dimensions. The whole paraphernalia of fractal geometry can thus be brought to bear on urban patterns and processes. Fractal dimensions, which determine the extent to which cities fill the space they occupy, are a means for classifying cities using their relative densities and accessibilities. Many of these ideas pertain to how cities develop in terms of compaction and sprawl, and they find immediate expression in terms of the network structures that tie land uses together and provide the glue for achieving the sort of agglomeration economies and scaling effects that define cities in the first place.

Interactions, Flows, and Networks

Interactions between different individuals rooted in time and space define the nature of a city. Cities, as Glaeser (2011) and Jacobs (1961) before him have argued so persuasively, are about "connecting people." In many ways, this is the leitmotif our science professes. The various processes that take place in cities, which bring people together to produce and exchange goods and ideas, define a multitude of networks that enable populations to deliver materials and information to support such endeavors. Physical and social networks tend to mutually reinforce one another as they develop. From an initial hub such as the origin of settlement, individuals are attracted usually in proportion to what already exists. This is the function of agglomeration that we described earlier—that is, for an existing population P_i at location i, the growth of the population is proportional to this size φP_i, where φ is the growth rate. Imagine that this is a hub or node in a network. Then the hub grows, assuming the rate of growth is greater than 1, by new links from other populations in the hinterland of the hub connecting to it, the number of such new links being in proportion to its size. Each of these other nodes that form the links attract links from other nodes in the same manner, with new nodes emerging randomly in

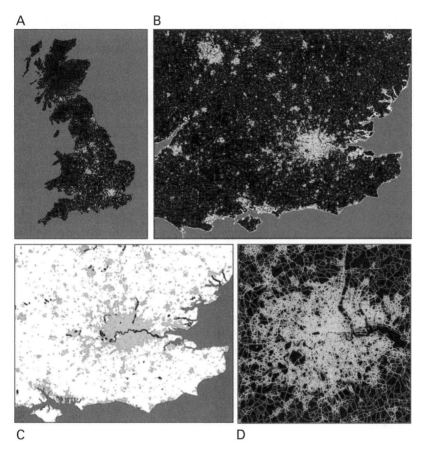

Figure 1.3
Self-similar urban morphologies from population, remotely sensed imagery, and street network representations. Top left (a) and right (b) show the urban morphology of the United Kingdom and the south of England from gridded population census data (from Batty and Longley, 1994). Bottom left (c) is an image taken from remotely sensed data for 2000, and bottom right (d) is from street network data for Greater London. Note there are clusters on all scales that accord to both the diffusive growth and rank-size scaling that we discuss in this and later chapters of this book.

the wider region. This is a model in which the rich get richer. It generates a cumulative causation and lies at the basis of how networks develop in many different domains. Barabasi (2002) calls it the "preferential attachment" model, and one of its striking features is that the size distribution of the nodes—which can be locations (cities), parts of cities, groups of individuals, institutions, and so on, for the model is generic—follows a scaling law. In short, the frequency of nodes of increasing size in terms of their links gets ever smaller in accordance with a power law, there being many small nodes and very few large ones. This is one of the basic signatures of this new science, for it reflects the way competition determines size and the way resources are bid for in a competitive environment. It is not only applicable to networks but also to city sizes as well as the sizes of different locations within cities, where it appears as Zipf's Law (Zipf, 1949). We will return to this below, first when we pull together the key signatures of spatial complexity that define this new science through the so-called laws of scaling, in the next two chapters where we examine flows and networks, and in chapters 4 and 5, where we will look at rank size and hierarchies.

How networks form in cities relates to the fractal patterns we showed earlier, with the physical imprint of these signatures forming the channels through which people and materials move. The flows in these networks also depend on the size of the locations between which links are established, but usually these flows are mitigated by a deterrent effect of distance—the friction of distance, as it is called—with d_{ij} defined between any two locations i and j. A particularly long-standing form of this flow relation, first proposed to model movements in human space almost immediately after Newton published the laws of motion in his *Principia* in the late seventeenth century, is the gravitational force defined as

$$T_{ij} \propto P_i P_j d_{ij}^{-2}, \tag{1.1}$$

where P_i, P_j are population masses and T_{ij} is the interaction between places or zones i and j. It is the so-called inverse square law used by Ravenstein to represent migration flows between cities and regions in Britain in the late nineteenth century (Tobler, 1995). This is also a scaling law, and from the 1950s, it has been used in many forms to model different kinds of traffic flow at a variety of scales. Indeed, it lies at the basis of much operational land use and transportation modeling (Batty, 1976), which still forms the basis of many applications of this science, as we will illustrate in chapter 9. We will exploit these ideas of interaction quite extensively in the next two chapters for both flows and their networks, which will form key components of this new science.

Social networks in which space is merely implicit also characterize cities and the way their individuals and groups interact to trade, exchange, build friendships, exercise decisions, manage other groups, and so on. Currently there is a great flurry

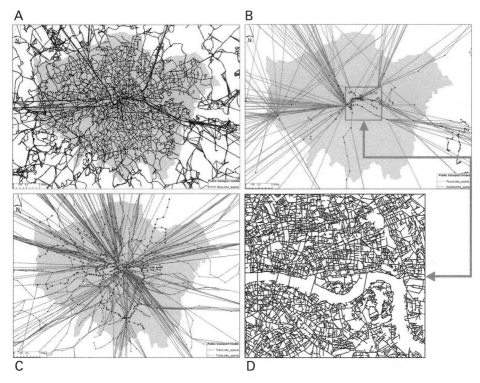

Figure 1.4
Coupled transport networks generating a convoluted dynamics of traffic. Top left (a) shows bus routes in London, with the longer straight lines being the routes of long-distance buses; top right (b) shows intercity coaches and ferries; bottom left (c) are long-distance and overground rail (gray) and tube (dark); and bottom right (d) is a sample of the network in Central London on which all road transport takes place.

of activity in modeling such networks, and the emergence of a new form of network science where the focus is on detecting patterns in networks—clusters called small worlds, shortest links, weak ties, bridges between communities, and hierarchies—is becoming ever more central to our science, as chapter 3 will illustrate. We show some of these networks in figure 1.4 to give the reader a sense of what they portray. The big challenge is in coupling networks, in developing ways in which both material (energy) and ethereal (information) networks are coupled to one another, ways in which such networks are cascaded into each other, and the way processes spread and diffuse over such links (Newman, 2010). The cutting edge is figuring out how such networks interlink and interlock and how the patterns of morphology that are the physical manifestations of the social and economic processes that define the way cities work build on such network representations. The key to understanding how networks fracture and split, how economies of scale and innovations are realized through the way different networks relate, and the ways in which prosperity and the creation of wealth are linked to these network effects are central questions our science needs to address. Scaling ties all of these ideas together.

Evolution and Emergence

So far our descriptions of cities—apart from identifying the key role of feedback in conditioning how dynamic processes build on existing patterns to reinforce size and to generate agglomeration economies—have not formally treated how cities actually develop through time. The patterns shown earlier, which are largely built from modules operating from the actions of individuals (or at least individuals acting for groups and institutions) from the bottom up at relatively small scales, evolve through time in such a manner that any snapshot at any cross-section shows an emergent order that is the product of countless individual decisions. Patterns emerge from the processes that spread the effects of these decisions spatially via various forms of segregation and diffusion.

Imagine that development proceeds in an orderly pattern in regular neighborhoods around some seed that motivates growth. If we define the urban landscape as a grid and a neighborhood as the eight cells around any given cell, the first of which is the seed, then development takes place around a cell already developed, in a certain unvarying local pattern. If a cell is developed already, we might fix a rule that a cell that is then developed around this cell must be as far as possible from the source. This would then lead to the development of four radial lines around the seed cell that spanned the space as four lines of development. If we relaxed the rule and specified that at each time period, a cell is developed in any of the eight positions if one is already developed in the neighborhood, then the process would lead to a diffusion of development around the seed cell, with all cells being eventually developed but in random order with respect to the chronology. We show some of the patterns typical of such diffusion in figure 1.5, where the patterns are constructed in a modular way, do not fill the whole space, and are clearly fractal. These are often referred to as cellular automata (CA), and they are central to the notion that hierarchical structure is produced by their operation in unceasing fashion, generating objects that are self-similar and hence fractal across many scales. They thus constitute the elements or "agents" of many models of physical urban growth developed during the last two decades (Heppenstall, Crooks, See, and Batty, 2012).

In contrast, we might consider an already developed system composed of two kinds of individual—let us say Republicans and Democrats—who are located randomly across a space. If the rule is that individuals are quite content to live side by side as long as their neighbors of a different kind are not in the majority, then we

Figure 1.5
Idealized urban patterns generated from the bottom up using modular rules for constructing development among nearest neighbors (from Batty, 2005).

might assume that the landscape is a checker board of Republicans and Democrats. However, if one of these individuals switches color arbitrarily so that one neighborhood is now composed of five Republicans and three Democrats centered around a Democrat who has switched his or her allegiance to be a Republican, the neighborhood will become more segregated. In fact, what happens is that the entire landscape unravels to produce highly segregated areas of Republicans and Democrats. This model was first proposed by Schelling (1969) and it produces a classic example of emergence in that, although individuals are quite prepared to live side by side if there is an equality of view—Republicans are balanced against Democrats—as soon as this difference shifts in favor of one or the other, the pattern begins to unravel and eventually what appears to be modest support for a balance of interests or views becomes extreme. Figure 1.6 shows how such segregation can take place from a set of initial conditions under which Republicans and Democrats are distributed randomly.

Size, Shape, and Scale: The Laws of Scaling

Relationships Based on Space, Distance, Size, and Frequency

What ties all these morphologies and processes together is the idea of scaling (Batty, 2008). When we say an object scales, we mean that the object resembles in some way a smaller or larger object of the same form, although these forms might be different in some distorted but regular way. For example, if an object scales with respect to its space, this means its spatial form might have the same or transformed proportions as a smaller or larger object (Bonner, 2006). In its strict form, this implies self-similarity, which is the fundamental premise on which fractal structures are based (Mandelbrot, 1982). In fact, often the object will sometimes have proportions that are distorted in a particular way due to the fact that as it changes in size, its proportions must adjust commensurately to conserve some critical functions of the object. This implies that scaling can involve qualitative change, in that such distortions still imply self-similarity but one that undergoes some transformation. This is sometimes called self-affinity. There are many features of cities that scale with size and imply transformations of shape, and we will extract these from our previous discussion before identifying a set of scaling laws that will appear over and over again in this book. In fact, these scaling laws will be the classic signatures of our new science.

Cities clearly change in shape as they change in size, and we usually call this allometry. It reflects agglomeration economies, which, when measured over all cities, we refer to as "interurban," and when measured within cities as "intraurban." The competition implicit in agglomeration is also reflected in the fact that there are many more small cities than big cities, with the asymmetry that to

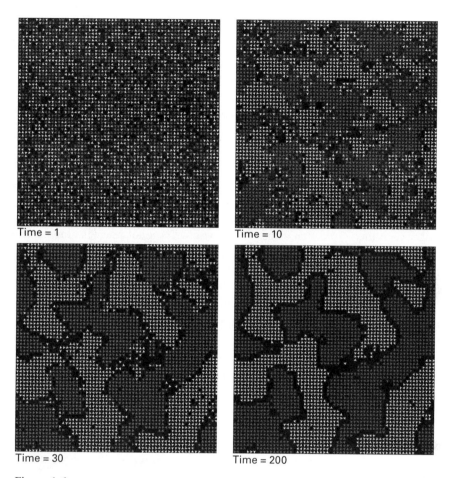

Figure 1.6
Segregation from a random spatial distribution t(time) = 1 to a highly polarized but stable distribution by t = 200.

become a big city you must first be a small city. This is reflected in the regularity of the distribution of city sizes, which we measure by ranking sizes. Cities are also distributed with respect to their size in such a way that small cities are nested in the hinterlands of bigger cities. This again implies that big cities are spaced more widely than small cities, and the hierarchical structure that results is codified in central place theory. Inside the city, populations appear to interact with one another more intensely in bigger than in smaller cities, although this is by no means a consequence of size per se. It may be a consequence of competition for space, in that densities tend to be higher in big than in small cities, offering the possibility for closer physical, and of course more frequent, interaction. As we noted earlier, treating cities in the wider landscape as point locations, the possible number of interactions increases in proportion to the square of the population, P^2. However, Dunbar (1992), among others, has suggested that the feasible number of potential friends and colleagues—in short, interactions—is on average about 150 for each person, implying that in towns with a population of less than 150, everyone knows everyone else. The limit must peak at a town of population 150, where there are $150P$ total interactions and where for every city above this size, the number of interactions varies linearly as $150P$. In fact, as the pressure to interact is greater in bigger cities, it is likely that total interaction will be slightly greater than linear, and this is implied in work relating to income and other economic quantities that we will note in later chapters.

It is also obvious that populations interact with one another less as the distance increases between them, and this of course is the gravitational law. It is implied in the direct analogy with the inverse square law in physics and lies in the trade-off between space and distance (or travel cost) in models of the urban economy dating from von Thunen (1826/1966). We might also argue that as cities get larger and occupy more space, notwithstanding increasing densities, distances between populations increase, and this acts as a deterrent on intraurban interactions, perhaps canceling out any of the pressures to interact that higher densities and competition tend to drive. Many other kinds of activities in city systems that diffuse over space fall off with distance, and in some instances, as cities grow in relatively undisturbed landscapes, it is possible to see distance and time as being equivalent as activity diffuses from some central market or hub.

Seven Laws of Scaling

As yet there is no formal synthesis of scaling in city systems, so it is worth drawing together these relationships before we discuss some of the most important in more detail. To this end, we can list these relationships as scaling laws, and somewhat casually, we will identify seven in all, with some observations that do not quite have the same status as laws, as yet. It might be argued that there are many more than seven, or indeed there is only one: that is, as all of these laws appear to follow from one another, there may be only one basic scaling law with respect to size and shape. We can formalize these a little more, but we must exercise some caution. They may not be laws in the accepted sense of the term in the physical sciences, but they reveal strong regularities that seem to persist in time and space. Thus, exercising the caveat of all other things being equal, *ceteris paribus*, we can state the following about cities:

- As cities grow, the number of "potential connections" increases as the square of the population. In computing, this is *Metcalfe's Law*, the network equivalent of *Moore's Law*, formalized when local area networks were first invented and which now dominates all network phenomena. A somewhat wilder speculation that information sharing is doubling every year has recently been made by Mark Zuckerberg, and some have referred to this as his law (Hansell, 2008).

• As cities get bigger, their average real income (and wealth) increases more than proportionately, with positive nonlinearity, with their population. This we might call the *Bettencourt-West Law* after their recent work on allometric scaling, but it might also be called *Marshall's Law* after the originator of the idea of agglomeration economies in cities at the end of the nineteenth century.

• As cities get larger, there are less of them. The frequency distribution of cities is highly skewed to the left, often being approximated by an inverse power law which can be transformed to a rank-size relationship. Its most elementary form is called *Zipf's Law* after Zipf (1949), who first articulated and popularized this relationship for city sizes and word frequencies.

• As a city grows in size around its original seed of settlement, which is usually the marketplace or its point of government, various densities, specifically the price of land—rent—and the density of occupation decline nonlinearly with distance or travel cost from its central point or central business district (CBD). This was pointed out by Clark (1951) for population density and by Alonso (1964) for rent, but it is essentially attributed to von Thunen (1826/1966), who first observed this in an agricultural context with respect to land use around the center of his estate in 1826. We should call this *von Thunen's Law*. Its inverse is the mass-radius relation, which is a fractal law implying that the population mass increases superlinearly with increasing distance from the CBD with an exponent (fractal dimension) that is between 1 and 2 (Batty and Longley, 1994).

• As cities get larger, interactions between them scale with some product of their size, as Marshall's Law implies, but interactions decline with increasing distance or travel cost between them. In one sense, this is the two-dimensional equivalent of von Thunen's Law, and in terms of gravitational models, it might be called the *Alonso-Wilson Law* after early work on spatial interaction, which generalized such models (Alonso, 1976, 1978; Wilson, 1970). However, in its most generic form, it has already been enshrined as *Tobler's Law* (strictly his first law of geography), which says that "Everything is related to everything else, but near things are more related than distant things" (Tobler, 1970).

• As cities get bigger, their central cores decline in population density and their density profiles flatten. This has been observed for many cities, and it tends to be due to land use displacing population at the center and land uses spreading to a wider hinterland on their peripheries. It was first noted for Paris by Bussiere (1970), and we will thus call it *Bussiere's Law*. We can generalize it as applying to include all activities where the law suggests, in fact, that densities rise in the center but fall at the periphery over time, a consequence of technology and increased wealth.

- As cities get bigger, they get "greener" in the sense of becoming more sustainable. This, of course, is a recent finding, and it is exactly the opposite of what happened as cities grew during the Western Industrial Revolution and as some but not all cities are still growing in the developing world. This we will call *Brand's Law* after Brand (2010), whose observations on the growth of cities and sustainability suggest that to maintain the quality of life that cities seek, higher densities are necessary for good social interaction and tend to make sustainable solutions to transport and construction much more cost-effective than lower-density sprawl.

There are many other more casual observations about how activities change in cities with respect to their size and space as measured through distance or travel cost. For example, as cities grow, their average density appears to increase, although there is no systematic body of evidence that has been compiled for this to date. Yet this is reflected in the fact that the average travel time within cities also increases, while more people tend to travel by public transport. There are many similar observations that might get elevated to the status of our "loose laws" as more evidence is systematically compiled.

The Scaling of Growth, City Size, Density, and Interactions

It is worth examining four of these laws in a little more detail to illustrate the similarities and differences between them, but first we need to state exactly what we mean by scaling. Imagine we have a function $y(x_i)$ that might be some attribute that varies spatially over locations or zones i with respect to some other attribute x_i. If we scale x_i by some "scalar" λ, then we say that a function "scales" if its scaled value is proportional to its previous value; that is, if $y(\lambda x_i) \propto y(x_i)$. The only function for which this occurs is a power law, which we can write as $y(x_i) = x_i^\alpha$, where the power α can take on any value. Then scaling this relationship by λ, we get

$$y(\lambda x_i) = (\lambda x_i)^\alpha = \lambda^\alpha x_i^\alpha = \lambda^\alpha y(x_i), \tag{1.2}$$

where it is clear from equation (1.2) that the "new" scaled distribution $y(\lambda x_i)$ is simply proportional to the old $y(x_i)$.

One of the most important scaling laws relates to the way attributes of cities change relative to their size and to one another. This is Marshall's Law, which is central to the economies of scale that emerge from agglomeration. It is also allometry, and it implies qualitative change, often being associated with changes in shape in biological populations where, to function, an animal or plant has to adjust its mass to its linear dimension. The easiest way to illustrate this is to consider how changes in the size of one geometric attribute of cities—the area, A_i, that they occupy or fill—relates to their population, P_i. For example, if cities expanded into the third dimension as well as into the other two, then we might consider cities increase in population as $P_i \propto d_i^3$, where d_i is a linear dimension. Area, of course, increases as the square of this dimension, $A_i \propto d_i^2$, and thus we can relate population to area as follows: $P_i \propto A_i^{3/2}$. This power law is an example of positive allometry where the power is greater than 1. If population were to rise proportionately with area, we would consider this to be isometry, and there is considerable evidence to think that the allometric coefficient—the power—is nearer to 1 than 3/2. If it is less than 1, then this is called negative allometry. Allometry is thus the study of changes in shape with size (Bonner, 2006). Clearly, population growing into its third dimension constitutes a change in shape with respect to what it is measured against (which is the flat plane), but it is more usual to consider changes in this power law to mirror economies or diseconomies of scale. There is a good deal of evidence, for example, to suggest that income Y_i grows more than proportionately with population, that is,

$$Y_i \propto P_i^{\alpha}, \tag{1.3}$$

where $\alpha > 1$. Physical infrastructures, such as road space, appear to grow less than proportionately with $\alpha < 1$, reflecting positive and negative allometry, respectively (Bettencourt, Lobo, Helbing, Kuchnert, and West, 2007).

Another law relating to size involves the frequency of objects of differing sizes, such as cities, firms, incomes, and related phenomena whose distribution is determined by competition for limited resources. The relationship between cities in a hierarchy of central places, for example, is one of scaling in that the frequency f_s of cities of size class s which have population P_s scales as a power law, which in generic form can be written as

$$f_s \propto P_s^{-\zeta}, \tag{1.4}$$

where the power ζ is always greater than 1. In its strictest form, it is $f_s \sim P_s^{-2}$. In fact, a much more convenient way to represent this frequency is to take the counter-cumulative distribution, which is the rank r_s, and to show that the typical rank-size distribution is represented as

$$P_s(r) \sim r_s^{-\zeta+1}. \tag{1.5}$$

If $\zeta = 2$, then $P_s(r) \sim r_s^{-1}$, which is the pure Zipf relation, first popularized in his book *Human Behavior and the Principle of Least Effort* (1949). This is the so-called rank-size rule, and it applies, of course, not only to city sizes and related phenomena such as incomes and firm sizes, but to the evolution of the size of nodes in what Barabasi (2002) calls "scale-free" networks. Mandelbrot (1982) defines it as a basic size relation that defines fractal phenomena, and with the strict Zipf Law, then it is easy to see why such a power law is scale-free. If we change the rank by doubling rank r to $2r$, our rule becomes $P_{s'}(2r_s) \sim (2r_s)^{-1} = 2^{-1}r_s^{-1} \sim 2^{-1}P_s(r)$. In short, the population size is a simple scaling. So the population of rank 2 is one half that of rank 1, and so on. We can generate these rank-size relations using various models: proportionate effect and preferential attachment, of course, but also by subdividing a large hinterland into mutually exclusive subdivisions in a modular and regular manner, making various assumptions about population densities, thus linking Zipf, von Thunen, and Tobler's Laws together. This ties this kind of scaling back to both central place theory in the interurban context and to urban economics in the intraurban context (Simon, 1955; Gabaix, 1999; Rozenfeld, Rybski, Andrade, Batty, Stanley, and Makse, 2008). Berry and Okulicz-Kozaryn (2012) present an excellent review and extension of these ideas, which shows the importance this kind of scaling continues to have on the development of our science.

In terms of distributions of activity within or between cities, scaling pertains to densities and interactions, which we have already anticipated in a previous section. Interaction effects, which in their purest form imply Metcalfe's Law, are one reason why creative pursuits, innovations, and even income scale more than proportionately with population. Consider again the "potential interactions" for a population P, which we can write as P^2. Only a fraction of these interactions might be realized, but as we have a population P where everyone interacts with him or herself, then the total number of interactions T for this population will be $P \leq T \leq P^2$. Assuming that self-interactions are excluded and we only count symmetric interactions once, then the upper limit is scaled down to $T = P(P-1)/2$. Of course, as the population of a city increases, then the area over which the population has to interact also increases. The furthest distance to travel to engage in interaction is in fact linear with distance—if d is the radius of the area, then we might write $T = P(P-1)/2d$, but if this is constrained by area, then as area is the square of distance, the potential interaction collapses to the gravitational form $T \cong (P/d)^2$. We might also argue that there is a constant fraction ξ of any population that can interact—Dunbar's (1992) number that we noted above—and thus the potential interaction will lie in the range from $\xi(P^2/d^2) \leq T \leq \xi(P^2/d)$. Now, it is likely for many creative pursuits that a large potential interaction does force greater interaction than simply the size of the population would imply, and thus the findings that income and related attributes scale superlinearly (or as positive allometry) with population would be borne out.

This logic leads directly to modeling interaction itself. We have already stated that interaction between two different places is usually modeled using gravitational force, but we will state it now in more generic terms as

$$T_{ij} = KP_iP_jd_{ij}^{-\phi}, \qquad (1.6)$$

where K is a constant that contains various other scalars that determine the interaction, and ϕ is a parameter that is usually greater than 1 and reflects the intensity of the friction that distance imposes on interaction (Wilson, 1970). Clearly this model is scaling, for if we double the friction as $(2d_{ij})^{-\phi}$, we scale the interaction as $2^{-\phi} T_{ij}$. This relation is much more generic, in that models of interaction such as this have been applied for many years (and continue to be) in land use transportation modeling. It is easy to see that we can derive von Thunen's Law from this in terms of density. If we assume one origin, the CBD for example, which we define as P_0, and many destinations defined by P_j at distances d_{0j} from origin 0, and we also normalize equation (1.6) by $P_0 P_j$, then we can write the density (of population) as

$$\rho_j = \frac{T_{oj}}{P_0 P_j} = K d_{0j}^{-\phi}. \tag{1.7}$$

We can tie some of these laws together, or at least provide the way forward to doing this. If we assume that population density varies as $\rho_j = K d_{0j}^{-0.5}$, and the frequency of these populations is given by the circumference around the circular city with center 0, which is $f_j = 2\pi d_{0j}$, then it is easy to show that the frequency of populations ρ_j is $f_j \propto \rho_j^{-2}$, which is Zipf's Law for the intraurban case. To extend this to a central place system, we need a rule to generate the hierarchy from the smallest places to the largest, and various nesting of populations ρ_j can be devised to generate consistent rank-size scaling. In terms of relating this to allometric scaling, we need to assume something about the various attributes of cities which vary superlinearly or sublinearly. Distance or road space is the obvious variable, and again with various manipulations, we can show that this varies sublinearly as population size increases, consistent with what we observe in real cities (Bettencourt, Lobo, Helbing, Kuchnert, and West, 2007). This is a synthesis in the making, but it is for others to develop a detailed and consistent argument. Our focus here is more applied and like many features of our science, we will leave this theory hanging for future work.

In chapter 2, we will develop these relationships in considerably more detail when we examine how we model flow systems in cities but for the moment, it is simply worth noting that all these scaling relationships can be linked to one another. Relating these, however, is still somewhat of a challenge, for they all mirror different perspectives on city systems. Our quest is not to develop a complete or even consistent theory of scaling for city systems, for this would involve us in explaining urban theory in ways that would divert us from our central task. But it is important to establish the fact that scaling laws appear everywhere in our science and are useful, indeed essential, in detecting the various signatures that give rise to structure in city systems.

Developing a Calculus for Urban Science

Jane Jacobs (1961) in her seminal book *The Death and Life of Great American Cities* argued that cities ". . . happen to be problems in organized complexity. . . .

They present situations in which a half-dozen or even several dozen variables are all varying simultaneously *and in subtly interconnected ways*" (her italics). Our laws of scaling in the previous section and the various approaches to complexity that we have described earlier mirror these interconnections and illustrate the kinds of subtleties that pervade our science. She continues, "Cities . . . do not exhibit one problem in organized complexity, which if understood explains all. They can be analyzed into many such problems or segments." This very much echoes the sentiments already spelled out here, in that there is no single approach that can ever be applicable everywhere and by everyone, yet there are still invariants and regularities that can be exploited in our concern for designing better cities. Moreover, the complexity sciences are sufficiently open to embrace many different approaches, for part of the very definition of complexity is the idea that no one approach is predominant (Miller and Page, 2007). Hence our insistence that what we are doing here is developing "a" science of cities, not necessarily "the" science.

We have quite forcibly defined our understanding and our design of cities in terms of science, and it behooves us to qualify our definitions in particular, since many approach the study of the city as an art form, while some see the wider context as one of humanism. As the reader already expects, our definition of science is somewhat catholic, in that the science we use here is one that embraces all systematically ordered knowledge. By science, we mean an organized body of knowledge produced using commonly agreed tools and methods that can be reproduced over and over again by different individuals. The knowledge that is produced may be interpreted differently, and different ways of applying the tools to produce or reproduce it may lead to variations, but the key criteria are explicitness and transparency. This is quite different from ideas and knowledge we consider art, since the production of art is individual and formed by an intuition that is personal. This is not to say that in science, intuition is not important; it clearly is at all stages, and it can make a difference in the production of knowledge, but its role is implicit. Cities are often seen as works of art, and the history of city planning has been dominated by a concern for "the city beautiful." This complicates our science, for it presupposes that good cities are a mixture of principles involving aesthetics, economic efficiency, and social equity, which in turn inevitably reflect a mixture of art and science. In fact, none of this is inconsistent with plotting a science of cities, for this science must be able to accommodate all viewpoints and provide a set of working tools in which any and every perspective can be accommodated.

The essence of the tools and methods that we will apply and outline in the following chapters lies in defining relationships—interactions—between the great array of components that we need to take into account in understanding the city. Our mantra is that interactions lie below locations, and that location is a function of interaction; indeed, that traffic is a function of land use was a key concept in city planning more than a half-century ago (Mitchell and Rapkin, 1954). To this end, we will introduce tools to explore and understand cities that are based on the dualism in interaction between flows and networks, which are different sides of the same coin. In the next chapter, we will begin by examining flows, which are somewhat independent of the networks that provide their physical structure. Our flow theory will be based on spatial interactions, and we will begin by laying bare the structure that has been developed for modeling transportation of different kinds. We will also describe how flows can be visualized and recount a history that identifies flows as a significant explanatory basis for understanding how cities function in terms of their local dynamics. This will set the scene for chapter 3, which takes a much more physical view of relationships that define the structure of cities. Here we will focus on networks, but before doing so, relate networks to systems of relationships between individual populations, locations, and their various attributes, showing how this multivariate view of the world leads naturally to the idea of networks. Once we have elaborated these ideas, we will have set the scene for part II of this book. There we will launch into applications to the many different phenomena that characterize the city, showing how models and methods to understand urban structure and dynamics can be seen as a natural consequence of flows and networks.

2

Ebb and Flow: Interaction, Gravity, and Potential

Moving elements in a city, and in particular the people and their activities, are as important as the stationary physical parts.
—Kevin Lynch, *The Image of the City* (1960, p. 66)

Analogies between cities and machines and cities and organisms provide immediate characterizations of the overall structure and dynamics of urban form. Machines illustrate an intricate clockwork composed of moving parts that click together to provide a synchronized harmony of form and process, one that is static in conception but built, as we illustrated in the previous chapter, in a strict hierarchical manner. In contrast, organisms based on connections that manifest themselves as networks to deliver flows of energy and sustain their moving parts cannot span the entire space in which the organism resides, but strictly self-organize the structure into hierarchically efficient systems of delivery. Analogies between the city and the human body, in which the city center is the heart that provides the motor to pump energy (in the form of traffic) on a hierarchically organized system of links that move goods and people back and forth to production sites go back hundreds if not thousands of years, certainly to the time of Leonardo da Vinci and thence to the Greeks. Victor Gruen's book *The Heart of Our Cities* (1964) is but one of many images of this metaphor, which builds on the notion of organism without specifically developing it in other than casual terms. He said: "I can visualize a metropolitan organism in which cells, each one consisting of a nucleus and a protoplasm, are combined into clusterizations to form specialized organs like towns" (p. 271). Pictures of cities as systems of connection that sustain this image provide excellent examples of the skeletal structure that dominates the way we will articulate our argument in this and the next chapter, where we will deal first with flows and then with both their physical and topological embodiment as networks.

The notion of flows and networks provides a dual conception that places each of these on different sides of the same coin (Lambiotte et al., 2011). In fact, there is somewhat of a disjunction in science between flows and networks, for each can be discussed without specific reference to the other. Although we will follow this distinction here in beginning to develop ideas about how cities are constellations of flows spanned by overlapping networks, many visible but an increasing number hidden from physical view, we will seek to synthesize flows with networks and vice versa in the many perspectives on the city that we develop in parts II and III. In our conception, flows are much more to do with processes that operate across spatial and temporal scales, and betray a dynamics of change that drives the functions of the city in the short term and transforms it over the long term. Networks in the first instance are generally the physical containers whose capacity constrains flows of energy and information, manifested as materials, people, or ideas, their physicality existing in real Euclidean space but often hidden from our immediate senses as in cyberspace or information space. Networks might be constrained to the geometry of the line, the plane, or the volume but equally, as in the case of social networks, they may have no such bounds. This complicates their role when it comes to cities, where our focus is still very much one of physicalism.

More at issue here, however, is that flows add up to what happens at locations: flows are summations or integrations—syntheses—of what happens at locations, and in their simplest form imply mere accounting but in deeper terms may also imply a whole that is "more than the sum of its parts." Flows are also forces whose summation or integration generates potentials that pertain to what goes on in locations. This is the paradigm we will exploit here, for it underpins the basis of spatial interaction modeling. A similar logic accords to networks in that the strength of their hubs or nodes is a simple function of their links, at least at physical or topological levels. In this chapter, we first provide a vocabulary and notation for representing generic flows between locations. In fact, this framework can be used to represent any set of distinct and unambiguously defined objects that define the system of interest, although our focus here will be on spatial locations. Our representations will provide the algebra for subsequent simulations, but it also enables us to generate various visualizations of the complexity provided by flow systems. These in fact reveal, as we noted in the last chapter, signatures of the complexity of city systems.

Once our notation has been fixed, we examine some of the history of visual representations, and this prepares us for exploring the simplest models of flows—gravitational models that form the basis of spatial interaction theory. For the last half-century, these models have formed the basis for simulating transport, migration, and other material flows that move people and ideas around cities and regions and larger entities such as the globe. We then describe the standard framework for generating consistent models, which can be derived using information theory, and this enables us to generalize and adapt these models to many different kinds of spatial systems. Cities, in fact, manifest an essential tension in terms of flow patterns—

pulling people and material and information into their cores (centralization) and pushing these same activities out to their edges in symmetric fashion (decentralization), but setting up their own asymmetry that we will exploit in examining flows, for which obvious asymmetric patterns are intrinsic. We then move on to examining these asymmetries in their own right, as well as adapting these models to show departures between symmetry and the asymmetry of the reality. We will conclude with some ideas about what these essential static conceptions of flows imply for longer-term change. This will set up our focus upon networks—in their most simplistic terms as the containers of these flows—which form the subject matter of chapter 3.

Representing Flows

Flows and Locations

We will first classify sets of places that are locations in terms of those defining flows that originate in places and those that are destined for those same or a different set of places. We call the volume of activity at the start of a flow the origins, O_i, which we define for a set $i = 1, 2, \ldots, I$, where I is the total number of such places and the volume of flow that is destined for a place D_j, where $j = 1, 2, \ldots, J$, and where J is the total number of destinations. Note that I need not be the same number (or set) as J, but in many applications $I = J$. In such cases, we will define the number of origins and destinations as N where the same set of locations is supposed, although the activity volumes are usually different. In our subsequent treatment and without loss of generality, we will assume $N = I = J$ unless we state otherwise, and thus when we come to sum flows with respect to locations, the range of summation will be implicit over N.

We define a set of flows between origins and destinations as T_{ij}. This notation reflects "trips" in transportation modeling but is much more generic in definition than this kind of routine transport. The most basic accounting relations can be stated as

$$O_i = \sum_j T_{ij}, \tag{2.1}$$

and

$$D_j = \sum_i T_{ij}, \tag{2.2}$$

where the total sum of flows T is formed from the general summation that links flows to origins and to destinations as

$$T = \sum_i \sum_j T_{ij} = \sum_i O_i = \sum_j D_j. \tag{2.3}$$

We made the point in chapter 1 that for any given pattern of locations, there are many trip patterns that might satisfy these; in short, this is the problem we face in cities when trying to explain a pattern of locations in terms of a pattern of flows when we have little data on the flow patterns themselves. In fact, as we will see, we will generate models of such flow systems that maximize the probability one pattern of flows rather than any other is likely to be the one we observe. The reason we shift from locations to flows is precisely because there are multiple explanations of location patterns in terms of flows, and we need to know which one is the most appropriate.

We can get some sense of this quantitatively from the fact that the representation we have set up in equations (2.1) to (2.3) is sometimes referred to as the "transportation problem." This problem is usually posed as one where we have costs of transport from origin to destination defined as c_{ij}, and we need to seek a trip distribution $\{T_{ij}\}$ that minimizes

$$C = \sum_i \sum_j T_{ij} c_{ij}. \tag{2.4}$$

We can work out that if all the flows are greater than zero and of integer value, then there are at least $N!$ feasible solutions—that is, assignments of T to the link pairs ij that meet the constraints on origins and destinations given in equations (2.1) and (2.2) (Doig, 1963). For example, in the small problem we illustrate below, in which we have flows from places of employment to places of residence for the $N = 33$ municipalities that make up Greater London, there are $N^2 = 1,089$ possible flows, including flows from a zone to itself, but there is of the order 10^{37} numbers of possible trip distributions that can give rise to the observed distribution of employment and residential population over these zones. This is as good an argument for focusing our science on flows rather than locations, on networks that deliver people, materials, and information to places rather than on patterns of locations themselves. Furthermore, although most of these possible distributions are likely to be implausible, there are still many that can give rise to the same pattern of locations. Our quest is to gain an understanding of cities that takes account of this kind of equifinality—that is, the same distributions can be arrived at in different ways, and when we look at cities as locations, there are many competing views of how such distributions might arise using quite plausible but different patterns of connection.

Flow Systems: Desire Lines, Network Flows, and Potential Fields
The conundrum about explaining flows in terms of locations extends immediately to ways in which we might understand their complexity. The usual method of visualizing such flows is by plotting directional vectors from origins to destinations or

vice versa, with the flow volume set equal to the width of the vector. These are referred to as "desire lines" in transportation modeling, but it is immediately obvious that the greater the number of origins and destinations, the number of lines that cross increases as approximately the square of the number of locations N^2, and our ability to discriminate pattern from such flow maps becomes increasingly difficult, with much attention having to be focused on presentation. To give an idea of this problem, we will use the 2001 journey from work to home matrix between the thirty-three municipalities comprising Greater London, setting the threshold for plotting these at different levels, thus successively thinning the plot with a view to extracting the dominant pattern. In fact, we form the aggregate matrix $T_{ij} + T_{ji}$, ignore the self-flows, that is T_{ii}, and plot the resulting aggregate pattern where the number of significant cells of the matrix is $N(N - 1)/2$. The maximum value in this matrix is 39,411 and the minimum 7. In figure 2.1, we show a selection of plots from $T_{ij} \geq T_{k\ell}(\text{min})$ to $T_{ij}(\text{max})$ for which we clearly mark the average position $T_{ij} \geq \overline{T} = 5505$, which is defined as

$$\overline{T} = 2 \sum_{i} \sum_{j>i} (T_{ij} + T_{ji}) \Big/ (N(N-1)). \tag{2.5}$$

The problem with this kind of diagram is that it is hard to draw so that salient patterns are easily and effectively revealed. They end up a mess, although animation (which we cannot show here, but is implied by figure 2.1) does help reveal hidden structure.

Here we avoid displaying more complex structures at higher levels of resolution for Greater London (which we do explore in later chapters) because these would become an even greater tangle of lines. If we were to try and represent two way flows—from i to j and back—the problem of visual representation gets even thornier, although there has been some progress, as we will show a little later (Wood, Dykes, and Slingsby, 2010; Wood, Slingsby, and Dykes, 2011). Historically, this may be the reason why this kind of flow map hardly appeared before the dawn of computers. The first known computer illustration was made in 1959 for the Chicago Area Transportation Study using a digital display device called the *Cartographatron* (Creighton, Carroll, and Finney, 1959).

Figure 2.2 shows some 9.9 million personal trips plotted in much the same manner as we show in figure 2.1, but it is clear this kind of visualization is simply impressionistic, not really revealing the more complex polycentric nature of Chicago, and instead simply emphasizing the monocentric nature of the city. Tobler (1987) remarks that the relative imprecision and fuzziness of this sort of display, due to the ability to defocus the device, is something that has generally been lost in more modern displays.

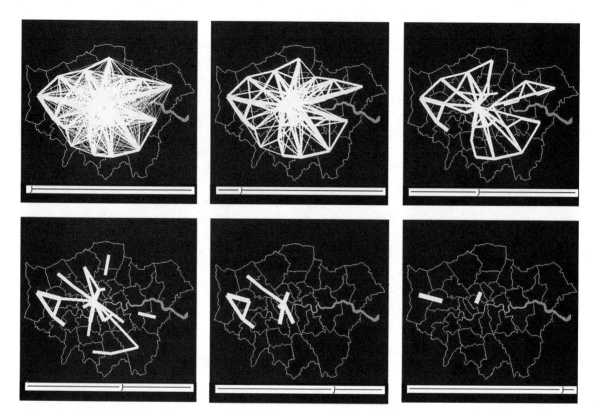

Figure 2.1
Visualizing aggregate flows by volume, from the full 33 × 33 matrix to the largest flows, with the upper middle frame set at the flow average.

Flow systems such as these can be visualized in many other ways, and there is an obvious simple extension that involves producing averages of flows at origins or destinations, or both. This involves producing a weighted average of flows in the form of a potential, which is then visualized as a vector whose direction is given by the average of all flow directions around the zone in question and whose magnitude is generated using the size of the flows. This vector field is strongly related to the notion of potential in that if the flows are seen as forces, then the average vector might be likened to the gradient of the potential field around the zone in question. To construct this for origin zones i and destination zones j, we first scale the flows T_{ij} and express them as a proportion P_{ij} of the total flow T as

$$p_{ij} = \frac{T_{ij}}{T}, \sum_i \sum_j p_{ij} = 1. \qquad (2.6)$$

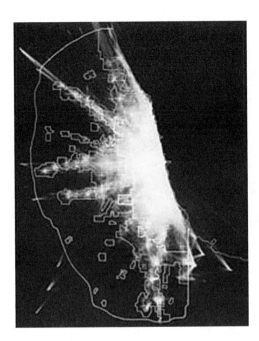

Figure 2.2
The earliest digital visualization of personal trips on the 1959 Cartographatron display device.

The coordinates of zones i, x_i, y_i, and j, x_j, y_j define the direction of flow, and we then produce a weighted average flow for each origin in terms of coordinate displacements:

$$\Delta x_i = \sum_j p_{ij}(x_j - x_i) \text{ and } \Delta y_i = \sum_j p_{ij}(y_j - y_i); \qquad (2.7)$$

and for destinations as

$$\Delta x_j = \sum_i p_{ij}(x_i - x_j) \text{ and } \Delta y_j = \sum_i p_{ij}(y_i - y_j). \qquad (2.8)$$

Our new vectors define directions of average flow for each origin and destination as $\vec{O}_i = [(x_i, y_i), (x_i + \Delta x_i, y_i + \Delta y_i)]$ and $\vec{D}_j = [(x_j, y_j), (x_j + \Delta x_j, y_j + \Delta y_j)]$. These are called interaction winds by Tobler (1976), who has done the most to exploit this kind of characterization. Several variants based on specifying different kinds of weights are possible, and we show these for the thirty-three locations defining Greater London in figure 2.3. One of the great advantages of compressing flow information into this form is that the various asymmetries in the system are revealed. In figure 2.3(a), where the origins are employment sites much more densely clustered than residential sites, the predominant movements are from the edge of the city to

A   B

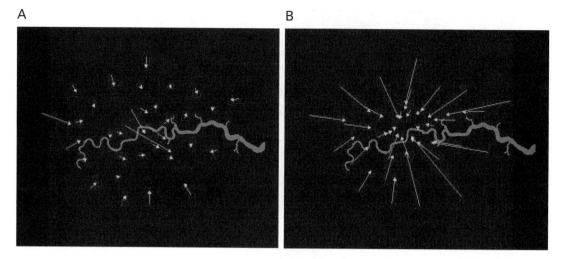

Figure 2.3
Directional flows from (a) origins to all destinations and (b) vice versa.

the center, whereas when we reverse the visualization to show movements from residential destinations to employment origins in figure 2.3(b), the average directions are somewhat different. But the structure is more complex than a simple reversal, as these pictures reveal.

A History of Representations

There are currently some dramatic advances in visual representation using various forms of digital media, but it is hard to escape the fact that information is inevitably lost when flows are reduced to the attributes of locations, as we did above in figure 2.3, and two-dimensional flow matrices are reduced to one dimension. There have been various attempts at simplifying flows without losing the notion of flow and direction. One method, popularized by Ernest Ravenstein (1885, 1889) who was one of the first to speculate on the simulation of movements between cities using migration, chooses only the most dominant flows across single boundaries to adjacent and contiguous regions, thus reducing the $N(N-1)$ flows (which exclude flows that exist within each region) to something on the order of $\sim 5N$, where we assume there is an average of four additional locations contiguous to a single one. We reproduce Ravenstein's map as figure 2.4; it is clear that that his "currents of migration," as he called the map, represent the dominant flow field that can be visually interpreted from the overall direction of flows. Such visualizations are useful where there is a dominant direction but are hard to use where there are countercurrents, vortices, and eddies in the pattern of movement. In this case, too, it is difficult to

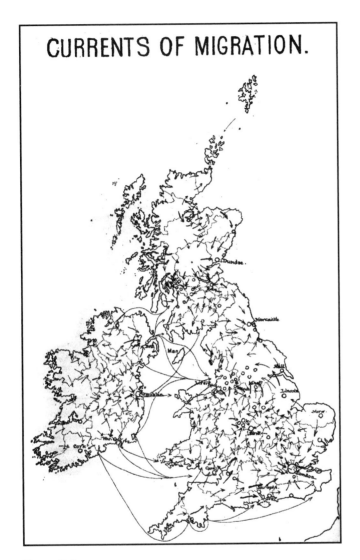

Figure 2.4
Ravenstein's (1885) Currents of Migration.

reconstruct precisely the logic Ravenstein actually used to produce his famous map (Tobler, 1995).

Two generations before Ravenstein, the German geographer Johann Kohl (1841), in a remarkable and prescient thesis on the function and structure of cities, speculated on the way urban functions were connected together through flows of activity that spanned space in a manner reminiscent of the fractal structures we introduced briefly in the previous chapter. The idea of activities flowing to the center and filling space in a parsimonious but efficient way dominates his approach, which is reflected in his many network diagrams of how transportation connects the city with functions and activities within its hinterland. Although we are mixing our spatial scales a little here, as well as mixing cross-sectional pictures of cities and regions with their flow dynamics, one of the best representations from Kohl's work is shown in figure 2.5. The

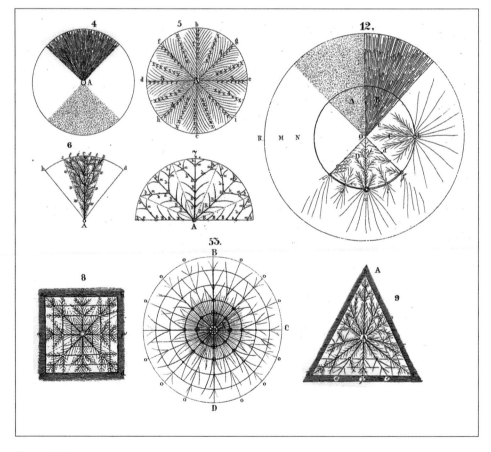

Figure 2.5
Kohl's (1841) flow fields.

parallel with force fields, migration fields, interaction winds, and the networks that dominate many of the ideas in this chapter and the rest of this book is striking, notwithstanding the fact that Kohl mysteriously left the reader to speculate on how such ideas were applicable to the way cities were structured. Thomas Peucker (1968), who first brought this work to our attention, says of Kohl that his aim, like Cerda and Geddes who came after him, was to establish a science of cities in geographical terms. Kohl hoped, says Peucker, "... no more and no less than that this book should alter the entire foundation of geographical science or rather that with it geography would gain a real basis and be marked as a true science" (1968). Despite the fact that his treatment might belong more to the following chapter on networks, the flow systems he established to illustrate his thesis appear synonymous with those that illustrate the flows and movements that form our focus here.

In fact, the dominant way in which flows have been and continue to be displayed involves aggregating them into directional fields akin to those proposed by Ravenstein, while providing more literal volumetric patterns. Such visualizations are akin to assigning the field of flows to the networks that contain them in geographical space, in much the same way that transportation engineers and planners, having predicted the distribution of trips between origins and destinations, assign these trips to the network. This, of course, is the way physical trips are bunched when the network is considerably more parsimonious (as it usually is) than the array of flows that represent our desires to travel. However, this is not possible for social networks, and thus collapsing flows to route assignment is not a general method for visualization in any case. Nevertheless, here we will provide a summary of the many flow visualizations that follow this path.

The first known map of such flows was developed by Henry Harness in 1837 for the British Army in Ireland for use in the Pale of Dublin and beyond (Robinson, 1955). In figure 2.6(a), we show one of these maps, which is based on the "passenger conveyance volumes" counted by the constabulary along the main highways. Harness was commissioned to do this work for the government, which was at the time considering the introduction of railways into Ireland, but what is noteworthy is that this appears to be the first known example of how traffic flows along physical networks. Following Harness, there was a flurry of such flow maps, such as Charles Joseph Minard's famous 1869 map of the decline and fall of Napoleon's army as it marched on Moscow from Paris between 1812 and 1813. Minard's map, which was popularized by Tufte (1983), is only one of many. In figure 2.6(b), we show another of his maps illustrating trade on a global scale, in which the network is implicit rather than explicit, as in the Irish maps.

Assigning flows to networks is one of the key stages in transportation modeling, where flow volumes from a distribution model—T_{ij}—are usually assigned to the shortest path between each origin i and its destination j, with volumes tending to

A   B

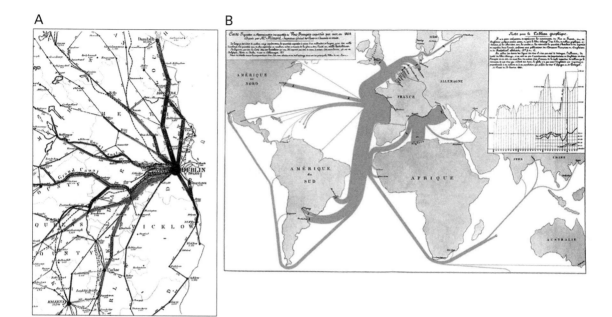

Figure 2.6
Flow volumes assigned to explicit and implicit networks. (a) Harness's 1837 Passenger Conveyance Map in the Dublin region. (b) Minard's 1864 Wine Flow Map: one origin (France) and many destinations collapsed into single streams.

build up as critical hubs of activity in the city (e.g., the CBD) are approached. These assignments can be complex, for they vary across the hierarchy of transport systems and vary in geometrical structure with mode, and contemporary methods tend to animate their function. As flow volumes usually vary during the working day, the capacity of the routes is often adjusted to show increases and decreases. As in all our representations, there are differing degrees of abstraction due to the routes on which flows occur and the scale of the volumes to be visualized. In figure 2.7(a), we show a typical flow map for peak-hour flows on the public bicycle scheme on the street system in Central London, where volumes have been assigned to the major street networks. Assignments to other modes such as bus, heavy rail, and subway are invariably more abstract and difficult to compare with one another in the same visual representation, but there are many related visualizations, such as compressing information into the origins of such trips in spatial tree maps as developed by Wood, Slingsby and Dykes (2011).

There are many flow maps currently being produced from a variety of spatial data. In an almost literal analogy with the flow of blood through the body, Cruz (2012) has developed an animation of the volume and speed of the flows for various traffic patterns in Lisbon; figure 2.7(b) shows a still from the animation.

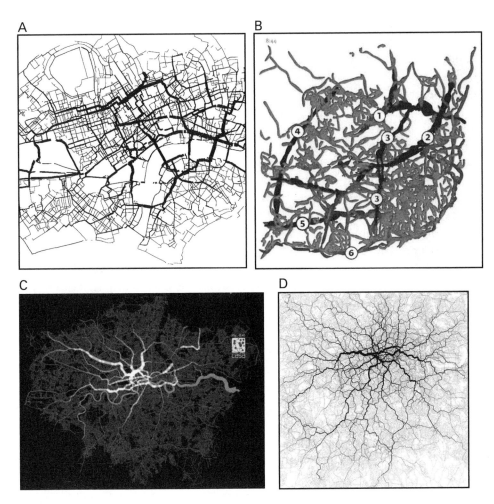

Figure 2.7
Assignments of flow to various network systems. (a) Assignment of public bicycles to the street network in Central London over a typical 24-hour period. (b) Lisbon's traffic in analogy to the flow of blood (from Cruz, 2012). (c) The flow of travelers on the London Underground system at the morning peak. (d) Pathways in Greater London based on location of tweets (from Fischer, 2012).

Reades (2012) illustrates the same kind of pulsing effects on the London underground using a similar animation, shown in figure 2.7(c). These pictures cannot do justice to the way in which such flow systems behave, and readers are referred to the Cruz (2012) and Reades (2012) websites (and to YouTube and Vimeo), where one can view these animations. There is now also interest in visualizing flows that can be captured using various digital sensing, in particular smart phone usage and location, as well as the transmission of contacts and messages from various computing devices. Locations of photographs, short text messages such as those associated with Twitter feeds, various kinds of geopositioning systems that track users and vehicles, data from smart cards that record travel and so on—all of these are providing new ways of capturing data about flows in the city. In figure 2.7(d), we show tracks from Twitter feeds in London generated by Fischer (2012) as part of a remarkable series of visualizations of such media in many large cities. These volumes are not sourced with any strong context of use, for the tweets that compose them cover everything that can be sent using short text messaging systems. Nor is the time period over which they are captured particularly significant, but as these kinds of flows come to dominate more and more of our knowledge of how people connect in cities, there will doubtless be many explorations of their usage for more considered purposes, for planning and design, management, and service delivery.

The assignment of flows to the routes they follow is often augmented by some systematic transformations or abstractions that clarify their complex interrelation. Figure 2.8(a) reflects our ability to morph such flow systems according to their volumes. The map of Budapest produced by Feles, Gergely, Bujdosó, Hajdu, and Kiss (2012) indicates how the importance of the street pattern can be systematically transformed according to the volume of the flow without losing the visual integrity of the set of relative locations. Cartograms are systematic map distortions that transform the map from the actual area to areas that reflect the density of points, making the areas proportional to the density. Dorling (2012) has explored many locational phenomena in this way, but he has also generalized this to interaction patterns. In figure 2.8(b), he shows how one might distort the map of daily commuting flows in England and Wales. Representations of flows do not have to follow the network system, and thus a bit of poetic license can be exercised. In figure 2.8(c), Austwick, O'Brien, Strano, and Viana (2013) color and smooth trip volumes associated with the morning peak network for the public bikes data that is shown in figure 2.7(a), following a similar method to that developed by Wood, Slingsby, and Dykes (2011).

There are many more examples of spatial flows that can be visualized using novel techniques, but we need to change tack now and consider how such spatial systems might be modeled as flow fields. Other visualizations that abandon the

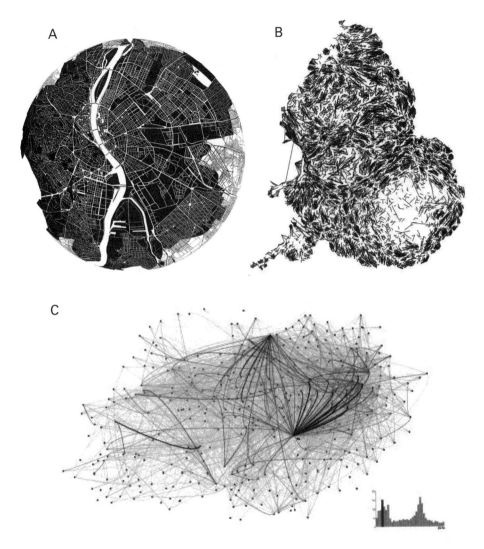

Figure 2.8
Distorting flows in various network systems. (a) The street network in Budapest (from Feles et al., 2012). (b) Daily commuting flows in England and Wales in 1981 in cartogram form. (c) Flows at the morning peak of public bikes in Central London.

spatial geometry of the map will be illustrated in later chapters as our focus changes toward social and organizational networks that describe ways in which cities function. In this chapter, we will return to these types of representations after we have explored how to model flows, for in modeling such systems, we are able to visualize more powerful concepts related to the change in such fields over space and time.

Elementary Models of Flows: Gravitation and Potential

The Basic Model

There is a simple model of spatial interaction that articulates flow as a force between the volumes of activity at any two locations, the origin i and the destination j, which is based on Newton's law of universal gravitation. We introduced this model in chapter 1 in equations (1.1) and (1.6), where we argued that gravitation is a good analogy to the way populations interact in diverse ways. In fact, Newton's law is a generalization of his second law of motion, which dictates that force is equal to the product of mass and acceleration, but with respect to two bodies i and j or nodes defined by their size (or population) P_i and P_j and their deterrence effect. This law can be restated from equation (1.6) as

$$T_{ij} = K P_i P_j d_{ij}^{-\phi}, \tag{2.9}$$

where, again, we define T_{ij} as the interaction or flow of energy between locations i and j, d_{ij} as some measure of deterrence such as distance between i and j, K is a scaling constant (the so-called gravitational constant in classical mechanics), and ϕ is a parameter that in the pure scaling case is usually 2, thus implying an inverse square law of distance. A variant of this model was first used by Ravenstein (1885, 1889) in formulating and testing his laws of migration.

One of the key points about this model is that it is symmetric in its structure if the distance or deterrence matrix $\{d_{ij}\}$ is symmetric; that is, $d_{ij} = d_{ji}$, $\forall ij$. In general, however, operational developments of these models have not detailed or exploited this symmetry, since to apply these models to real flow systems, distortions posed by multiple attributes of the nodes (for the nodes are actually geographical locations for the most part—points or areas) need to be directly considered. The most complete statement of these models is the family of spatial interaction models developed by Wilson (1970). He defines four variants, of which the first and most general is the model in equation (2.9). These variants constrain the predicted flows to various external locational information pertaining to the total observed activity in the system—the total interactions T; the total interactions associated with out-degree nodes, which are the origins i, O_i; and the totals associated with the in-degree nodes, which are the destinations j, D_j. These totals are defined as three different constraints on the matrix of flows $\{T_{ij}\}$, which we stated above in equations (2.1) to (2.3), and at the risk of belaboring the point, restate for the case of populations at origins and destinations as

$$P_i = \sum_j T_{ij}, \tag{2.10}$$

$$P_j = \sum_i T_{ij}, \tag{2.11}$$

and

$$T = \sum_i \sum_j T_{ij} = \sum_i P_i = \sum_j P_j. \tag{2.12}$$

Four generic types of model can be defined. First, the *unconstrained model* in which there are no constraints on origins or destinations—equations (2.10) and (2.11) do not apply—but the overall interaction activity is constrained by equation (2.12). In this case, there is only one constant for the entire system, which we will call $K(= K_i = K_j, \forall ij)$, and the gravity model in equation (2.9) is thus recovered. The second and third variants are sometimes called "location models," which are origin-constrained or destination-constrained with either equations (2.10) or (2.11) applying, respectively. The models that are derived have constants for either origins or destinations K_i or K_j, but not both, and these ensure that equations (2.10) or (2.11) are met. These models are defined by Wilson (1970) as *origin-constrained* or *destination-constrained* and are part of the broader class of "singly constrained" models. The fourth model in the family is one where both origin and destination constraints in equations (2.10) and (2.11) apply and both sets of constants K_i and K_j appear in the model. This is a called a doubly constrained or an *origin-destination-constrained* model. Note that in this family of models, in the first unconstrained model, flows on each link ij, flows originating at out-degree nodes i, and flows destined for in-degree nodes j are all to be predicted from the model. For the singly constrained models, one of these sets of nodes is constrained, while for the origin-constrained model, the activity flowing out to various destinations is constrained but not that flowing into the destinations, which is to be predicted by the model; and vice versa for the destination-constrained model. In the case of the doubly constrained model, there is no prediction of activities at origins or destinations, as both of these are given, and thus the focus is entirely on the distribution of trips. This latter model has traditionally formed the basis of trip distribution in traffic flow models prior to assignment of trips to the network, which is what we visualized in the previous section.

The Unconstrained Model: The Symmetric Baseline

The unconstrained model we will now write, following equation (2.9), as

$$T_{ij} = KP_iP_jd_{ij}^{-\phi} = T\frac{P_iP_jd_{ij}^{-\phi}}{\sum_i\sum_j P_iP_jd_{ij}^{-\phi}}. \tag{2.13}$$

This includes the constant K, which ensures that the overall constraints on total flows T in equation (2.13) are met. If the distance matrix is symmetric, as we will assume, then the model is symmetric in its flows and in the activities originating at each origin node and destined for each destination node. We can state this as follows by summing the flows explicitly over destinations, then origins, and this gives

$$s_i = \sum_j T_{ij} = KP_i\sum_j P_jd_{ij}^{-\phi} = KP_iV_i, \tag{2.14}$$

$$s_j = \sum_i T_{ij} = KP_j\sum_i P_id_{ij}^{-\phi} = KP_jV_j. \tag{2.15}$$

V_i, V_j are in fact population potentials—a little like potential energies or summations of forces (Stewart, 1941). In spatial systems, these reflect the relative propinquity, or nearness, of an origin to all destinations or of a destination to all origins—that is, accessibilities (Hansen, 1959)—with of course these being identical in the symmetric case, which is that implied by equations (2.14) and (2.15). They are also defined as accessibilities in land use transport planning and used to judge the potential of locations. As such, in comparison with the actual populations at origin or destination locations, they are a measure of how the real system differs from its symmetric idealization; that is, how s_i differs from P_i and s_j from P_j. In this sense, we define these potentials as the ratios $V_i = s_i/KP_i$ and $V_j = s_j/KP_j$.

We will examine the implications of the baseline symmetry in a later section and return to this in chapter 3, but it is worth noting that if we define the conditional probability

$$p_{ij} = \frac{T_{ij}}{P_j}, \text{ and } \sum_i p_{ij} = 1, \tag{2.16}$$

it is easy to see that the destination potential—equation (2.15)—can be written as a linear function of the origin potential. This is

$$\begin{aligned} s_j &= \sum_i p_{ij}s_i = \sum_i \frac{P_id_{ij}^{-\phi}}{V_i}KP_iV_i = KP_j\sum_i P_id_{ij}^{-\phi}, \\ &= KP_jV_j \end{aligned} \tag{2.17}$$

and the same holds for the reverse process where the conditional probability matrix is defined with respect to destinations rather than origins. These are, in fact, Markov processes, and we will exploit these ideas in chapter 3, while simply noting there that the link between these ideas and gravity modeling is explicit in the work of Smith and Hsieh (1997) and Bavaud (2002).

The Generic Flow Model

We now turn to the generic constrained model, which distorts the baseline to embody the correlations and idiosyncrasies of the real system. In what follows, we will focus on showing how close this doubly constrained model and its variants are to the baseline, because one very obvious extension of spatial interaction modeling is to consider how close the actual system is to some idealized form or structure or to some null hypothesis about the operation of such systems. We write the generic model as

$$T_{ij} = K_i K_j P_i P_j d_{ij}^{-\phi},$$

(2.18)

and it is immediately clear if we compare this to equation (2.9), our symmetric equivalent, that the factors K_i and K_j replace K, ensuring that the origins, destinations, and distance between them are distorted to enable the origin and destination constraints on flows to be met. In short, it is useful to think of any of the members in the family of spatial interaction models as being distortions or deviations from this symmetric equivalent, which we take as the baseline model.

We will now use constraint equations on origins and destinations in equations (2.1) and (2.2), which use distinct origin and destination volumes O_i and D_j to define the constants K_i and K_j from

$$\sum_j T_{ij} = O_i = K_i P_i \sum_j K_j P_j d_{ij}^{-\phi},$$

(2.19)

and

$$\sum_i T_{ij} = D_j = K_j P_j \sum_i K_i P_i d_{ij}^{-\phi}.$$

(2.20)

These are explicit functions of the origins and destinations, that is

$$K_i = \frac{O_i}{P_i \sum_j K_j P_j d_{ij}^{-\alpha}},$$

(2.21)

$$K_j = \frac{D_j}{P_j \sum_i K_i P_i d_{ij}^{-\alpha}}.$$

(2.22)

It is worth noting immediately that as we have differentiated origin and destination activity O_i and D_j from each other and from the population P_i or P_j, then the mass variables or attractors in the basic gravity model equation that are populations are transformed to the origin and destination activities by their respective constants. If we assume $O_i = D_i = P_i, \forall i$, then these variables cancel in the definition of the constants in equations (2.21) and (2.22) to give

$$K_i = \frac{1}{\sum\limits_j K_j P_j d_{ij}^{-\alpha}} \quad \text{and} \quad K_j = \frac{1}{\sum\limits_i K_i P_i d_{ij}^{-\alpha}}, \tag{2.23}$$

and the model can be written in terms of the origin-destination terms O_i and D_j in equation (2.9), noting of course that these are the same as the populations. There is one critical point which we will take up later, and that is that if the distance matrix is symmetric, using populations for origins and destinations, the symmetric model in equation (2.9) is recovered and the constants become $K_i = K_j = K, i = j$. All other results from the symmetric model follow, and we will return to this below.

Reverting to the nonsymmetric generic form, we can write this doubly constrained model in direct probability form rather easily if we note that

$$T_{ij} = O_i p_{ij} \quad \text{where} \quad \sum_j p_{ij} = 1, \tag{2.24}$$

and

$$T_{ij} = D_j q_{ij} \quad \text{where} \quad \sum_i q_{ij} = 1. \tag{2.25}$$

These equations then become

$$D_j = \sum_i T_{ij} = \sum_i O_i p_{ij} = \sum_i O_i \frac{K_i P_i d_{ij}^{-\phi}}{\sum\limits_k K_k P_k d_{ik}^{-\phi}}, \tag{2.26}$$

and

$$O_i = \sum_j T_{ij} = \sum_j D_j q_{ij} = \sum_j D_j \frac{K_j P_j d_{ij}^{-\phi}}{\sum\limits_k K_k P_k d_{jk}^{-\phi}}. \tag{2.27}$$

Before we move, later in this chapter, to examine the relationship between the generic and symmetric models, we will deal with the case of the singly constrained model explicitly, as this has some interesting simplifications from the generic case. Then taking only the origin-constrained model, which we write as

$$T_{ij} = K_i P_i P_j d_{ij}^{-\phi} = O_i \frac{P_j d_{ij}^{-\phi}}{\sum\limits_j P_j d_{ij}^{-\phi}} = O_i p_{ij}, \tag{2.28}$$

the constant K_i can be stated directly from equation (2.21), where we set $K_j = 1, \forall j$, giving

$$K_i = \frac{O_i}{P_i \sum_j P_j d_{ij}^{-\phi}}. \tag{2.29}$$

The singly constrained model in equation (2.28) can now be used to predict the activity locating in the destination zone that we call D_j'. In general, this will differ from the observed activity D_j because it is a prediction from the model. The transition relation can thus be written as

$$D_j' = \sum_i O_i p_{ij} = \sum_i T_{ij}, \tag{2.30}$$

implying an asymmetry in the reversible relation, which can be written directly from equations (2.26), (2.27) and (2.30) as

$$O_i = \sum_j D_j' q_{ij} = \sum_j \frac{K_i P_i d_{ij}^{-\phi}}{\sum_k K_k P_k d_{kj}^{-\phi}} D_j' = \sum_j \frac{T_{ij}}{\sum_k T_{kj}} D_j'. \tag{2.31}$$

We can, of course, develop the same structures for the case of a singly constrained model where we estimate the origin activities O_i' for given destination activities D_j.

Generating Consistent Flow Models

Our treatment of these models is already based on deriving consistency in allocation through strict accounting of interactions; that is, ensuring that flows sum to known totals, which act as constraints that the particular model must meet. Wilson's (1970) family of interaction models is structured in this fashion. However, there are more powerful frameworks for generating such models, and these involve setting the flow problem in an optimization framework where constraints and objective functions can be chosen accordingly. We have already had a glimpse of this at the beginning of this chapter when we examined the transportation problem and counted the number of possible solutions to variants of that problem.

One of the most widely used functions that describes variety in complex systems is the measure of information (or entropy) developed by Shannon (1948), which measures the degree of uncertainty in the probabilities of different components that make up the system in general. If there is a probability of each event occurring, and we define this probability of a flow occurring between an origin and a destination as p_{ij}, we can take the logarithm of each probability $\log p_{ij}$ and form the weighted average H as

$$H = -\sum_i \sum_j p_{ij} \log p_{ij}, \sum_i \sum_j p_{ij} = 1. \tag{2.32}$$

If each probability p_{ij} is the same as any other, and then one of these events actually occurs, this gives a maximum of information, and it can be easily confirmed that, in this case, $H = \log M$, where M is the total number of trip probabilities and, in this notation where the number of origins and destinations are the same, $M = N^2$. In the case where only one of these probabilities is positive and equal to unity and all the rest are zero, if the event occurs then no information is gained and $H = 0$. The function in equation (2.32) satisfies these limits, and thus it is a suitable measure for describing the structure of systems that can be fashioned in these terms.

We can now generate the family of spatial interaction models by placing constraints on the origins, destinations, total flows, and other relevant information, such as the cost of travel. We will develop these models using this probability formalism rather than explicit constraints associated with flow volumes, but we can recover all our earlier models by noting flows T_{ij} can be generated by multiplying the appropriate probability by the total flow volume T. The most constrained system is where we assume that all the interactions originating from any zone i must sum to the probability p_i of originating in that zone, and all interactions destined for zone j must sum to the probability p_j of being attracted to that destination zone. The constraint that these origin and destination probabilities sum to 1,

$$\sum_i \sum_j p_{ij} = \sum_i p_i = \sum_j p_j = 1, \tag{2.33}$$

is consistent with the entropy measure in equation (2.32), but equation (2.33) is redundant with respect to the origin and destination normalization constraints, which can be stated explicitly as

$$\left.\begin{array}{l} \sum_j p_{ij} = p_i \\[2mm] \sum_i p_{ij} = p_j \end{array}\right\}. \tag{2.34}$$

The most critical constraint involves generating models that scale in some manner with distance or travel cost c_{ij}, and the usual way to achieve this is given as

$$\sum_i \sum_j p_{ij} c_{ij} = \bar{C}, \tag{2.35}$$

where $\bar{C}$ is the average travel cost.

The model derived from the maximization of equation (2.32) subject to equations (2.34) and (2.35) is

$$p_{ij} = K_i K_j p_i p_j \exp(-\gamma c_{ij}), \tag{2.36}$$

where K_i and K_j are normalization constants associated with equations (2.34), and γ is the parameter on the travel cost c_{ij} between zones i and j associated with equation (2.35). It is easy to compute K_i and K_j by substituting for P_{ij} from equation (2.36) in equation (2.34), respectively, and simplifying. This yields equations

$$\left. \begin{array}{l} K_i = \dfrac{1}{\sum\limits_j K_j p_j \exp(-\gamma c_{ij})} \\[2em] K_j = \dfrac{1}{\sum\limits_i K_i p_i \exp(-\gamma c_{ij})} \end{array} \right\}, \tag{2.37}$$

which are formally identical in structure to the doubly constrained gravity model constants in equations (2.21) to (2.23). A particular consequence of this maximization is that the function of travel cost or distance is expressed not as a pure scaling law, as we suggested in chapter 1, but as a negative exponential, which is a more adaptable form of function, as we will see elsewhere in this book.

As already noted, these models can be scaled to deal with real trips or population simply by multiplying these probabilities by the total volumes involved, T for total trips in a transport system, P for total population in a city system, Y for total income in a trading system, and so on. This system, however, forms the basis for the family of interaction models that can be generated by relaxing the normalization constraints; for example, by omitting the destination constraint, $K_j = 1, \forall_j$, or by omitting the origin constraint, $K_i = 1, \forall_i$, or by omitting both, where we need to invoke the total normalization in equation (2.33) to provide an overall constant K. It is worth noting that these models can also be generated in nearly equivalent form using random utility theory, in which they are articulated at the level of the individual rather than the aggregate trip-maker and are known as discrete choice models (Ben-Akiva and Lerman, 1985).

We will examine only one of these models, a singly constrained model where there are only origin constraints. This might be a model used to predict interactions from work to home, such as the one we will explore in chapter 9, given that we know the distribution of work at the origin zones. Then, noting that $K_j = 1, \forall_j$, the model is

$$p_{ij} = K_i p_i p_j \exp(-\gamma c_{ij}) = p_i \frac{p_j \exp(-\gamma c_{ij})}{\sum\limits_j p_j \exp(-\gamma c_{ij})}. \tag{2.38}$$

The key issue with this sort of model is that not only can we predict the interaction between zones i and j, but we can predict the probability of locating in the destination zone p_j', that is

$$p'_j = \sum_i p_{ij} = \sum_i p_i \frac{p_j \exp(-\gamma c_{ij})}{\sum_j p_j \exp(-\gamma c_{ij})}, \qquad (2.39)$$

from which we can derive the total flows in the destination zone as $D'_j = T p'_j$.

If we were to drop both origin and destination constraints, the model becomes one that is analogous to the traditional gravity model we began with in equation (2.9). However, to generate the usual standard gravitational form of model in which the "mass" of each origin and destination zone appears, given by P_i and P_j respectively, then we need to modify the entropy formula, thus maximizing

$$H = -\sum_i \sum_j p_{ij} \log \frac{p_{ij}}{P_i P_j}, \qquad (2.40)$$

subject to the total normalization on probabilities in equations (2.32) or (2.33) and a constraint on the average "logarithmic" travel cost $\overline{\ln C}$,

$$\sum_i \sum_j p_{ij} \ln c_{ij} = \overline{\ln C}. \qquad (2.41)$$

The model that is generated from this system can be written as

$$p_{ij} = K \, P_i P_j c_{ij}^{-\phi}, \qquad (2.42)$$

which is formally equivalent to the model we began with in equation (2.9). The effect of travel cost/distance is now in power law form, with ϕ the scaling parameter.

Exploiting the Asymmetry of Flows

Restating the Baseline Model
We will return to our discussion of symmetry in flow systems. This is an important aspect of the way movement occurs in a city system, as we have seen in patterns such as the journey to work that we explored earlier in this chapter. In fact, all spatial interaction models of the kind we have presented here are quasi-symmetric. Returning to the earlier formalism of explicit trips, attractors based on population, and deterrence based on Euclidean distance, we can normalize the generic doubly constrained model in a symmetric equivalent as

$$A_{ij} = \frac{T_{ij}}{K_i K_j} = \frac{P_i P_j}{d_{ij}^{-\phi}}. \qquad (2.43)$$

The way we have formulated this model enables us to see any flow matrix $\{T_{ij}\}$ in relation to any other through its various constants (and hence constraints), which are imposed on the basic symmetry. Equation (2.43) is symmetric not only in its summations over its origins and destinations but in its flows; that is, $A_{ij} = A_{ji}, \forall ij$. As the model is always symmetric independently of the value of the constants K_i and K_j, any model in the family can be seen in terms of the original symmetric model where potentials V_i and V_j occur quite naturally in terms of the origins and destinations. This is made clear using the definition of $\{A_{ij}\}$ in equation (2.43), from which

$$\sum_j A_{ij} = s_i = \frac{1}{K_i} \sum_j \frac{T_{ij}}{K_j} = P_i \sum_j P_j d_{ij}^{-\phi} = P_i V_i, \tag{2.44}$$

and

$$\sum_i A_{ij} = s_j = \frac{1}{K_j} \sum_i \frac{T_{ij}}{K_i} = P_j \sum_i P_i d_{ij}^{-\phi} = P_j V_j, \tag{2.45}$$

and is the most straightforward demonstration that the weighted potential values for origins and destinations are symmetric. To generate the relevant flow models in the family, all we need do is note that if the global constant is required for the unconstrained model, then $K_i K_j = K$, and if the origin (or destination) constraints are required for the singly constrained models, then $K_j = 1$, and thus $K_i K_j = K_i$ (or $K_i = 1$ and $K_i K_j = K_j$).

The fact that the unconstrained gravity model in its basic form is symmetric has long been regarded as a problem in that most, if not all, flow matrices that are observed in real spatial systems are asymmetric. In fact, the purpose of much spatial interaction theory is to provide a rationale through information or constraints that are known or hypothesized in determining how such asymmetries are to be reflected in the model formulation. There are at least two other strategies for dealing with asymmetry, and we will note these here. The first is to make any asymmetric flow matrix symmetric using various plausible definitions, while the second is to decompose the asymmetric matrix into its symmetric and asymmetric equivalents, treating the symmetric one using some of the methods presented here. Much of the analysis that follows is due to Tobler (1976, 1981, 1983), who has produced a series of modified gravitational models that exploit these asymmetries. We will follow his example here.

Tobler's Models
The first method is to convert an asymmetric flow matrix $T_{ij} \neq T_{ji}$ into a symmetric one. Tobler (1976) argues that the asymmetry is likely to be due to asymmetries in perceived distances between origins and destinations in the basic gravity model. To resolve this, he suggests that the modeled flows be first added as

$$T_{ij} + T_{ji} = 2K P_i P_j d_{ij}^{-\phi}, \tag{2.46}$$

and then averaged in estimating a generic flow $\hat{T}_{ij} = \hat{T}_{ji} = [T_{ij} + T_{ji}]/2$ as

$$\hat{T}_{ij} = \frac{T_{ij} + T_{ji}}{2} = K P_i P_j d_{ij}^{-\phi}. \tag{2.47}$$

This is of course equivalent to the flow system that we visualized for Greater London earlier in this chapter, pictured in figures 2.1 and 2.3.

The averaged flow which is now symmetric can be explained by a symmetric model, but considerable information is lost in this process. Thus a wider logic for this averaging is based on a second method of decomposition. If we premise that the distortion to symmetry is associated with the distance d_{ij} in such a way that we can write a new distance, with r a global correction factor and ρ_{ij} a specific local correction to the actual distance, we can write the unconstrained gravity model as

$$T_{ij} = K \frac{P_i P_j}{d_{ij}^{\phi}} (r + \rho_{ij}), \tag{2.48}$$

where the value of the flow between j and i can now be written as

$$T_{ji} = K \frac{P_j P_i}{d_{ij}^{\phi}} (r - \rho_{ij}) = K \frac{P_j P_i}{d_{ji}^{\phi}} (r + \rho_{ji}), \tag{2.49}$$

noting the symmetry of distance and the fact that we assume $\rho_{ji} = -\rho_{ij}$. If we divide equation (2.49) into (2.48), we get

$$\frac{T_{ij}}{T_{ji}} = \frac{r + \rho_{ij}}{r - \rho_{ij}}, \tag{2.50}$$

from which it is easy to compute the local correction factor ρ_{ij} as

$$\rho_{ij} = r \frac{T_{ij} - T_{ji}}{T_{ij} + T_{ji}}. \tag{2.51}$$

Using equation (2.51) in (2.48) gives

$$T_{ij} = K \frac{P_i P_j}{d_{ij}^{\phi}} \left[r + r \frac{(T_{ij} - T_{ji})}{(T_{ij} + T_{ji})} \right], \tag{2.52}$$

and from equation (2.52), the baseline symmetry can now be written as

$$\frac{T_{ij} + T_{ji}}{2} = K \frac{P_i P_j}{d_{ij}^{\phi}}. \tag{2.53}$$

Using this in equation (2.52) produces the unique decomposition of an observed flow into

$$T_{ij} = \left[\frac{T_{ij} + T_{ji}}{2} \right] + \left[\frac{T_{ij} - T_{ji}}{2} \right], \tag{2.54}$$

its symmetric and skew-symmetric components. This provides a different basis from the one we sketched earlier for evaluating the degree of distortion in any real system from its symmetric baseline in the unconstrained gravity model.

We can devise various measures from this result. For example, based on squared deviations between respective flows, as $\Phi = [\Sigma_i \Sigma_j (T_{ij} - T_{ji})^2 / 2]^{1/2}$. As Tobler (1976) argues, the essence of this decomposition is to remove the symmetric part of the model from the data and to focus on explaining the residuals, which in this case is the skew-symmetric part. This follows the long-standing but largely unknown "method of residuals" articulated by Coleman (1964), who argued that it is the residuals that should be modeled in spatial interaction, not the actual flows (Batty and March, 1976).

There are other models that exploit symmetry in various ways. Dorigo and Tobler (1983) suggest that flows might be modeled as a linear sum of push and pull effects, which in our notation would be

$$T_{ij} = K \frac{P_i + P_j}{d_{ij}^{\phi}} = K \frac{P_i}{d_{ij}^{\phi}} + K \frac{P_j}{d_{ij}^{\phi}}. \tag{2.55}$$

This is clearly symmetric, with the added feature that it can be separated into specific, nonsymmetric push and pull effects, which we also show in equation (2.55). Dorigo and Tobler (1983) explore the properties of this model in some detail, particularly focusing on the separability of the push and pull effects, as well as suggesting forms that have nonlinear effects. This model can be subjected to all the routine analysis presented earlier, but few additional insights seem to come from this simplification. Lastly, it is worth noting that Alonso (1976), in his quest for a general theory of movement, proposed an interaction model that potentially has a symmetric baseline but for which he hypothesized that the attractivities, often taken to be the population mass terms or the origin or destination terms or some other locational variables, could be formulated as power functions. His model can be written as

$$T_{ij} = A_i^{\varphi} B_j^{\varsigma} t_{ij}, \tag{2.56}$$

where the terms A_i^{φ} and B_j^{ς} are combined attractivities that can also contain constants that ensure various constraints are met, and the parameters φ and ς are elasticities that reflect the agglomeration or scale economies implicit in the push-pull effects of spatial interaction. The interaction effect t_{ij} might also be a deterrent, and if it is a symmetric distance, then the symmetric baseline can be incorporated into this model structure. Various extensions to his model have been suggested (Hua, 2002; de Vries, Nijkamp, and Rietveld, 2000), but these are only concerned with proposing different explanations of the model attractors, rather than focusing on asymmetry per se.

Flow Potentials

When we examined the potentials in the form of interaction winds in an earlier section, we essentially produced weighted vectors averaging the directions of flow in the system with respect to origins and destinations. We demonstrated that where origins and destinations were very different, as for example in flows based on the journey to work, then the predominant directions of flow mirrored these differences, which in the case of strongly monocentric cities such as London led to flows away from the center with respect to employment origins and to the centers with respect to residential destinations. These were illustrated in figure 2.3. If we were to assume that origins were the same as destinations and that the flow matrix were symmetric, then this should generate identical patterns for origin and destination vector flows. In fact, our desire line flow maps in figure 2.1 assume this, as does Tobler's symmetric baseline gravity model in equations (2.46) and (2.47).

To demonstrate this average pattern based on symmetry, we form the symmetric "averaged" flow matrix as

$$\hat{T}_{ij} = \hat{T}_{ji} = [T_{ij} + T_{ji}]/2, \tag{2.57}$$

and then convert this average into a weighted probability flow matrix as in equation (2.6), which is now

$$\hat{p}_{ij} = \frac{\hat{T}_{ij}}{\sum_i \sum_j \hat{T}_{ij}}, \quad \sum_i \sum_j \hat{p}_{ij} = 1. \tag{2.58}$$

The coordinates of zones i, x_i, y_i, and j, x_j, y_j, which define the direction of the symmetric flow, are generated using displacements that have the same form as in equation (2.7), which we repeat here for origins,

$$\Delta \hat{x}_i = \sum_j \hat{p}_{ij}(x_j - x_i) \text{ and } \Delta \hat{y}_i = \sum_j \hat{p}_{ij}(y_j - y_i), \tag{2.59}$$

and for destinations as

$$\Delta \hat{x}_j = \sum_i \hat{p}_{ij}(x_i - x_j) \text{ and } \Delta \hat{y}_j = \sum_i \hat{p}_{ij}(y_i - y_j). \tag{2.60}$$

Using equations (2.57) and (2.58), it is easy to show that origin and destination vectors based on these displacements are identical; that is,

$$[(x_i, y_i),(x_i + \Delta \hat{x}_i, y_i + \Delta \hat{y}_i)] = [(x_j, y_j),(x_j + \Delta \hat{x}_j, y_j + \Delta \hat{y}_j)], \tag{2.61}$$

as is clear from the map computed for Greater London from these equations illustrated in figure 2.9(a). It is instructive to note the predominant symmetric direction of flow, which is probably dominated by residential rather than employment

A

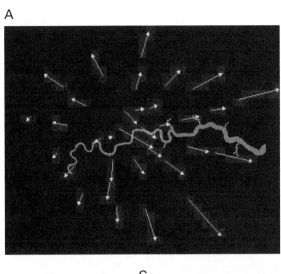

B 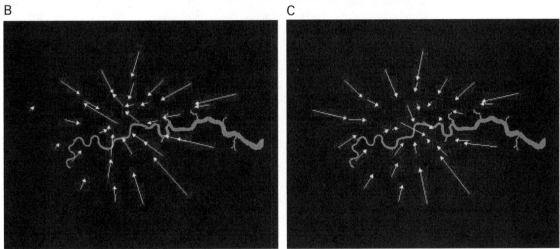 C

Figure 2.9
(a) Directional symmetric flows. (b) Asymmetric flows from origins. (c) Asymmetric flows from destinations.

locations, mirroring our intuition that the major direction of flow in London is from suburb to center.

Tobler's (1981) second model, which we will introduce briefly, takes the opposite approach and works with the asymmetric flow matrix formed from

$$\tilde{T}_{ij} = -\tilde{T}_{ji} = T_{ij} - T_{ji}, \sum_i \sum_j T_{ij} = 0. \tag{2.62}$$

Normalizing $\tilde{T}_{ij}$ is more difficult because these asymmetries can be both positive and negative, although this difference in sign is best seen as one of direction rather than absolute difference. Therefore we can normalize equation (2.62), forming the weight, not the probability, as

$$w_{ij} = \frac{\tilde{T}_{ij}}{\sum_i \sum_j |\tilde{T}_{ij}|}, \sum_i \sum_j w_{ij} = 0. \tag{2.63}$$

The coordinates for origin and destination zones now differ and are generated using displacements that have the same form as in equation (2.7), which we write as

$$\Delta \tilde{x}_i = \sum_j w_{ij}(x_j - x_i) \text{ and } \Delta \tilde{y}_i = \sum_j w_{ij}(y_j - y_i), \tag{2.64}$$

and for destinations as

$$\Delta \tilde{x}_j = \sum_i w_{ij}(x_i - x_j) \text{ and } \Delta \tilde{y}_j = \sum_i w_{ij}(y_i - y_j). \tag{2.65}$$

We have computed the vectors for both origin and destination asymmetries in the greater London flow matrix, and these are shown in figures 2.9(b) and (c), respectively. The gradients for the origins are away from the center, while for the destinations are toward the center, as we might expect from this kind of monocentric city system.

It is worth noting Tobler's (1981) specific model of asymmetric flows, which he argues should be modeled using the traditional symmetric gravity model in equation (2.13). This simulates the flow T_{ij} as

$$\tilde{T}_{ij} = T_{ij} - T_{ji} = KP_iP_jd_{ij}^{-\phi} - KP_jP_id_{ji}^{-\phi}. \tag{2.66}$$

If the model is accurate, then there is no asymmetry and $\tilde{T}_{ij} = 0$. If we sum equation (2.66) over the zones j, we get

$$\Delta_i = \sum_j \tilde{T}_{ij} = \sum_j T_{ij} - \sum_j T_{ji} = K(P_iV_i - P_iV_i). \tag{2.67}$$

Now as we might expect, the potentials may not be equal to one another with respect to origins and destinations and the net difference at origins (or destinations) Δ_i is not equal to zero. Tobler (1981) argues that we can approximate equation (2.66) with a new but symmetric attractor for origins and destinations that we call A_i and A_j, which we can write as

$$\tilde{T}_{ij} = G\left(A_i d_{ij}^{-\varphi} - A_j d_{ji}^{-\varphi}\right). \tag{2.68}$$

If we now sum equation (2.68) over j (or i), we can write the net difference Δ_i as

$$\Delta_i = \sum_j \tilde{T}_{ij} = G(A_i \sum_j d_{ij}^{-\phi} - \sum_j A_j d_{ji}^{-\phi}). \tag{2.69}$$

It is possible from the net differences to solve for the attractors A_i and to then plot these attractors and examine their gradients. This is a considerably more formal extension of the notion of interaction winds than the one we have illustrated here, but in general, it is possible to approximate the kind of directional vectors that we have computed from data in figures 2.3 and 2.9 using solutions to equation (2.69). Although we have not calibrated our models to the data in this chapter, once we do so—as in chapter 9—we will then be able to generate directional vectors from predicted flows based on variants in the family of models presented extensively in this chapter.

From Flows to Networks

Flows are the raw material on which activities at locations are built in the sense that in cities, they imply ways in which energy and information in their widest sense are distributed to sustain the functioning of the urban fabric. In this chapter, we have presented the theory of spatial interaction, which is the best-developed theory of flows we currently have for cities. However, it is a theory about how flows take place largely across space, often at discrete cross-sections in time. Although flows always imply a dynamics of change, for it takes time for a flow to occur, our focus here has been very much on space, and insofar as the flows in these models might pertain to time in terms of migration and trade, for example, the focus on their dynamics is implicit rather than explicit.

One factor that will continue to dominate our approach is the fact that flows (and networks) are difficult to represent unambiguously. Plotting flows and networks always leads to a clutter that at worst simply obscures meaningful patterns and at best tends to remain ambiguous, even for the most innovative kinds of visualization. In fact, representing flows as averages in some sense, as attributes of locations that might be suitably weighted aggregates of interactions, is a favored strategy we have adopted here, and there are many others that are currently being explored. For example, Wood, Dykes, and Slingsby (2010) favor tree maps that compress many interactions into small multiples that can be nested in terms of the original map, whereas Rae (2009) favors aggregates of flows to make out differences in locations, letting the user interpolate meaningful information from such collections of data. Despite some ingenious methods of unraveling graphs that have been developed in network theory, examples of which we will introduce in the next chapter, progress has been slow, and when flows are added to networks, methods for representation become unwieldy. However, there is considerable progress in this field of visualization at present, and new insights are being developed all the time. We will avail ourselves of some of these as our science develops in subsequent chapters.

In the next chapter, we will switch our viewpoint rather abruptly, forgetting flows but examining the networks on which such flows invariably take place. In one sense, this disjuncture is part of the way science has developed, and insofar as there has been any convergence of flow and network theory, then it is through the notion that flows might represent the weights on the arcs of a graph. In fact, there has been very little concern in seeing how networks might be embedded into spatial interaction flow theory, largely, one suspects, because the articulation of flows has been slightly more abstract than the level at which networks are represented. Desire lines, for example, although embedded in Euclidian space, do not have physical presence when they are plotted, generated, or modeled, but only when they are assigned to a network, something that takes place in transportation planning late in the process of simulation.

In what follows, we will first stand back from networks and examine the more generic problem of connections and relations that define how the different elements of cities relate to one another. Indeed, as this has rarely been pursued explicitly, we will interpret much multivariate and pattern analysis as a particular case of relational analysis, and we will thus focus on how such relations come to be defined to show how elements that make up the city are related. These may or may not have explicit flow and network characterizations, but in developing our science this way, we will focus much more on the notion of connections and correlations, correspondences and coordinations than on networks per se, only introducing networks once we have developed this approach. When we have achieved this, we will be in a position to say how these basic tools and approaches can be used to unravel the functioning of cities and generate insights into their structure and dynamics.

3

Connections and Correlations: The Science of Networks

And if this age, *the connected age*, is to be understood, we must first understand how to describe it scientifically; that is, we need a science of networks.
—Duncan Watts, *Six Degrees: The Science of a Connected Age* (2002, p. 14)

The implication that flows and networks are opposite sides of the same coin, that one cannot exist without the other, needs to be finessed, for they imply rather different perspectives on how we might represent and simulate connections and interactions within cities. In short, each have a very distinctive structure. In the last chapter, we assumed that flows could exist between any two locations, whereas this is rarely the case with networks, largely because their spatial patterning in cities implies some kind of embeddedness in physical space. While the structure of flows depends largely on how they constitute activity at locations, the structure of networks depends more on how locations are connected (or not) rather than on the size of the nodes that constitute their origins and destinations. We will begin at the point where we left off in the previous chapter, first illustrating how flows can be underpinned by networks, then changing our means of representation and introducing rather different measures of structure. This will open up a collection of methods for representing relations between the basic objects that comprise our cities. Here we will seek to develop a generic approach to relations and connections that enables us to build different kinds of networks between urban objects or components, of which we will define many types.

Throughout this book, we define cities in terms of elements or components that we regard as *irreducible objects* that compose urban structure and are endowed with behaviors that incorporate the way the city functions. These objects are sometimes referred to as "agents" and, for the most part, we will consider these as locations, activity types, individuals, or aggregates of population, all of which have some distinct purpose. Such objects will have attributes that make them distinct, and relations or connections that exist as flows or networks linking objects to one another, often with respect to their different attributes. Flows and networks can be rooted in physical spaces as sets of interactions or transactions between locations, or as links between different nodes, with planar graphs often defining their associated networks. Or they can be more abstract, not rooted in Euclidean space but in social space, where their topology defines the relevant graphs that are used in their representation. Relations can be within a single set of objects (as *simple graphs*) or between two different sets (as *bipartite graphs*), with higher-level relationships between three or more sets of different objects usually being reduced to networks between two sets of differing objects linked in sequential fashion. All of these ideas will be clarified in this chapter, where our focus on networks is underpinned by our desire to express relations or connections as an essential basis for our understanding of cities.

We will begin by showing how flow systems such as those portrayed in the last chapter have a natural equivalent in terms of networks (Lambiotte et al., 2011). We will then argue that such flow systems can be approached through ideas about relations and correlations, and this will provide us with a logic for constructing networks, rather than simply observing that they exist independently in their own right. They do, of course, and many of our examples will start from this assumption, but the notion that we might construct and derive networks from sets of relationships will provide us with a powerful way of thinking about cities. Having introduced the key concept of the bipartite graph, we will then examine the simple statistics of such graphs in terms of their type and connectivity. This will lead us to defining processes on graphs, in particular looking at the way the states that define the hubs or nodes—or origins or destinations of graphs (and flow systems)—define processes in which such states change as a function of the flows or actions that define links between the system's elements. We will then introduce several simple ideas about how cities are structured in terms of planar graphs, and conclude with their generalization to topological graphs. In no sense is this intended to be a comprehensive treatment of networks, for we will pick and choose what we need to progress our argument. Readers are encouraged to explore the field of network science more generally, which is basic background to the ideas presented here (Newman, 2010), but we do not offer any substitute for this.

From Flows to Networks

We will begin with the flow matrix T_{ij}, which records movement and which in the case of planar graphs might be embedded in two-dimensional Euclidean space, between origins i and destinations j. Let us throw away space for the time being and consider these origins and destinations as objects of interest, such as individuals, groups, or activities, that may be spatially grounded but in general are not. Just as we plotted these movements as desire lines in figure 2.1, we illustrate them in the same way here, but without their location in space. We will do this using the convention that the nodes or hubs—origins and destinations—can be represented as points around a circle, with flow and network connections defining the particular circular mesh. Such a representation only works, however, when the origins are the same set of objects as the destinations, for if they are not, then we need to represent these using a bipartite graph where they are distinct. This we will do a little later, but for now in figure 3.1, we present the flows between all thirty-three municipalities in London based on the journey to work data that we plotted at different thresholds in figure 2.1.

In fact, we plot the symmetric flows, for it is difficult to get a good pictorial representation of the directed flows T_{ij} and T_{ji} for networks with more than about ten nodes. Frankly, most networks of any size are hard if not impossible to disentangle into effective-looking relationship graphs, and thus although our flow graphs built on the circle are somewhat arbitrary, in the absence of any very strong hierarchical or other clustering, we will represent them in this way. Thus figure 3.1 is based on symmetric flows S_{ij}

$$S_{ij} = T_{ij} + T_{ji}, \tag{3.1}$$

where we have imposed flow thresholds on the representation. We begin with the complete flow matrix plotted in figure 3.1(a); what we call a strongly connected matrix in figure 3.1(b), where we only illustrate flows greater than 11,000; and a disconnected matrix in figure 3.1(c), where the threshold is chosen to disconnect the hubs from one another, thus forming two or more separate systems. We are still working with flow systems, but the binary graphs that underlie these flows are simply plots of connections between the hubs, which we show to the right of the flow patterns in figure 3.1. Because the matrix is symmetric, these graphs are *undirected*. Were we to plot each individual flow, then the underlying graph would be *directed* and called a *digraph*. We will have recourse to use such graphs later in this book.

The three graphs shown in figure 3.1 are classified in the order shown as *completely connected, strongly connected,* and *disconnected* (Harary, Norman, and Cartwright, 1965). A completely connected graph is one in which every flow emanating from an origin or destination reaches every other directly. In a strongly connected graph, such flows reach every origin or destination directly or indirectly—that is, through intermediate nodes—and in a disconnected graph, at least one flow emanating from a node cannot reach another, effectively breaking the graph into two or more subgraphs. There is one other type we need to note, called a *weakly connected* graph. Such a graph does link every node with every other, but it is not possible in such graphs to find a directed path from every node to every other. In this sense, a weakly connected graph is something less than strongly connected, but for the most part in the models that we introduce here, our focus will be on

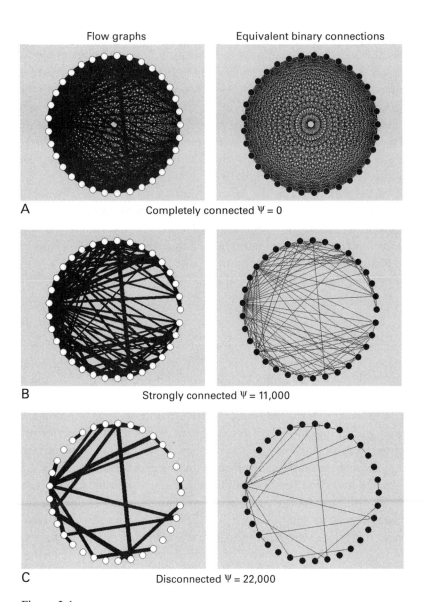

Flow graphs   Equivalent binary connections

A   Completely connected $\Psi = 0$

B   Strongly connected $\Psi = 11{,}000$

C   Disconnected $\Psi = 22{,}000$

Figure 3.1
Flows, connectivity, and networks at different flow thresholds.

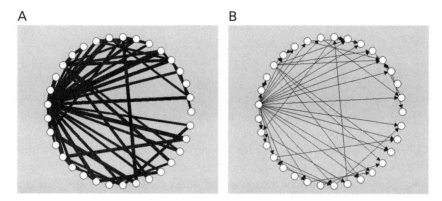

Figure 3.2
Directed flows and graphs illustrating a weakly connected system.

symmetric and strongly connected systems, largely because this kind of science is in its infancy and it is often difficult to be precise about the direction of connection between nodes.

From our example, we can threshold the flows to reveal a system that is weakly connected. This is based on trial and error; figure 3.2 shows an example. The graph is quite sparse, and the arrowheads show direction. We have set the threshold at about 38% of the maximum directed flow, and at this point all nodes are still part of the system but the number of paths through the graph is quite small. In fact, this is indirect evidence that this flow system is highly asymmetric. As we will see, flow systems of this kind have rarely been examined as graphs, thus pointing to the disjunction between flows and networks, which we will try to address as we continue to elaborate this science.

Our focus on connections as a more generic way of looking at movement and interaction immediately raises the question of how connected the sorts of structures we have been examining are. As we will see, there are many such measures, but it is worth examining an aggregate index to give some sense of how flow systems and their graphs change as they become less dense. The simplest measure is to simply count the links that exist in the graph and form the ratio with the total possible links. We will define a graph generically as $G(N, g)$ with nodes N and links g. This is in contrast to its associated flow system T, where g_{ij} exists if there is a flow T_{ij} from node (or origin) i to node (or destination) j. If we set the flow threshold at Ψ, then we define a binary relation on the graph as

$$g_{ij} = \begin{cases} 1, & if\ T_{ij} \geq \Psi, i \neq j \\ 0, & otherwise \end{cases}.$$ (3.2)

The connectivity $C(\Psi)$ is then defined as

$$C(\Psi) = \frac{\sum_i \sum_j g_{ij}}{N(N-1)}, \tag{3.3}$$

where it is clear that we are not counting in this particular measure the self-flows T_{ii}; hence, $g_{ii} = 0, \forall i$. The total maximum number of links is the sum of all the other possible links between the nodes, which depends on the number of nodes N. This measure, $C(\Psi)$, varies between 0 and 1, with 1 occurring when there are connections everywhere, as in the flow matrix in figure 3.1(a), and 0 occurring when there are no flows and the system is entirely disconnected—that is, when no node is linked to any other, the trivial case.

This is very simple, but it yields an immediate structural consequence that relates to the evolution of networks, and in fact to the dynamics of networks (Watts, 2002). If we gradually relax the threshold Ψ, the connectivity gradually reduces as fewer and fewer links in the underlying graph compose the system. We can identify points at which the system moves from a strongly to a weakly to a disconnected mode, and the path that this takes can be somewhat surprising. We will plot two measures for the London data, starting from a situation in which all links are positive in the completely connected graph—that is, $C(\Psi) = 1$—and knocking out flows one by one in order from the smallest and ending with only one flow, the largest. We have done this for the symmetric flow matrix in figure 3.1, and in figure 3.3 we plot two of these measures. First we simply count the number of nodes $n(\Psi) \leq N$ that remain as part of the connected structure—that is, strongly or weakly connected—and we immediately see that for a long time as links are taken out, the system remains strongly connected, but suddenly it becomes disconnected. We will not do this here, but were we to trace the shortest paths through the resulting graphs, these paths would be of infinite distance for a disconnected structure but would suddenly become finite and then would converge very quickly on a fairly stable value. Simply adding more links once this threshold has been passed would not decrease the path length very much. This is an example of a phase transition, in that the system moves quickly from disconnection to connection. We have examined this problem elsewhere, but it is quite basic to the structure of networks (Batty, 2005). More usually this is demonstrated not by taking out links but by gradually adding them into the system, building up the system to the point where it suddenly becomes connected, and at that point passes a critical threshold where many actions and movements are possible that are not possible in a disconnected system. In one sense, this kind of transition is a simple model of how cities might develop from disconnected parts that suddenly realize economies of scale. Figure 3.3 in fact illustrates this transition for the London

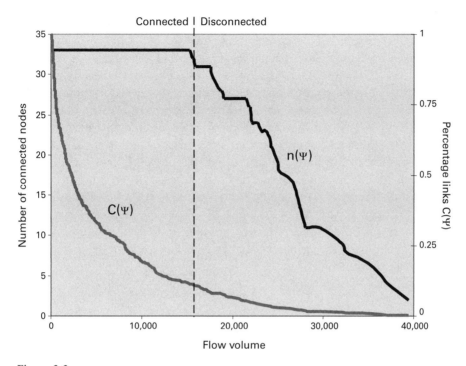

Figure 3.3
The transition from complete connection to disconnection.

example, although this is not a model of how the flow system in the metropolis actually developed.

Observing and Predicting Network Structure

So far our analysis of flows has largely been from the perspective of measuring and observing movements, such as trip making, migration, the flow of information, and so on. The London journey to work matrix, for example, is taken from observed flows that are recorded from the universal decanal census of population, and although we may be able to predict such flow patterns using the gravitational models of the last chapter, the underlying networks are assumed to be given. However, in more general terms, when we examine connections between objects of interest, we also focus on how to explain such connections in terms of other objects or their attributes. Such a focus on how to derive connections and their patterns lies at the heart of multivariate analysis, and it is possible to approach correlations between objects in terms of network structure. We will explore this here, for it represents a powerful, if oblique, way of looking at networks, giving us the opportunity to derive different types of networks by changing our perspective on how connections between objects can be formulated.

We will begin by defining a set of objects or elements $\{Y_i, i = 1, 2, \ldots, N\}$ and a set of attributes $\{X_k, k = 1, 2, \ldots, M\}$, from which we observe that each object i is associated with an attribute k through a set of connections or associations defined by the $N x M$ matrix $\{A_{ik}\}$. We can generate various comparisons between objects and their attributes by comparing or counting the number of common elements with respect to attributes through the comparison of objects, or with respect to objects through comparison of attributes. Objects might be locations, and thus common elements between locations could be seen as flows. In terms of attributes that might be characteristics of locations, the commonality is more likely to be in terms of how these characteristics compare across locations. Let us assume that instead of counting common elements, the comparison is with respect to their distribution. We will effect this by multiplying the elements together. First we can form the matrix F_{ij} of comparisons between the objects as

$$F_{ij} = \sum_k A_{ik} A_{jk}, \tag{3.4}$$

where each element F_{ij} is a multiplicand of the distribution of objects i and j across attributes k. It is clear from equation (3.4) that the matrix is symmetric; that is, $F_{ij} = F_{ji}$; and this means the marginals, the row and column sums are equal:

$$\sum_j F_{ij} = \sum_i F_{ij} = \sum_i A_{ik} \sum_k A_{jk} = \sum_k A_{ik} \sum_j A_{jk}. \tag{3.5}$$

A similar but opposite comparison of the attributes across locations can be made by forming the attribute comparison matrix $H_{k\ell}$ as

$$H_{k\ell} = \sum_i A_{ik} A_{i\ell}, \tag{3.6}$$

which is also symmetric with equal row and column sums. The general idea is that by making such comparisons through these connections, networks can be derived or generated, and we will use these ideas extensively in later chapters. We will not introduce matrix notation at this point, but will do so a little later in the effort to make these ideas a little clearer.

There are two variants of this kind of comparison. First, the multiplicative comparisons may need to be normalized in some way with respect to their size, and we will note this below. But it is also possible to slice the relationship matrix $\{A_{ik}\}$ into integer or binary form and then achieve the comparisons, and this can be seen as a kind of normalization. Once achieved, the matrix $\{F_{ij}\}$ could also be sliced as we did with the flow matrix earlier, and then various graph analyses initiated in the search for patterns and processes that are associated with such connections. The same, of course, can be done for the alternative matrix $\{H_{k\ell}\}$ with such operations providing a wide range of possible structures for flow and network analysis.

The most obvious comparison we can develop in this manner is to compute the various correlations between the objects or their attributes. To illustrate the standard measure sometimes called the Pearson product-moment correlation, for the distribution of attributes for any two objects i and j, which we call ρ_{ij}, we deduct the mean values $\bar{A}_i$ and $\bar{A}_j$ from the relevant observations, forming the covariance of each object and then dividing by the standard deviation of each. The form is

$$\rho_{ij} = \frac{\sum_k (A_{ik} - \bar{A}_i)(A_{jk} - \bar{A}_j)}{\sqrt{\sum_k (A_{ik} - \bar{A}_i)^2 \sum_k (A_{jk} - \bar{A}_j)^2}}. \tag{3.7}$$

This is much better behaved than the generic form in equation (3.4), and it can be sliced and diced to form a variety of different flow systems or networks. Of course, the correlation varies between −1 and +1, and if the sign of association is not required but only the strength, then the coefficient might be squared as ρ_{ij}^2 (although this gives yet another interpretation in terms of the variance explained). Other forms of correlation exist, but the point we make here is simply that much multivariate analysis based on extracting pattern and structure from data often has a network interpretation, which we can exploit easily and effectively in these terms. Once we explore bipartite graphs below, where the basic object-attribute matrix is really the flow matrix associated with such a graph, we will begin to give substance to these relationships.

Bipartite Graphs: Generating Flow Graphs and Their Duals

A bipartite graph is based on the links between any two sets of objects or their attributes that are intrinsically different from one another. Sometimes, the objects and attributes are referred to as "modes," with the bipartite graph a two-mode system (Borgatti and Everett, 1997). In unweighted form, they are derived from flow matrices that have the form of the NxM association matrix $\{A_{ik}\}$. These kinds of relations are not, strictly speaking, flows in the sense of movement that our spatial interaction patterns imply, but it is convenient to think of them existing on a substrata that is the equivalent graph, which we now call Γ_{ik}. We can define this without loss of generality as a binary graph formed by thresholding the matrix $\{\Gamma_{ik}\}$ using the value Ψ in much the same way we did for the flow matrix $\{T_{ij}\}$ in equation (3.2). Then

$$\Gamma_{ik} = \begin{cases} 1, & \textit{if } A_{ik} \geq \Psi, \\ 0, & \textit{otherwise} \end{cases}. \tag{3.8}$$

Just as we formed the flow matrix by counting the associations or their differences across objects i or attributes k, we can form the higher-level flow matrices (which reduce two modes to one) between these sets. For comparisons between objects in terms of attributes, the weighted graph Θ_{ij} can then be formed as

$$\Theta_{ij} = \sum_k \Gamma_{ik}\Gamma_{jk}, \tag{3.9}$$

and what we will now call its *dual* can be formed from

$$\Phi_{k\ell} = \sum_i \Gamma_{ik}\Gamma_{i\ell}. \tag{3.10}$$

Θ_{ij} and $\Phi_{k\ell}$ are not, in fact, binary graphs even though they are formed from binary bipartite graphs, but they can be reduced to such forms if they are thresholded once again as

$$\bar{\Theta}_{ij} = \begin{cases} 1, & \textit{if } \Theta_{ij} > 0, \\ 0, & \textit{otherwise} \end{cases} \text{ and } \bar{\Phi}_{ij} = \begin{cases} 1, & \textit{if } \Phi_{ij} > 0, \\ 0, & \textit{otherwise} \end{cases}, \tag{3.11}$$

where the appropriate slicing values are set above zero.

These flow and binary bipartite systems appear in many kinds of problems, and the first applications go back more than seventy years (Davis, Gardner, and Gardner, 1941). In later chapters, we will use them extensively to represent and model street systems in terms of their segments and intersections, linking this to space syntax methods (Hillier and Hanson, 1984). We will also employ them to represent social networks between actors involved in conflict resolution concerning problems of urban design, which we focus on in part III of this book. They appear in representing and analyzing social relationships in Q-analysis (Atkin, 1974) and in Coleman's (1973) theory of social exchange. In network science, Watts (2002) refers to such bipartite representations as affiliation networks. To illustrate their use at this point, however, we will consider the networks based on spatial interaction flow systems that form our earlier examples and the focus of the last chapter. Such interaction matrices between origins and destinations can be treated as unipartite graphs, but, strictly speaking, they are bipartite because origins and destinations are different sets of objects.

To demonstrate their use, we have thus have selected five employment zones as origins and seven residential zones as destinations from our Greater London flow data set. The origins are large inner employment zones, and the destinations the residential zones of which they are a part as well as two other inner-city zones. Where they are is of no consequence to this argument, but what the flow matrix

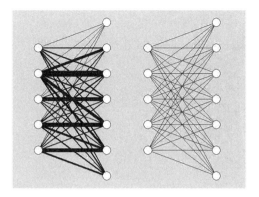

Figure 3.4
The bipartite flow graph and its binary equivalent.

$\{A_{ik}\}$ formed from a subset of $\{T_{ij}\}$ actually means is relevant. The rows of this matrix are the flows—trips from employment to residential zones—and if we then form the correlation or association between each pair of origins in terms of their distribution across destinations, then the matrix formed as $\{F_{ij}\}$ in equation (3.4) is a measure of how close any two employment origin zones are in terms of the pattern of trips they generate to residential areas. It is useful to show this flow matrix in bipartite form with its equivalent thresholding to binary form, as in equation (3.8). We illustrate this in figure 3.4; we have not drawn any arrows on the links in the graph, but clearly the underlying bipartite graph is directed. If two rows of this matrix were the same, this would generate a correlation of 1. In the same way, if we compare any two destinations with respect to how trips flow to their origins, then the dual matrix formed as $\{H_{k\ell}\}$ in equation (3.6) is a comparison of how close any two destinations are in terms of the trips that flow into their different employment sites. The first matrix thus measures how closely correlated are employment sites, and the second, population sites, but both in terms of their relationships to the other. If the profile of trips was the same for every origin to all its destinations, then both of these matrices would show correlations of 1.

Now we can conveniently show how the flow matrix and its dual can be formed. We have used equations (3.4) and (3.6) to compute these rather than equations (3.9) and (3.10), which have already been sliced. We present these in figures 3.5(a) and 3.5(b) (and their bipartite graphs are obvious by inspection in that the flows are simply reduced to lines). What these show immediately is that if we count or compare these flows from origins to destinations and back to origins in figure 3.5(a) and vice versa in figure 3.5(b), then these relationships are symmetric. In fact, when we compute the correlation matrices based on equation (3.7) for $\{\rho_{ij}^2\}$ and their dual equivalents $\{\rho_{k\ell}^2\}$, these provide obvious examples of this symmetry, shown in

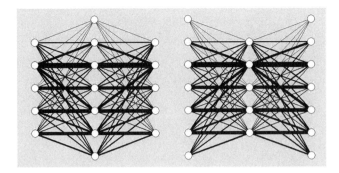

Figure 3.5
Joining bipartite graphs (a) through destinations and (b) through origins.

A   B

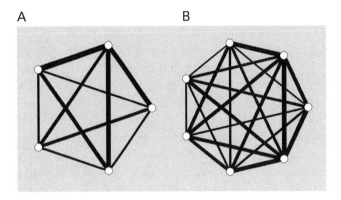

Figure 3.6
The correlation flow graph (a) based on origins and (b) its dual based on destinations.

figures 3.6(a) and 3.6(b). They provide the basis for an analysis and subsequent explanation of these patterns, but note that they are focused entirely on the pattern of interactions, not simply on the pattern of locations.

In fact, there is little analysis of urban interactions in these terms, and to illustrate how we might progress this, we will now briefly examine the full trip matrix $\{T_{ij}\}$ between all thirty-three London municipalities (boroughs). To develop these comparisons, we will compute the relative squared differences between elements in the basic bipartite flow matrix $\{A_{ik}\}$, which in fact is the original trip matrix, now focused on the origins being very different from the destinations in object type. These squared differences are not normalized for size, since we consider that the large values from one origin (or destination) compared to large values from another are more similar than large compared to small. The distribution, of course, is also accounted for by these differences, which we can define for origins and destinations as

$$\varphi_{ij} = \sqrt{\sum_k (A_{ik} - A_{jk})^2} \text{ and } \xi_{k\ell} = \sqrt{\sum_i (A_{ik} - A_{i\ell})^2}. \tag{3.12}$$

These association metrics are essentially indices of dissimilarity that show how dissimilar one object is from another. They are too messy to present graphically as flow graphs, but we can cluster the objects, gradually relaxing the degree of similarity and adding more and more objects into the classification. The tree charts or dendrograms that result show how close different origins are to one another in figure 3.7(a) and how close are destinations in figure 3.7(b). In fact, the clusters that define each of these systems are quite similar for both origins and destinations, thus suggesting that location is a significant structuring feature in conditioning how people travel to and from work. Geographically near places seem to cluster, and from a casual analysis, places with similar demographics are close to one another. In essence, what these compositions reveal is that if you work in a place that is similar to another, then these places are also similar with respect to where you live. This is a complex issue that needs considerable thought in terms of its implications for how spatial interaction systems are organized.

Connectivity, Clusters, and Small Worlds

Many of the tools and models we use to describe flows and networks are applicable to both kinds of system, but flow systems tend to be much more homogeneous in structure than networks, and thus most of the structural measures that we introduce in this section apply to graphs. In short, structure in graphs relates to the presence or absence of links, whereas in many flow systems, there are flows along every link, as we illustrated in the examples above. To an extent this is also a scale problem, but the measures we will use tend to focus on connectivity and clusters, which only become relevant when a subset of the possible links between nodes are absent. Aside from structural measures, recent developments in network science have generated many statistical models of the density of nodes and links, and we will complete our brief survey of structural measures with some analysis of statistical distributions.

We have not been very rigorous in defining the elements that compose a graph, largely because the terminology is highly variable with few strong conventions. Nodes, hubs, and vertices are used to define the elements or basic components of the graph, and links, arcs, or edges are used to specify relations between the elements. We will use nodes and links with the graph defined as $G(N, g)$, where N is the number of nodes and g_{ij} is a generic link between node (or origin) i and node (or destination) j. This is consistent with our more informal treatment earlier. We

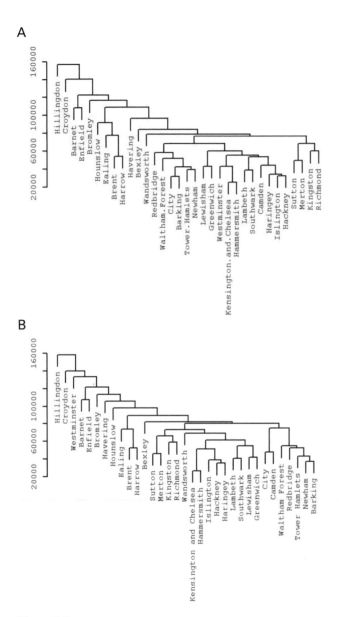

Figure 3.7

Hierarchical clustering of differences between (a) origins based on destination flows and (b) destinations based on origin flows.

have already defined a basic measure of connectivity for the whole graph in equation (3.3) as $C(\Psi)$, which is the ratio of the positive binary links defined above the threshold Ψ, if appropriate, to the total number of possible links, not counting self-links. Of course, this measure could be generalized to apply to different subgraphs, and in one sense, any partitioning of the graph into clusters pertains to such connectivities. The most obvious measures of structure involve the nodes, which are defined in terms of their degrees. The in-degree of a node d_j is the number of links that are destined for that node, defined as

$$d_j = \sum_i g_{ij},$$

(3.13)

and the out-degree d_i the number of links originating from the node

$$d_i = \sum_j g_{ij}.$$

(3.14)

These are measures of nodal density and relate to the connectivity as

$$C(\Psi) = \frac{\sum_i d_i}{N(N-1)} = \frac{\sum_j d_j}{N(N-1)},$$

(3.15)

which varies from 0 to 1. It is easy to see that the average density (or average in-degree) calculated as $NC(\Psi)$ varies between 1 and $N-1$.

A component of a graph is either a node or a group of nodes that are completely connected to one another and thus can be averaged as a sort of super node. To an extent, when we decompose or compose a graph into a hierarchy, we are assuming that at each level the components are connected in this way, although they may not be, for usually a criterion of similarity is applied, and the hierarchy builds up as the similarity between nodes and groups of nodes decrease. In the case of our hierarchies in figure 3.7, these are in fact based on completely connected weighted graphs, which are flow systems with the criterion of similarity relating to the strength of connection, not the fact that there is or is not a connection. This leads us immediately to a whole series of measures that pertain to subdividing the system. The most usual one is a measure of clustering, where a cluster is defined as a completely connected subgraph based on a particular node i. For a directed graph, we count all the links in its neighborhood based on g_{ik}, for $\forall k \neq i$, and we form the ratio of this number with the total possible links that could exist in the neighborhood when all links are positive. Then the clustering coefficient first defined by Watts and Strogatz (1998) is

$$C_i = \sum_{k, i \neq k} \frac{g_{ik}}{d_i(d_i - 1)},$$

(3.16)

where the average cluster coefficient for the entire graph can be calculated as

$$C = \sum_i C_i / N. \tag{3.17}$$

Note that C differs from the total connectivity index $C(\Psi)$, but they might be expected to co-vary as they are measuring similar intensities of links.

Central to the notion of a graph are walks, paths, and cycles. A walk is a sequence of nodes that are not necessarily distinct, whereas a path is a walk with distinct nodes. A cycle is a walk that is a loop, and it may also be a path if its nodes are distinct. In fact, graphs are connected—either completely, strongly, or weakly if there is a path from every node to every other node, either directed or undirected. A straightforward intelligible introduction to the field is provided by Jackson (2010), who draws on ideas that go back much earlier, before the development of network science (Harary, Norman, and Cartwright, 1965). Much of graph theory deals with the analysis of paths through graphs, but here we will not exploit this very much, since our focus is more on flow systems in part II and on processes on graphs in part III. There is, however, a growing area within network science that deals with urban networks in these terms, and we will note this below.

A related concept that involves the size of the graph relates to the average distance that any node is from any other; this feature has constituted the basis of defining different types of graph in terms of size and connectivity. Imagine that each node has d neighbors and each neighbor is at a unit distance from every other neighbor to whom it is connected, as in a binary graph. It is quite easy in simple graphs of this kind to compute the average distance to all neighbors who are 1, 2, 3 and up to n steps or distance away from any node i, as all the nodes have the same number of neighbors. Note that we are assuming that n is a unit distance. Then for 1 step away, there are $d - 1$ neighbors, for 2 steps away $d(d - 1)$ neighbors, for 3 steps $d^2(d - 1)$ and in general for n steps, $d^{n-1}(d - 1)$. To an order of magnitude, then, the number of neighbors encountered is $N \sim d^n$, and we can thus compute the number of steps or the typical distance as $n = \log(N)/\log(d)$. If each node has, say, 100 neighbors, then we can compute the typical step distance to reach, let us say, 10 billion people, as 5. This is, in effect, the diameter of this graph, and if this were a good model of how the world's population was configured in terms of friendships, then it shows immediately that there is a small number of steps to reach everyone, indeed less than the so-called six degrees of separation, that characterizes what has been called the "small world" problem (Milgram, 1967; Watts, 2002). However, this is not a good model because people group into clusters, and thus the typical distance is likely to be somewhat greater.

The opposite extreme to a graph made up of densely connected clusters is a random graph. Such graphs are formed by selecting arcs that link different nodes chosen at random from the total set of N, with the requirement that no more than one link can occur for every pair of nodes. In such graphs, the average distance between any pair of nodes is small relative to the number of nodes, but the cluster coefficient is also small. Chung and Lu (2002) show that the average unit distance or step length n in such graphs depends on the degree distribution, but in graphs with power-law degree distributions, it can be as small as $\log[\log(N)]$. In a graph of clusters, the cluster coefficient is large, but the average distance is large too. In fact, many social systems appear to have relatively small path lengths and relatively large cluster coefficients, reflecting some sort of compromise between random and clustered graphs. Watts and Strogatz (1998) call these kinds of graphs "small worlds," and they show that the average distance or steps n in such graphs increases as $\log(N)$. There has been intense activity in recent years to unravel the structure of many networks in different fields that show evidence of small worlds (Watts, 1999, Newman, 2010). In fact, planar graphs—which we will examine later and which apply to spatial networks—do not contain small worlds, because their planarity means that links rarely cross each other due to the nature of Euclidean space. Hence they do not reveal dense clusters.

A considerable focus in network science is not on the structure of graphs, at least directly, but on their statistical properties, which emphasize the distribution of links that originate or are destined for different nodes. The probability p of any node having d links is binomial, and if the probability is small and the number of nodes is large, then the distribution is approximately Poisson. In short, many such degree distributions approximate the normal. The other class of networks that to an extent lie at the other extreme are those in which the node degrees are distributed in skewed fashion. A very few nodes have many links or very high degrees, whereas most have a very low number of links. These networks can be formed by a process of preferential attachment, with networks developing in such a way that the rich nodes get richer and the poor stay poor. This process is accredited to Barabasi and Albert (1999), although it goes back in terms of networks to de Solla Price (1965), and before that in terms of stochastic systems to Yule (1925). We will not say much in this book about the formation of networks in terms of their dynamics, but in chapter 5, we will explore how these networks generate skew distributions of their node degrees, which follow power laws and are scaling. This picks up on the laws of scaling introduced in chapter 1, but only in the next two chapters will we broach these issues in earnest, first for locational distributions and then for networks.

We will introduce two more important properties of networks in this chapter. First we will look at processes defined on networks, and then we will examine ways of embedding networks into the spatial systems, showing how planar graphs differ from spatial graphs. Processes on networks may be of many kinds, but all have in common the idea that networks act as communication channels for flows to disseminate or diffuse between nodes that are directly connected. Networks can thus generate cascading flows where disruptions can be traced if nodes or edges are damaged in some way, but processes can also be defined to illustrate how information or ideas or opinions can be pooled—how conflict between the nodes can be resolved—as agents exercise different degrees of influence or power over one another. This raises the notion of interaction networks reflecting transactions, trade, or exchange, something we will take up in detail in part III. In some senses, diffusion is central to all these kinds of processes, but in the next section, we will focus on how communication in networks leads to conflict resolution which implies steady states or equilibrium in processes acting on graphs. This, as we will see, is key to centrality in street network systems, which we invoke in chapters 6 and 7. The idea of opinion pooling using Markov processes on networks is central to the ideas we develop in part III, where our focus will turn to methods for modeling how key agents and stakeholders might engage in reaching consensus over city plans.

Processes on Networks: The Dynamics of Flows

Defining a Basic Probability Process

We will first define the conditional probabilities p_{ij} for the movement of an object from i to j from the unipartite graph or flow matrix F_{ij} as

$$p_{ij} = \frac{F_{ij}}{\sum_i F_{ij}}, \sum_j p_{ij} = 1, \tag{3.18}$$

where we assume the matrix is strongly connected, that is, where it is possible for every origin to reach every destination and vice versa, either directly or indirectly. This is thus a transition matrix that we can use to show how objects might move around the system between the common set of origins and destinations. Imagine an element of the flow or a single walker $o_i(t = 1)$, which we define as $o_i = [0, 0, 0, \ldots, 1, 0, \ldots, 0]$ which means that the probability of the walker starting at $t = 1$ in state i is 1. Then the probability of the walker ending up in any destination j at time $t = 2$ is based on the transition probability equation,

$$o_j(t + 1) = \sum_i o_i(t) p_{ij}. \tag{3.19}$$

We can reiterate this process for $t = 3$, which leads to

$$o_k(t + 2) = \sum_j o_j(t + 1) p_{jk} = \sum_i \sum_j o_i(t) p_{ji} p_{ik}. \tag{3.20}$$

It is much easier to write this as a matrix equation where $\mathbf{P}$ is the $N x N$ matrix of transition probabilities and $\mathbf{o}(t)$ is a $1 x N$ vector of probabilities that the walker is in state i. Equation (3.20) can be written as

$$\mathbf{o}(t + 2) = \mathbf{o}(t)\mathbf{P}^2, \tag{3.21}$$

which implies the recurrence relation

$$\mathbf{o}(t + n) = \mathbf{o}(t)\mathbf{P}^n. \tag{3.22}$$

Now, for a strongly connected transition matrix, this converges to a steady-state matrix,

$$\mathbf{Z} = \ell i m_{n \to \infty} \mathbf{P}^n, \tag{3.23}$$

which is a matrix where each row is identical. Using equation (3.23) in (3.22), it is easy to show that the steady-state vector $\mathbf{o}$ giving the probability of the walker entering each state is

$$\mathbf{o} = \mathbf{o}\mathbf{Z} = \mathbf{o}\mathbf{P}. \tag{3.24}$$

This is a classic first-order Markov chain, and we will explore it in more detail in later chapters. It is an excellent but simple process model for the dynamics of change—for diffusion—on a graph or network, and it has many applications. If we consider that it is a singly constrained spatial interaction process, then the first iteration is the basic gravity model, which can be written as

$$D_j(t + 1) = \sum_i O_i(t)p_{ij}. \tag{3.25}$$

If we continue this iteration with $O_i(t + 1) = D_i(t + 1)$, then this could be seen as a migration process, or indeed as a process of search across the location space. However, most of the applications of this model in this book will be on social and political networks rather than on spatial networks, and we need to keep this in mind for future applications. In this sense, the process is generic.

Forward, Backward, and Reversible Processes

We have demonstrated the model for an asymmetric flow matrix F_{ij} or strongly connected binary equivalent, which can either be a direct specification of the flows or as the product of a bipartite graph, as in equations (3.4) or (3.6). If we use a basic symmetric gravity model, as in equations (2.9) and (2.13) in chapter 2, or the product (or its dual) of a bipartite flow matrix or graph, then $\{F_{ij}\}$ is symmetric; that is, $F_{ij} = F_{ji}$. This has important implications for the Markov chain. This symmetry is reflected in the fact that the in-degree of any node is the same as its out-degree; that is, that $s_i = s_j$, $i = j$, where we now define these degrees as

$$s_i = \sum_j F_{ij} \qquad \text{and} \left.\begin{array}{l} \\ \\ \\ \end{array}\right\}.$$
$$s_j = \sum_i F_{ij}.$$

$$(3.26)$$

There are two processes that can be defined on these symmetric networks, which we might rather casually term the *forward* process and the *backward* process. The forward process involves the transition matrix p_{ij}, which we define again as

$$p_{ij} = \frac{F_{ij}}{s_i} = \frac{F_{ij}}{\sum_j F_{ij}}, \quad \sum_j p_{ij} = 1. \tag{3.27}$$

This has the steady-state equation in equation (3.24), which can be written as

$$o_j = \sum_i o_i p_{ij}, \sum_i o_i = 1. \tag{3.28}$$

It can be easily demonstrated by substitution that these weights or probabilities are proportional to the in-degrees and out-degrees; that is,

$$s_j = \sum_i s_i p_{ij} = \sum_i F_{ij} = \sum_i \left(\sum_j F_{ij} \right) \frac{F_{ij}}{\sum_j F_{ij}}. \tag{3.29}$$

This is a very simple but rather remarkable result we will exploit very heavily in subsequent chapters, for it implies that such symmetry is a form of equilibrium in which the degree of a node is directly reflected in the importance of the object, activity, or actor. In later chapters, we will say more about this process and how it can be formulated in terms of its eigenvalues, which in fact reflect the process of convergence to the steady state and the ultimate probability weights.

The backward process involves one of averaging, which we define as follows. Starting from a vector of, say, positive values $c_j(t)$ at time t, we can then form an average of these values over each of the links that node i is linked to as

$$c_i(t+1) = \sum_j p_{ij} c_j(t). \tag{3.30}$$

We could work this process through time, but it is enough to show that the process converges to a limit or consensus or average value $c = c_i, \forall i$ (for strongly connected symmetric matrices), defined as

$$c = c_i = \sum_j \frac{s_j}{\sum_k s_k} c_j(t), \forall i. \tag{3.31}$$

This can be easily demonstrated by performing the recursion on equation (3.30) and noting that the resulting matrix converges to the stochastic matrix **Z** with all its rows equal to **o**, which, as we have shown, is proportional to the in-degree or out-degree of each node in equations (3.28) and (3.29).

These two processes—the forward, which is essentially a random walk, and the backward, which is an averaging or consensus-generating process—are representative of various systems and their applications that we explore in later chapters. However, it is also worth noting that the transition probabilities defined in the steady state in equation (3.28) specify a reversible process or reversible Markov chain, which can be stated as

$$\left. \begin{array}{l} o_i p_{ij} = o_j p_{ji} \\ s_i p_{ij} = s_j p_{ji} \end{array} \right\}. \tag{3.32}$$

This follows directly from the symmetry of F_{ij}. This means that the process defined on the out-degrees is the same as that defined on the in-degrees. If we thus state the reverse process as

$$s_i = \sum_j q_{ij} s_j, \tag{3.33}$$

where the transition matrix q_{ij} is defined as

$$q_{ij} = \frac{F_{ij}}{s_j} = \frac{F_{ij}}{\sum_i F_{ij}}, \quad and \quad \sum_i q_{ij} = 1, \tag{3.34}$$

then it is obvious that this is exactly the same as the process defined on the transition matrix $\{p_{ij}\}$, for in essence the in-degrees are identical to the out-degrees—which is the key requirement for symmetry when it comes to defining processes focused on the nodes or components of the system.

Related Symmetric Structures

Our basic structure is a symmetric matrix $\{F_{ij}\}$, which we define as the *default symmetry*. It is assumed that this matrix is not the identity matrix or the unit matrix, and thus its elements indicate some structure in which the nodes and their arcs are correlated with one another. If the matrix were the identity matrix, then no node would be correlated with any other, and if the unit matrix, then all nodes would have the same set of links to one another. This default symmetry might also be augmented through some attribute information $\{z_i\}$ of the nodes that define

$$Q_{ij} = z_i F_{ij} z_j. \tag{3.35}$$

It is obvious that the symmetry of the initial matrix is maintained from

$$\left.\begin{aligned} z_i' &= \sum_j Q_{ij} = z_i \sum_j F_{ij} z_j \\ z_j' &= \sum_i Q_{ij} = z_j \sum_i F_{ij} z_i \end{aligned}\right\}, \tag{3.36}$$

with the summations on the right-hand side of equation (3.36) being transposes of one another due to the symmetry in $\{z_i\}$ and $\{F_{ij}\}$. Note that as part of the same default class, we could take other functions that ensured the matrix was positive definite, such as z_i^β, thus forming Q_{ij} from $z_i^\beta F_{ij} z_j^\beta$ or suchlike.

However, a more basic symmetric structure, which we refer to as the *elemental symmetry*, is one where the nodes are not correlated with one another through their interactions but where these interactions are formed from some symmetric information about the nodes; that is, from the independent products of attributes of the nodes $\{x_i\}$. Then

$$F_{ij} = x_i x_j, \tag{3.37}$$

where it is clear that the matrix is partitioned into nodal elements such that

$$\left.\begin{aligned} \sum_j F_{ij} &= s_i = x_i \sum_j x_j = x_i X \\ \sum_i F_{ij} &= s_j = x_j \sum_i x_i = x_j X \end{aligned}\right\}. \tag{3.38}$$

The structure of this matrix is dependent on the product of the in-degrees and out-degrees, which is tantamount to implying that any random walk beginning at any in-degree or out-degree generates probabilities of first-order visits in proportion to the in-degrees and out-degrees of the node visits (Lambiotte et al., 2011).

There is a third structure, which might be defined as a *generated symmetry*, that is formed by taking an asymmetric network $\{W_{ij}\}$ and applying nodal information to the in-degrees and out-degrees separately, defined respectively by $\{x_i\}$ and $\{y_j\}$. The symmetric matrix is thus defined as

$$\left.\begin{aligned} F_{ij} &= x_i W_{ij} y_j \quad \text{where} \\ \\ s_i = \sum_j F_{ij} = x_i \sum_j W_{ij} y_j &\quad \text{and} \quad s_j = \sum_i F_{ij} = y_j \sum_i W_{ij} x_i \end{aligned}\right\}. \tag{3.39}$$

Equation (3.39) provides another way of exploring symmetry in network systems, and it resonates strongly with some of the ideas that we introduced in the last chapter, where we focused on how space could be distorted in flow systems as departures from some symmetric baseline. In fact, the idea of the symmetric baseline in spatial flow and network systems is a powerful idea in that it transfers attention onto the way in which space in cities might distort locations and interactions, thus directing attention to the residuals that occur when such symmetric structures are considered as the background canvas on which spatial behaviors take place. Throughout this book, we will have this generic idea in mind.

There are several similar ideas that have preceded the more recent discussion of these kinds of structures in network science. In the mid-1950s, French (1956), working in the psychometric and sociometric tradition, developed a model of consensus that essentially was based on the averaging procedure, which we called the backward process above in equation (3.30). Harary (1959) explored the same model but cast this into a Markov framework, unwittingly, perhaps, involving the forward process model as well. We used this as a method for averaging maps that implied conflicting values or opinions between actors who were generating a plan—which we associated with implying a consensus (Batty, 1974a). This is explored in depth in part III of this book, where we extend this to a more fully fledged model of decision-making (Batty, 1984), drawing on arguments made by de Groot (1974) and others such as Kelly (1981). More recently, Jackson (2010) has reviewed the de Groot model in a similar network context. He says: "The de Groot model is simple and tractable, and has a number of nice properties that make it a useful benchmark both in terms of positive and normative features," and he goes on to suggest the "influence that each agent has on the final consensus depends in very intuitive ways on the network structure ... where agents place equal weights on each of their friends, the influence of an agent is proportional to his or her degree." As Lambiotte et al. (2011) note, there are several other extensions of the backward model in the sociological tradition (for example, see Hegselmann and Krause, 2002) although most of these do not depart very much from the basic model. This is a burgeoning literature that we will note again in part III.

Spatial Networks and Planar Graphs

Our focus in this book is not simply on networks or graphs that exist in Euclidean space, but on sets of relations that have different degrees of embeddedness in space. These are networks whose nodes are fixed at specific locations but whose links are not, such as airline, wireless, and networks that relate to line of sight, all the way to social networks whose nodes and links might be continually moving but are definitely not associated with fixed points in space. However, all the networks we will examine have some spatial association in that they always pertain to cities and their physical planning. In this section, we will concentrate on the strictest of these spatial networks, those whose nodes and links are embedded in space, such as street networks. Invariably these pertain to the infrastructure that might carry the sorts of flows that we introduced in chapter 2, but note that the way we examined those networks, and those earlier in this chapter, tended to assume that although their nodes were fixed, their links were abstracted above the level of actual physical space itself. To an extent, this is a matter of the way we treat these systems rather than pertaining to their actual operation in two- (or three-) dimensional space.

In essence, the graphs or networks we will present here are referred to as *planar* where, without any loss of generality, we will assume that such graphs are undirected. Informally, these are graphs where there are no intersections between any of the links, were we to draw them in two-dimensional space. More formally, there is a conservation rule that says that the number of nodes N in such a graph less the number of links or edges E plus the number of faces F, where a face is defined as an area bounded by edges, is equal to 2. That is,

$$N - E + F = 2. \tag{3.40}$$

The simplest planar graph is a tree, which has no faces and is simply a string of edges with no loops. A more complex one is a set of triples of nodes such that there is a connected component for every triple $g_{ij}g_{jk}g_{ki} = 1$. Now imagine a graph being drawn by starting with two nodes g_i and g_j with a link $g_{ij} = 1$ between, and then the construction of another node g_k with new links $g_{ik} = 1$ and $g_{jk} = 1$, completing the triple. We then add another node g_ℓ and connect this to two of the nodes so far, so that the links do not cross. In this way, we can imagine the graph extending as a series of triples, building up a hexagonal-honeycomb structure. It is intuitively obvious that if we consider the distance from one node to another as $d_{ij} = 1$ when $g_{ij} = 1$, then the graph will grow no more than two units of distance each time a new node is added. That is, the total distance $D(N)$ in the graph can be defined as

$$D(N) = \sum_{\substack{\forall i, j \\ where\ g_{ij} = 1}} d_{ij} \propto N. \tag{3.41}$$

It is clear, then, that the average distance in the graph converges to a constant, and the average number of in-degrees (or out-degrees) per node also converges as $N \to \infty$. These results generalize to other kinds of planar graphs, where it can be shown that the upper bound on the *average* in-degrees per node is less than or equal to 6, and that the degree distribution is much more like a random graph; that is, there is no power-law degree distribution and very large hubs are extremely unlikely (Barthelemy, 2010).

This makes sense for networks such as streets, where it is physically difficult for intersections that are nodes to have more than 6 or 7 links entering or leaving the intersection. In a grid pattern, of course, the number of links is on average 4, and the range of possible in-degrees for most street and road networks is likely to follow a skewed distribution to the left, whose mean is likely to be about 4. For light and heavy rail networks, the average number of in-degrees is much less; in fact, for a series of subway systems in world cities, the average in-degrees of the densest core areas within their circle lines are of the order 2.5 (Roth, Kang, Batty, and Barthelemy, 2012). In fact, in all such networks, planarity tends to be an idealization in that roads sometimes do not intersect but cross through tunnels and bridges, and clearly the same is true of rail networks. Even utility networks cross in the same way, perhaps more so as their locational positioning is more flexible than for the movement of people and goods. For information networks, so far we do not have a clear perspective on their morphology or even topology, but it is probably the case that we would discover very substantial crossings if we were able to plot them in two-dimensional space. Moreover, when networks are planar in this sense, clusters are smaller and more uncommon than in the case of networks generated using preferential attachment and that show power-law degree distributions, which we will consider in chapter 5.

Figure 3.8 shows a dense networks of streets in central and inner London that is within a 6-mile square block (36 square miles in all, which is about 93 square kilometers). We simply represent each street intersection or junction as a node and each street segment between nodes as a link, and from this, we note the following

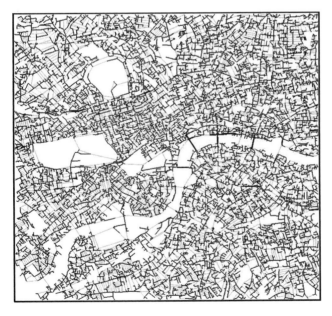

Figure 3.8
A planar graph of streets in Central London with the minimal spanning tree (which is most of the graph) imposed as dark lines on the graph.

key properties of the planar graph shown in the figure. The original graph is directed with one-way streets in places, many end nodes, and nodes dividing segments at traffic lights, crossings, and so on. In fact, the actual network representation is far from a pure planar graph, but as we are interested only in whether there is a connection between any two nodes, we disregard direction. Using these simplifications, from the N = 22,260 nodes, there are some 52,448 links or edges. This gives a very low average connectivity of the entire graph from equation (3.3) as 0.000106, which implies an average degree (noting the graph is symmetric and in- and out-degrees are the same) of $NC(\Psi)$ = 2.356. The distribution of degrees is extremely simple, with a maximum number of links of 6, of which there is one node with the distribution from 1 to 6 links as [4,615, 6,313, 10,130, 1,194, 7, 1].

The total distance in the graph is clearly the number of edges E, which from equation (3.41) is D = 52,448. If we use actual distances computed from the map itself, then this total is 3,179 kilometers, and we can compute the average real distance per segment as 63 meters. The average distance associated with any node—that is, distances to and from the node's nearest neighbors averaged over the number of all such distances—is 143 meters. In figure 3.9, noting the nodes are in arbitrary order, we plot the cumulative distance associated with each node as the number of nodes increases (where the nodes have to be connected), which shows that this distance simply scales linearly with the number of nodes, and with the area over which the nodes are located. The last feature of this network worth noting is that we show the minimal spanning tree in figure 3.8 as the thick lines. This represents

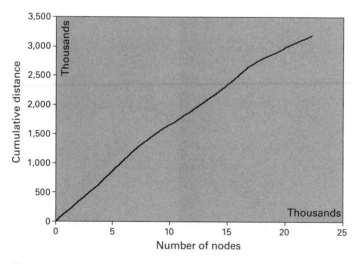

Figure 3.9
Cumulative distance in the planar graph.

a planar graph without any loops, which connects all the nodes. It is thus a tree, and it represents a much smaller total distance traveled in traversing the graph (that is, visiting every node). Of course, it is not an efficient system for delivering materials to the nodes in this dense part of London, but if all material had to be delivered to just one of these nodes from every other, it would be, as our analysis in terms of the morphology of the entire city, which we develop below, suggests.

To conclude our analysis of planar graphs, we have computed the same measures for the networks in Greater London illustrated in figure 3.10, where we show in (a), (b) and (c) this network disaggregated into three levels of hierarchical importance, which roughly relate to capacity, and in 3.10(d) the tube (subway) with the overground rail network imposed on it. Various measures from these networks are shown in table 3.1, and it is immediately clear that there is substantial convergence in terms of the road indices, but differences with the two rail networks are clear in

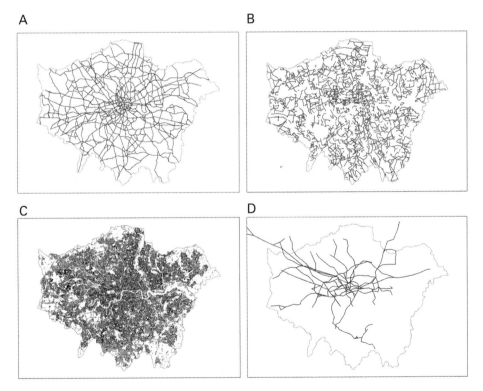

Figure 3.10
Hierarchical transport network graphs in Greater London. (a) The motorways and major A trunk roads within the GLA (Greater London Authority) boundary. (b) B and minor roads. (c) Local roads and paths within the GLA area. (d) Tube and overground rail networks.

Table 3.1
Properties of different physical network types in Greater London

Network type	Nodes N	Edges E	Connect $C(\Psi)$	Av degree $NC(\Psi)$	Total Real$D$ km	AvReal$D$ Segment km
Major trunk roads	19321	50484	0.000135	2.613	3436	0.068
Minor roads	21694	55577	0.000181	2.562	4305	0.077
Local streets	96477	327386	0.000035	3.393	25685	0.079
Extended subway-tube system	589	813	0.00235	1.380	996	1.225
Overground rail system	58	115	0.0348	1.983	173	1.502

that these networks are much more reminiscent of tree graphs. These rail networks have much higher average distances between the nodes than the road networks, and they also have higher connectivity, in that the number of segments is large relative to the possible number of segments in comparison with the road systems. In fact the road systems, despite being at three different levels of hierarchy, are very similar to one another, with very low connectivity and all three average distances between nodes less than 100 meters. This is clearly an artifact of the way these networks intersect with one another, implying that at the major road level, which consists of motorways and trunk roads, there is very little grade-separated construction since most major roads have evolved as an integral part of the generic road system, thus forming a key element in local transportation.

What this analysis also suggests is that these differences probably have some impact on the behavior of travelers, who switch from one network to another in making trips. The analysis of multimodal switching in this fashion, and the implications of this for coupled networks, are some of the great challenges in network science, and we do no more than raise them here. Note too that in the calculations in table 3.1, the tube network is different from that used in chapter 2 and in the section below, with the network in figure 3.10(d) being based on links between lines through common platforms rather than the topological version used earlier. These networks suggest that we require more powerful methods for linking them together, but as yet we do not have robust methods for analyzing how they intersect and function together. Later in part III, we will look at networks coupled in sequences that involve resolving conflicts, but it would be a mistake to assume our science has even begun to properly broach an understanding of the "networks of networks," which is what such coupling implies.

Centrality, Accessibility, and Allometry in Spatial Networks

Measures of Graph Centrality

The equivalent to accessibility or potential in the flow systems introduced in chapter 2 is usually defined as centrality in network systems. In-degrees or out-degrees are a clear measure of access to nodes but these only take account of direct links and, unlike measures of flow potential, do not reflect any distances greater than those that are immediately adjacent to the nodes in question. In the case of binary graphs, these measures of degree are simply a count of the number of links into or out of any node—numbers of adjacencies, not actual distances, although if the graph is weighted by actual distances, then more conventional measures of accessibility can be computed. We will repeat the degree equations (3.13) and (3.14) as

$$
\left.
\begin{aligned}
d_j &= \sum_i g_{ij} \\
d_i &= \sum_j g_{ij}
\end{aligned}
\right\}, \tag{3.42}
$$

and if actual distances d_{ij} were imposed on the binary links g_{ij}, this would provide measures relating to the two-dimensional Euclidean space. If all possible links were considered, then the measures for each node would reflect the total distances in the system from all nodes to the node in question, and it would be necessary to invert these measures in some way to produce a relevant index of accessibility. There is a measure related to this called closeness centrality that is used in social network theory (Jackson, 2010; Prell, 2012), and this is generated by first computing the total adjacent distances as D_i:

$$
D_i = \sum_j d_{ij}. \tag{3.43}
$$

We show the measure solely for out-degrees since the in-degree measure follows directly, but this is really a measure of decentrality and is usually inverted as L_i, which is thus defined as

$$
L_i = K D_i^{-1} = K \left(\sum_j d_{ij} \right)^{-1}. \tag{3.44}
$$

K is a constant sometimes set as the number of nodes less one, $N - 1$, and d_{ij} is measured as the shortest route from i to j.

A much more satisfying measure, particularly for binary symmetric graphs, is betweenness centrality C_k, first defined by Freeman (1979) as

$$
C_k = \sum_i \sum_j \frac{\sigma_{ikj}}{\sigma_{ij}}, \tag{3.45}
$$

where σ_{ikj} is the number, not the distance, of shortest routes from node i to j but passing through node k. This clearly picks up how a node is related to all other nodes, but it has to be weighted by the fact that there may be several shortest paths of similar length from i to j. If there is more than one, then each path is set as a fraction of the total such paths σ_{ij}. In the case of two paths, the count is thus 1/2 for each path; for three paths, 1/3 for each path, and so on.

There are many variants on these three measures based on degrees, closeness, and betweenness that depend on how distances are defined, but to illustrate their use, we have chosen a real but relatively simple spatial network—the London underground (tube) system. This is configured in more or less radial treelike form focusing on London's center, but with a complicated polycentric set of cross-links within the circular line that bounds the extended CBD of the metropolis (Roth, Kang, Batty, and Barthelemy, 2011). There are 307 nodes and 353 edges in this system, with connectivity $C(\Psi) = 0.0077$. The density, which is the number of links divided by possible links excluding the self-links, is $\Sigma_i \Sigma_j g_{ij} / N(N - 1) = 0.0038$, which is half the value of the connectivity (due to the symmetry of the system). We will not show the three measures for the system, only the betweenness centrality illustrated in figure 3.11, but we have computed the correlations between them. The correlation between the degree distribution and closeness centrality (noting that the in-degree and out-degree distributions are the same due to the symmetry of the network) is quite high at 0.727, but the correlation between degrees and betweenness centrality is lower at 0.513, as is the relationship between closeness and betweenness, which is 0.492. In short, these measures do pick up rather different kinds of accessibility, and there is potential that they might be extended to deal with flow systems, thus enriching the accessibility and potential measures introduced in chapter 2.

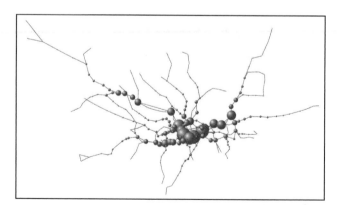

Figure 3.11
Betweenness centrality at the nodes defining stations on the London tube network.

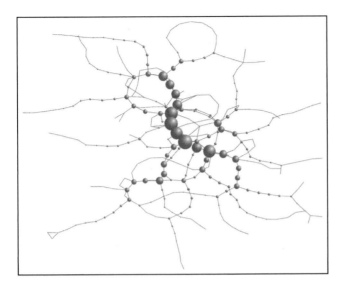

Figure 3.12
Betweenness centrality for the London network presented using the Harel-Koren force-directed algorithm (anchored at Heathrow Airport, bottom left).

Figure 3.11 shows the betweenness centrality indices as spheres, illustrating the relative simplicity of this network system (in comparison with most social networks with the same number of nodes). But it also illustrates how difficult it is to present graphs with many nodes and links even where, in this case, the graph is largely (but not entirely) planar, hence embedded in geographical space, and is of a treelike form. The geographical embeddedness is important in this case to appreciate the clustering of nodes in the CBD, but other graph representations are still useful for spatial networks. The class of graphs that involve displaying the nodes and edges in such a way that the least number cross and the edges are as equal in length as possible is a useful way of revealing structure, and in figure 3.12, we show how the "force-directed" algorithm can untangle relationships in a way that provides real insight.

The particular version we have used to generate figure 3.12 works by assigning forces to the edges as if they were springs and to the nodes as if they were charged particles, and these forces are applied by pulling the nodes together and forcing them apart in such a way that the graph ends up in a layout that makes each edge as equal in length as possible. This is achieved iteratively until some equilibrium is established. There are many variants; the one that generates figure 3.12 is from Harel and Koren (2002). The effect is to expand the central area and define the loops in the system such that the complicated nature of the polycentric hubs in the CBD become clearer. There are many such packages now that enable these kinds of visualizations to be produced, and we will use these occasionally to present networks in the rest of this book.

Spatial Allometry and Network Scaling

A very different but complementary approach to spatial networks relates to how networks scale, not with the number of nodes or distance but with the area over which they are defined. We can form the relation between R, the amount of road space, the population of each individual at its origin center, which is the total number of nodes N, and the average distance traveled $\bar{d}$ by a typical traveler from origin to destination as

$$R = N\,\bar{d}. \tag{3.46}$$

The average distance is the average radius of the area A, which is defined as $\bar{d} = A^{1/2}$, from which equation (3.46) can be written in generic terms as

$$R = N\,A^{1/2}. \tag{3.47}$$

In fact, if we define the population density as $\rho = N/A$, or $N = \rho\,A$, then equation (3.47) can also be written as

$$R = \rho\,A^{3/2}, \tag{3.48}$$

which is the classic example of superlinear scaling or positive allometry with respect to area. This relation is equivalent to the usual metabolic rate equation based on 1/4 power scaling for organisms that exist in three dimensions, where the similar relation is $V(R) = \rho V^{4/3} \propto V^{(D+1)/D}$, and D is now the dimension of the space in which the object exists (West, Brown, and Enquist, 1997).

Systems in which the network forms a tree spanning the entire space, with many origin nodes and one destination node, have the archetypal morphology for the monocentric city where all travel is to the center for purposes of work, exchange, and so on. This centralized system leads to equations (3.47) and (3.48) which we picture in figure 3.13(a) for a strict treelike fractal hierarchy of movement to a central node. However, most systems tend to be more decentralized than this, to the extreme shown in figure 3.8, which is more like a grid with many origins and destinations. Samaniego and Moses (2008) suggest that in such a system, the average distance of travel is inversely proportional to the density as $\bar{d} = \sqrt{1/\rho} = 1/\sqrt{\rho}$, which results from assuming that the inverse of density gives a unit of space, from which we derive its radius. Then equations (3.46) and (3.47) lead to

$$R = N\,\rho^{-(1/2)}, \tag{3.49}$$

which simplifies to

$$R = (N\,A)^{1/2}, \tag{3.50}$$

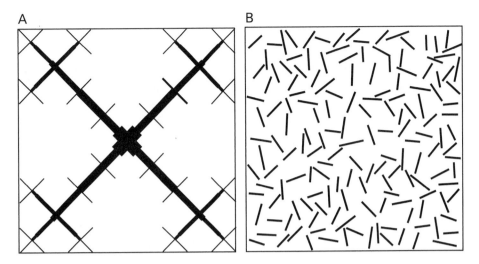

Figure 3.13
Network morphologies in (a) centralized and (b) decentralized cities.

where it is clear that the scaling is now linear with area when we note that $R = \rho^{1/2}A$. This makes sense when we examine figure 3.13(b), which is a random placement of lines between origins and destinations distributed in the city space.

Artifacts, Tools, and Visualizations

In concluding the first part of this book, which has introduced models, methods, and tools to be used in building our science of cities, we have described urban structure and dynamics not as artifacts of cities in their own right but as exemplars of how our methods and tools can be used to simulate and manipulate their form. In the rest of this book, we will focus much more on the substantive meaning of flows and networks as the underlying components of urban processes and morphologies, building on ideas that have occasionally surfaced so far in terms of location and interaction. In this chapter, which has introduced tools for representing and exploring networks, we have been at pains to emphasize that networks exist across the many dimensions of the city, which, although ultimately rooted in space and location, may be intrinsically aspatial or social without any explicit spatial dimensionality. In fact, the idea of a spatial network as the underlying skeleton on which the locational structure of cities is built has only been explored in general terms in terms of flows earlier in this and the previous chapters, and latterly in terms of the scaling of networks with respect to area, as illustrated in figure 3.13. Locations in cities built on flows and networks range from highly centralized to decentralized structures, from monocentric to polycentric, from strict hierarchies to overlapping lattices, and in part II we will explore these possibilities in many applications.

In visualizing complex systems, networks are regarded as being the key exemplar in that nodes and links are assumed to be visually understandable in two-dimensional space. Social networks that are not embedded in such space are the easiest to visualize since there is no requirement that links be separable in their own space; that is, they do not intersect one another, as many of the networks in this chapter do. But despite this limitation, two-dimensional visualization (or even three-dimensional) can become quite cluttered as the number of nodes increases to fifty or more. This is the case even for networks as simple as the London tube, which is largely a tree structure, as illustrated in the previous section. Moreover, given the fact that the nodes and their linkage patterns embody their own dimensionality, it is increasingly difficult to identify patterns visually in networks that have anything more than a handful of nodes. Thus visualization has very significant limits, despite the fact that we will use such pictures quite extensively in what follows. However, hierarchical summaries, as shown in figure 3.7, conventional plots, and an array of other devices such as the trajectories that we plot in the next chapter are as important as networks in making sense of how locations are built from interactions in determining urban form and structure.

There are several other kinds of network that we have not introduced so far. Specifically, there are connectionist approaches that involve neural networks, which tend to be part of multivariate analysis (Gurney, 1997) but are based on the default model that "everything is related to everything else," a crude summary of Hegel's dictum. We will note these briefly in part III, where we examine conflict relations between actors and factors that are an intrinsic part of the way cities are designed, but it is worth stressing that networks tend to be simplifications of the way components relate to one another in any case. In many respects, the way we approach cities tends to suggest that the phenomena of interest cannot be separated into clear causes and effects and that any partitioning into specific discrete, often binary, one- or two-way relation is an inevitable oversimplification. To an extent, this pitches us back into the world of flows rather than networks, or at the very least, weighted, fully connected networks.

Lastly, our focus here is on both positive and normative applications of these tools. It is perhaps too great a distinction to see part II as pertaining to a science of cities and part III a science of city planning, for we wish to fuse these more closely. But in part II, our perspective is clearly on representing and understanding cities and then simulating their form and process in ways that pertain to what currently exists, whereas in part III, this changes to consider ways in which cities might be designed and manipulated in terms of their locations and interactions. Although we will not invoke the term optimization very much in this book, for we see planning and design as more an issue of compromising than optimizing, or of "satisficing" to use Simon's (1956) classic term, our use of networks in part III is clearly focused on using these to enable normative processes of change. These themes are woven into our treatment in diverse ways, and as our argument develops, we will construct a science that comfortably contains both positive and normative perspectives as well as articulates a vision that extends to design, and vice versa.

II

The Science of Cities

In presenting our science, I am acutely aware that I will never know the background knowledge about cities that readers already have and that they will use to make sense of the ideas and methods already introduced. This is made all the more problematic because we are not only abstracting from the casual knowledge of cities that every city dweller has, but providing a particularly focused, formal perspective on how we should approach an understanding of cities. This treatment thus draws from many sources. In one sense, even the most general reader might find something of interest about cities here, but providing a general perspective that all readers might embrace is difficult. In developing a new formalized science built largely through methodological considerations, and one that includes a plurality of perspectives, this is a challenge a book such as this can only begin to address. I have chosen to develop these ideas by defining foundations based on methods and tools that assume cities are sets of actions, interactions, and transactions. We will then use these tools in part II to explore different aspects of cities in terms of their growth, size, scale, shape, and the processes that generate differences in these attributes (Batty, 2008). In part III, we apply similar tools to show how those charged with changing cities through design—planners, policy analysts, politicians, and so on—can effect a consensus about their future form.

This part of the book thus applies the tools and methods introduced in part I to six different perspectives on how we might think about cities. First, in chapter 4, we will explore systems of cities through their size and growth, focusing on regularities in city-size distributions that occur across the system, but pointing to the real puzzle that these distributions remain extremely stable while the cities that compose them change in size in quite volatile ways. We invoke the scaling laws introduced in part I quite extensively here, and this leads us to our second theme in chapter 5: how the system of cities and their parts scale with respect to the hierarchies that exist both between and within them. We thus begin to unpack individual cities, examining networks of relations that underpin city growth and showing how not only locations scale systematically, but also how sets of relations or networks linking populations to their urban cores and to other cities scale in similar ways.

This takes us to our third theme in chapter 6, where we move to a much finer scale involving networks. Here we invoke one of the most important ideas in this book: that since networks provide us with abstract conceptions of how cities function, they really need to be predicted rather than observed. We enable this by thinking of networks as relations between two sets (expressed as bipartite graphs) from which we can derive networks of relations between objects in either of these sets. We are thus able to predict their form and structure from a deeper concern for how different sets of objects in a city are linked to one another. We give substance to these notions using the street network and the way streets intersect with one another, thus introducing primal and dual problems of how we might think of networks. This lets us introduce a kind of space syntax that has been used to represent streets and their relations between one another. We illustrate a means of articulating the importance of streets and their intersections, thus enriching ideas about connectivity and accessibility that tend to be based on simple measures of what happens at particular nodes or locations. Accessibility is the subject of our fourth theme in chapter 7, which we develop on the back of this syntax. Besides defining more abstract networks of relations on what are physical networks such as streets, we also show how distance, hence shape and morphology, can be reconciled with networks of this kind. We thus introduce urban form into the argument, opening the door just a little to the study of coupled networks, the cutting edge of network science at the time of writing.

We switch tack a little with our fifth and sixth perspectives in chapters 8 and 9, which deal with different simulation approaches to cities. Chapter 8 focuses on bottom-up hierarchical processes of growth and development and introduces ideas about fractals linking back to hierarchical networks, to scaling, and to self-similarity. We show how such ideas can be operationalized using cellular automata models that grow cities as physical structures from the bottom up, and we illustrate how such ideas provide us with a handle on urban dynamics, something that is largely implicit in much of the material introduced so far. Chapter 9 presents another set of models, those based on spatial interaction, where our concern is to simulate city systems at the level of flows between fine spatial zones. Here we dwell on notions of equilibrium once more but also focus on developing tools that inform the debate about urban futures, which we argue should be structured in highly visual ways. Although we visualize our models and processes in quite rich detail in presenting these ideas, this focus is implicit apart from our excursion into land use transportation modeling and its use in exploring different "what if?" scenarios in the last chapter in this part of the book.

There are many ways of classifying these different approaches to cities. In part I, we were largely concerned with representations, albeit abstract and largely network-based. In part II, our focus is on modeling generic processes that generate physical forms, whether they be locations or networks. In part III, as we will see, we switch yet again to examine design and policy processes that take place in cities but are explicitly about changing the form of the city. In all these perspectives, we are concerned with the notion that locations and hubs, network links, and actors and agents who define what happens in cities exist as nodes in networks. One way of ordering these relative contributions' position, or importance if you like, is by working out their connectivity or accessibility to one another. In fact, much of the logic of this approach to cities is reflected in different notions of connectivity and accessibility and the tools of network and flow analysis introduced in part I are key to making these ideas applicable and practical.

One thing I have not done, which is extremely tempting but is far too tall an order here, is to present the reader with a more integrated history of what a science of cities in this vein has accomplished so far. The preamble hinted at this, but because there are so many different perspectives available, it is quite impossible. However, it is possible to provide some sense of how previous researchers have fashioned such a science. In fact, the architecture of this field is largely built around different sectoral perspectives on how cities function via the application of economic, geographic, engineering, environmental, and social disciplines to different sectors such as housing, industry, transport, and so on. In economics, macroregional analysis and micro-urban theories of how cities and regions function have been a major preoccupation, with the price mechanism being the major arbiter of how land use and their transport links configure themselves to form the cities we observe. Social physics has enriched this tradition somewhat and led to operational, practical tools, while social theory, often from a geographic perspective, has produced theories of how cities are structured and segregated, consistent to a degree with deductions from urban economic theory. Regional science and urban economics are the disciplinary foci that have come to define these approaches, supplemented by transport modeling and various approaches from demography. But there has been little emphasis on processes of energy exchange or even information exchange, nor has there been any strong ecological foundation to thinking about how cities function. To date, the theories and models that have existed—prior to complexity theory, that is—have been largely descriptive and often superficial. This presents us with an enormous challenge in knowing how to fashion a new science that is clearly required if we are to design future cities that embrace all this evident complexity.

In the chapters that follow, I will attempt to provide such insights, but those looking for an integrated science that is nicely packaged and available to apply immediately and without qualification to city problems will be disappointed. No such package exists, and it probably never will. Like physics, it might seem as though the field should aspire to an integrated theory, to a final theory if you like, but as in physics too, this is a mirage. It is more likely that a new science of cities will be composed of a series of approaches that adopt and adapt various methods and tools, and insofar as it is integrated, this can only occur through a consistent philosophy one might describe as complexity. The six chapters that follow are quite closely linked in some parts, but they simply define markers on the road to a better understanding, one that will have many twists and turns in the quest for designing more sustainable cities.

4

The Growth of Cities: Rank, Size, and Clocks

I will tell the story as I go along of small cities no less than of great. Most of those that were great once are small today; and those that in my own lifetime have grown to greatness, were small enough in the old days.
—Herodotus, *The Histories* (quoted in Jane Jacobs, *The Economy of Cities* (1969))

Settled agriculture emerged from mankind's beginnings as hunter-gatherers some 8,000 years before the common era (BCE), but there is controversy about when the first cities were actually established. Many assume that settled agriculture provided a surplus of wealth above basic subsistence and provided the spark for a division of labor that led to more urban pursuits, and in turn to various social and techno-logical innovations (Bairoch, 1988). Others such as Jacobs (1969), however, have argued that cities are intrinsic to mankind and that they date back to before the agrarian revolution, or at least were coterminous with it, existing in parallel to nomadic as well as agricultural activities. In fact, it is clear that by 4,000 BCE, cities were in existence in Sumeria, but the fraction of the population that was urbanized remained low for many thousands of years until the Industrial Revolution began in the West in the seventeenth and eighteenth centuries.

In chapter 1, we established the idea that as cities get larger, they get wealthier, and argued that this is due to agglomeration economies of various kinds that out-weigh diseconomies generated through crowding. Positive allometric scaling, in which the average income of urban population increases as cities get bigger seems to be an incontrovertible fact at any snapshot in time as well as through time as populations grow. This statement is, of course, conditional upon context but appears quite clear with respect to the growth of cities over the last 200 years, for which we have reasonably good estimates of income and city size. In fact, although world population has grown more than exponentially since prehistory, the proportion of the population living in cities has remained low and relatively stable until quite recently. We do not have good data on this proportion but from Chandler's (1987) data, which record the population of the largest fifty cities from about 500 BCE, the ratio of the sum of these city populations to world population hovered around 2%–3% until the beginnings of the Industrial Revolution. It then took off dramatically and currently (in 2012), it stands at about 12%. Forecasts suggest that we will all be living in cities by the end of this century—that is, we will all be urbanized— but at the same time, world population will probably stabilize at around 9 billion and to all intents and purposes growth as we have experienced it for the last 100,000 years will be over. The implications of this transition for cities are hard to fathom (Batty, 2011).

However, this history of urbanization and city building has been far from smooth. Morris (2010) argues that civilizations have always come up against what he calls a "hard ceiling," beyond which cities of more than one million or so in population cannot increase in size, effectively limiting urbanization, at least until the breakthrough of the Industrial Revolution. Urban civilization and its largest cities have waxed and waned across the epochs. However, as we will see in this chapter, although there are strong regularities in the system of cities with respect to the distributions of different city sizes, at all levels and through all times, there is also continual change. This is due to fierce competition for resources, which may well get greater as world population stabilizes after the demographic transition from high to low fertility rates passes. In this chapter, we will attempt to understand these changes; first, how cities sort themselves out into different sizes and, second, how volatility in the size of an individual city is transmitted up and down the urban hierarchy. We will begin with an almost nihilistic model of how populations grow that is simply based on random growth but is proportionate to size, which incorporates positive feedback. In this model, the rich get richer and the poor get poorer, but there is enough randomness to ensure that the rich and the poor have the possibility of swapping roles. We will then develop a variant of this model that explicitly considers resources, and then launch into several explorations of various city size distributions. We will contrast these with other urban size distributions—firm size and building size—and then steer our argument to questions of hierarchy and scale, which we will explore in the next chapter.

Population Growth and City Size

The Simplest Model
The basic premise for any kind of growth is the component of change between any two time periods t, and $t + 1$ is proportional to the size of the object that is the subject of growth. This is a positive feedback effect that occurs as growth compounds over time. If we assume the net rate of growth defined as λ remains constant over time, then the change in the size of the object, which we will consider to be a population $P(t)$, is

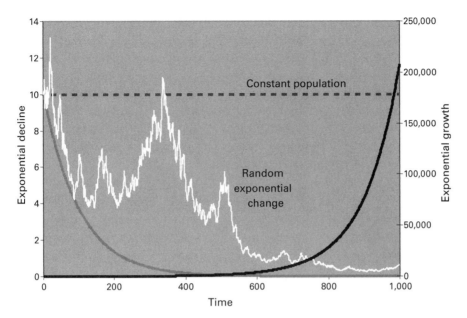

Figure 4.1
Exponential change. Continuous growth, black; continuous decline, light gray; decline, continuous constant dashed line; random change, white.

$$P(t + 1) = (1 + \lambda)P(t). \tag{4.1}$$

A simple recursion on equation (4.1) leads to a positive feedback or compounding expressed as

$$P(t + T) = (1 + \lambda)^T P(t), \tag{4.2}$$

which is the generic growth equation. It is easy to see that if $\lambda > 0$, then the population grows exponentially; if $\lambda < 0$, the population declines (negative) exponentially; and if $\lambda = 0$, net growth is zero, and the population remains the same, that is $P(t + T) = P(t)$. We show these graphs in figure 4.1, which makes clear the nature of the feedback. However, if the net rate of growth were to be chosen randomly at each time period within the range $-\xi < \lambda < \xi$, with a net mean growth rate of zero, this rate would compound and, starting with a positive population, the rate of growth would be likely to fall toward zero, as would the ultimate population. This, of course, depends on the sequence of random growth rates that vary around zero, for if the random growth rate were always positive (as is increasingly unlikely) through time, then the population would grow exponentially. This indeed holds the key to what happens when we consider a very large number of population objects, such as those comprising a system of cities, and we will explore this next. For a typical sequence of growth rates chosen randomly, we also show the population generated in figure 4.1 falls toward zero. To enable a population to oscillate randomly within a fixed range around its initial value, we need an additive, not multiplicative, model where the growth at each stage is independent of previous growth rates.

If we now think of our population model as pertaining to many different population objects such as cities, and we consider that the population is always positive, when we apply the growth rate chosen randomly between given limits to each of these populations, the proportionate effect would generate an increasingly small number of large positive populations (where the random growth was always greater than unity) and an increasingly large number of tiny positive populations (where the random growth rate was less than unity). We can now state the model for each population object or "city" $P_i(t + 1)$ as

$$P_i(t + 1) = (1 + \lambda_i(t + 1))P_i(t). \tag{4.3}$$

Recursion on equation (4.3) leads to the long time equation after T time periods, which is

$$P_i(t + T) = (1 + \lambda_i(t + 1))^T P_i(t)$$
$$= \prod_{\tau=1}^{T} (1 + \lambda_i(t + \tau))P_i(t). \tag{4.4}$$

The implication is that the population sizes would be distributed in some fashion resembling a lognormal distribution. We have simulated such a system for 10,000 population objects, beginning with each object having 1 unit of population and applying a randomly chosen growth rate $\lambda_i(t)$ to each object i between the limits $-0.01 < \lambda(t)_i < 0.01$. We show the distribution in figure 4.2(a), where it is clear that because of the zero limit on population size, the distribution is highly skewed to the left and follows what appears to be a lognormal form or even a power law. The maximum population size after 10,000 time periods is 12,119, with a mean population size of 9. In fact, this interpretation of the model is due to Gibrat (1931), who in studying firm size coined the term "law of proportionate effect" for the process of generating growth in this manner. Readers are referred to more rigorous demonstrations of the fact that this process always leads to these forms of distribution (Sornette, 2004; Saichev, Malevergne, and Sornette, 2010).

Blank and Solomon (2000) were among the first to demonstrate that if a lower constraint is placed on this process, which in this case consists of ensuring that a population is never able to decrease below, say, 1 unit, in the limit the distribution converges to a power law. This is clear from figure 4.2(a), where if we fit a distribution to $f[P_i(\tau)]$, we are able to show that

$$f(P_i(\tau)) = K P_i(\tau)^{-\alpha}, \tag{4.5}$$

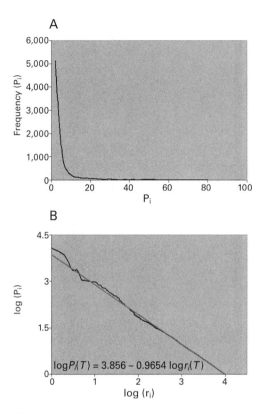

Figure 4.2
Proportionate growth: (a) the frequency distribution up to size 100 but noting that the maximum size is 12,000 and (b) the rank-size distribution over its entire range.

where K is a constant of proportionality and α is the inverse power of the distribution. We can fit equation (4.5) by minimizing the sum of the squared deviations from the predicted values by linearizing equation (4.5) to

$$\log f(P_i(T)) = \log K - \alpha \log P_i(T), \tag{4.6}$$

and finding the parameters K and α using regression analysis. If we consider equation (4.5) in continuous form as $f(P) = K\,P^{-\alpha}$ and we integrate this from some lower bound to its upper limit to form the counter-cumulative $\int_r^\infty KP^{-\alpha+1}dP$, this is equivalent to the rank r. In discrete terms for a time slice τ, this gives the rank-order of population $P_i(\tau)$ as

$$r_i(\tau) = KP_i(\tau)^{-\alpha+1}, \tag{4.7}$$

which we can linearize as

$$\log r_i(\tau) = \log K - \beta \log P_i(\tau), \tag{4.8}$$

where $\beta = \alpha - 1$ (Adamic, 2002). Equation (4.7) is in fact the rank-size rule or law popularized by Zipf (1949) that we introduced in chapter 1, and we can write this in alternate form as

$$
\begin{aligned}
\log P_i(\tau) &= \frac{\log K}{\beta} - \frac{1}{\beta} \log r_i(\tau) \\
&= \log Z - \frac{1}{\alpha - 1} \log r_i(\tau)
\end{aligned}
\tag{4.9}
$$

We graph equation (4.9) in figure 4.2(b) and it is quite clear that for most of its length, the relationship is linear, hence providing the classic signature of scaling based on a power law. In fact, this simulation produces an almost classic power law with a β value (power) of 0.965, with over 99 percent of the variance explained. This is extremely close to Zipf's Law, the pure rank-size rule where $\alpha = 2$ and $\beta = \alpha - 1 = 1$ (Zipf, 1949).

An Alternative Model Based on Swapping Resources

The basic reason why the population sorts itself out into a distribution that follows a pure power law is due to the interaction between the asymmetry of the boundary condition, which prevents the population from becoming negative, and the increasingly low probability that any population gets larger and larger. In one sense, the mechanism cannot be explained any further, and thus this can only be an artifact of the mathematical process used. As we will see in the next chapter, this probably does not accord to the precise timing of any real-world process of city growth, despite its plausibility in terms of the distributions produced. Before we move to examine empirical evidence, deeper explanations are required; we will explore one here.

Imagine that the population of a system of cities at time t is always fixed at a total population $P(t)$, the sum of N individual populations $P_i(t)$; that is,

$$
P(t) = \sum_i P_i(t), \forall t.
\tag{4.10}
$$

The only way cities can change in size is through migration. We assume that the migration associated with any city in any time interval $[t \rightarrow t + 1]$ is defined as $\Delta P_i(t)$, where $0 < \Delta P_i(t) < P_i(t)$. If we assume the simplest process of migration in which two cities j and k are chosen at random at time t, then the migration consists of adding an increment of population $\Delta P_j(t)$ to the population of the first randomly chosen city $P_j(t)$ and deducting the same amount from the second randomly chosen city k, which is

$$
\left.
\begin{aligned}
P_j(t + 1) &= P_j(t) + \Delta P_j(t) \\
P_k(t + 1) &= P_k(t) - \Delta P_j(t)
\end{aligned}
\right\}.
\tag{4.11}
$$

It is easy to show that the conservation constraint in equation (4.10) is met as

$$\sum_i P_j(t+1) = \sum_{i \neq j,k} P_i(t) + P_j(t) + \Delta P_j(t) + P_k(t) - \Delta P_j(t) = P(t).$$ (4.12)

Now, this is a process akin to atoms colliding randomly, gaining and losing energy (population) in such a way that total energy (total population) is conserved. For a system comprising, say, two cities with populations $P_1(t)$ and $P_2(t)$, then the probabilities of these two population events $\rho(P_1(t))$ and $\rho(P_2(t))$ when combined multiplicatively are equal to the probability $\rho(P_1(t) + P_2(t)) = \rho(P_1(t))\rho(P_2(t))$. It follows that the only function that meets this criterion is the Boltzmann-Gibbs distribution $\rho(P_1(t)) = C\exp[-P_1(t)/\overline{P}(t)]$, and $\rho(P_2(t)) = C\exp[-P_2(t)/\overline{P}(t)]$, where $\overline{P}(t)$ is the mean population per city and C is a scaling constant (Dragulescu and Yakovenko, 2000).

This result generalizes to many events, many cities, where the steady-state distribution is

$$\rho(P_i(t)) = C\exp[-P_i(t)/\overline{P}(t)].$$ (4.13)

This negative exponential distribution can be fitted directly from probability distribution, or it can be considered as a rank-size distribution if we integrate this in counter-cumulative form, linearize it and use least-squares regression to find its parameter. Assuming for the moment a continuous form for equation (4.13), we integrate $\int_r^\infty \rho(P(t))dP(t)$, and passing back to the discrete form, the rank-size distribution for the negative exponential can be written as

$$r_i(t) = K\exp[-P_i(t) / \overline{P}(t)] = K\exp[-\lambda P_i(t)],$$ (4.14)

which we can linearize as $\log r_i(t) = \log K - \lambda P(t)$. In fact, to make this consistent with our previous definition of the rank-size relation, we actually fit

$$P_i(t) = \log K - \frac{1}{\lambda}\log r_i(t),$$ (4.15)

from which it is clear that the parameter $\lambda^{-1} = \overline{P}(t)$.

We have generated such a distribution using the simple exchange model in equation (4.11) with the following parameters: $N = 500$ cities, total population 10,000, with the total number of time steps 1 million. We begin with a uniform distribution where each city has 20 persons and we fix the increment/decrement at $\Delta P_j(t) = 2$, $\forall j, t$. When we operate the model we ensure that no population $P_i(t)$ can fall below the zero threshold. If we choose two cities where this condition is violated when we make the swap, we ignore it and continue to choose two cities at random until the condition is met. The distribution of city sizes eventually converges to the Boltzmann-Gibbs distribution in equation (4.14). We show this as a probability distribution in figure 4.3(a) and in its rank-size logarithmic form in figure 4.3(b), where it is clear

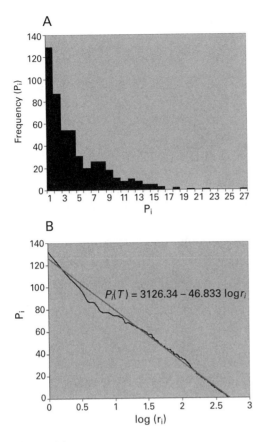

Figure 4.3
The additive exchange model generating the Boltzmann-Gibbs distribution. (a) The frequency distribution up to size 27. (b) The rank-size distribution over its entire range.

that the fit to the theoretical distribution is extremely good. The probability distribution is not as good a fit as the rank size, but this is an artifact of the presentation, and if averaged over a number of runs up to the limit of 1 million, this would generate a near-perfect distribution, as Dragulescu and Yakovenko (2000) show. The average population per city is also generated as the slope of the rank-size distribution, and this accords to the parameters and initial conditions of the swapping model. This, however, is not a power law, but it is possible to glimpse ways in which such processes of swapping—migration—do lead to distributions that are power laws by relaxing some of the conditions and using the same kinds of algorithm.

There are many models of exchange that generate scaling-like laws with respect to the population sizes that result from the process, with multiplicative rather than additive models tending to generate distributions that are nearer to power laws. However, such models are sensitive to rates of exchange and to the volume of exchange, and invariably they do not conserve the total amount of population in the system. A model that is similar to the one specified in equation (4.11) consists of choosing cities at random, adding a random proportion of the city's population to itself, and taking this random proportion from the population of the other randomly chosen city. Clearly symmetry is broken in this process, as the following equations imply—that is, the process cannot be reversed if the choice of random cities at the next iteration is the same but with the city that lost out in the previous round being the winner in the new round. Then

$$P_j(t+1) = P_j(t)[1 + \xi_j(t)] \\ P_k(t+1) = P_k(t) - \xi_j(t)\Delta P_j(t) \Bigg\} , \tag{4.16}$$

where $\xi_j(t)$ is a randomly chosen rate of change such that $0 < \xi_j(t) < 1$. Each time the swapping process is initiated, an exchange is not made if either of the populations in question have a value less than 1; that is, the swap is abandoned if $P_j(t)$, $P_k(t) < 1$. If we operate this process a large number of times, the probability distribution of city sizes appears to converge, but it is unclear what form this particular distribution has. In the example we show here, we start with the same parameters as in the Boltzmann-Gibbs model above, that is $N = 500$ cities, total population 10,000, total time steps 1 million, and a uniform distribution where each city has 20 persons. Then, using randomly chosen rates of change always less than 0.5, the distribution that emerges is near a power law, as we show in its rank-size form in figure 4.4. However, other distributions that are nearer the exponential emerge for lower growth rates, and an entire new class of models is generated if the population is allowed to become negative, as is the case when population might be considered

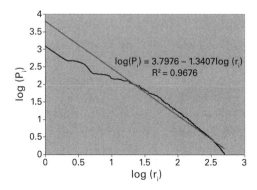

Figure 4.4
A multiplicative exchange model generating a power-law-like distribution.

as capital or wealth. These applications have been explored elsewhere (Ispolatov, Krapivsky, and Redner, 1998).

Scaling as the Signature of Complex Systems

Throughout this book, we will argue that the pervasive signatures of complex systems are distributions that vary regularly with the size of the components or objects comprising the system. Whether these are scaling in their strict form as power laws is of less consequence than the fact that regularity persists across different scales, which in the case of city-size distributions relates to the influence over space that cities at different levels of hierarchy reveal (Pumain, 2006a). We demonstrated this consequence in chapter 1 for different patterns that make up the city, but here we need to examine the extent to which the simplest organization of cities by their sizes reveals an ordering that explains their regularities in space and time. We have already introduced the means to map these signatures, and this consists of examining their size distributions using their rank order following Zipf (1949). If the distribution follows a power law, this rank-size ordering is as formalized in equation (4.9); if it is a negative exponential, the equivalent relation is as in equation (4.14) (or 4.15).

The simplest models that "explain" these distributions are those introduced in the last section, and these are based on either multiplicative growth or additive growth, the first leading to power-law distributions, the second to exponential distributions. In both cases, the way cities of different sizes are generated is due to cities being able to grow without limit but decline only to a lower threshold that is a hard boundary, below which a city ceases to exist. In these models, cities that reach this lower limit—which might be regarded as a subsistence limit—are in decline, but are then thrown back into the pool of candidate cities and are able to grow again. Models of random proportionate growth generate power-law distributions at least in their upper tails, but a more general form of distribution is the lognormal. Only when such models have a hard lower threshold does a power law emerge in the steady state (Simon, 1955; Gabaix, 1999; Blank and Solomon, 2000; Sornette and Cont, 1997).

In the rest of this chapter, we will examine how distributions of cities that form our main concern here change over space and over time, identifying the key issue: these distributions appear to be stable in time, showing little sign of change in their scaling over tens or even hundreds of years. However, when we examine the detailed dynamics of how their ranks shift in time, there is considerable volatility, with the objects in such distributions often persisting for no longer than fifty to a hundred years. To explore this kind of micro-volatility, we introduce a number of measures of rank shift over space and time and visualize size distributions using the idea of the "rank clock" (Batty, 2006a). We use the example of changes in the population of Italian towns between 1300 and 1861 to introduce these ideas and then compare this analysis with city-size distributions for the World from 430 BCE, the United States from 1790, the United Kingdom from 1901, and Israel from 1950. The morphologies of growth and change displayed by these clocks are all quite different. When we compare these to the distribution of US firms from 1955 in the Fortune 500 and to the distribution of high buildings in New York City and the World from 1909, we generate a panoply of different visual morphologies and statistics. This provides us with a rich portfolio of space-time dynamics that adds to our understanding of how different systems can display stability and regularity at the macro level with a very different dynamics at the micro level.

The models we have introduced to explain scaling are as parsimonious as they can be. Indeed they are almost nihilistic, in the sense that competition is implicit in their processes and cannot be unpacked in terms of why such multiplicative growth should be appropriate. In short, the assumptions do not extend to any rationale as to why competition between cities takes place in this way. Our extension to additive (and other multiplicative exchange models) hardly explains the kind of ordering that emerges. In the next chapter, we will extend these to network effects, in which the size of a node in a network grows using proportionate effect to add network links, from which such competitive effects emerge (Batty, 2006b). Indeed, Barabasi's (2005) models, which depend on preferential attachment, can be seen as a subclass of these kinds of growth models, all falling under a wider class of models that emphasize cumulative mutual advantage.

There are several types of growth processes that can be modeled using this generic mechanism. In terms of cities, populations that are measured by their size can grow or decline. Cities grow from the smallest settlements to the point where they "become" cities. Their growth is thus asymmetric, in that to be a large city, one must first be a small city. This is characteristic of all growth processes, of course. Because of difficulties in defining what a city is at the lowest population level, but more because of a lack of data, we will only work with the largest cities, taking the top 100, 200, and so on, thus establishing immediately that the analysis and visualization is relative to the size of the system in terms of numbers of objects chosen. In such cases, births of cities occur when they enter the top ranks—say, the top 100—and die when they leave this ranking. Cities are, of course, regularly spaced according to a spatially competitive hierarchy originally formulated by Christaller (1933/1966) under the banner of central place theory, and the random proportionate growth model appears to be consistent with this logic. However, what we do not do here (although we indicate how we might extend all this in the next chapter) is explore much more elaborate socioeconomic spatial models of the urban and regional system based on how individuals and firms operate spatially in economic markets. Here we prefer to simply argue that the basic mechanisms of proportionate and additive growth are sufficient to provide a substrata for interpreting how city systems evolve with respect to their size distributions.

There are, of course, difficulties in defining the spatial extent of cities, but generally these can be dealt with. It is much harder to ensure that firms are defined appropriately, since mergers and acquisitions can destroy any size grouping immediately in the absence of constraints, whereas in cities, spatial extent determines size. In terms of buildings, although these have generally grown taller as cities have also grown, and certainly over the last one hundred years, since skyscrapers were first constructed, buildings do not usually grow per se. They are usually demolished and constructed afresh, and only a small fraction of buildings are modified and grown or reduced in size *in situ*. It is easier to see how firms and cities compete to increase in size than it is to see this in buildings. Nevertheless, bigger buildings usually do occur in places where there are already high buildings, and it is easy to consider the random proportionate growth model being adapted to take account of copycat-like behaviors as ever higher buildings are proposed and constructed. In the case of buildings, then, we might expect a rather different dynamics from that which is characteristic of city and firm sizes. In the period in question—the last one hundred years or so—we will disregard the very small number of tall buildings that have been demolished, concentrating only on those that have been constructed anew, do not grow per se, and are still within their planned lifetimes.

The Space-Time Dynamics of Cities

We begin with a modest example involving the growth of cities in the Italian peninsula from the fourteenth to the nineteenth century (Bosker, Brakman, Garretsen, De Jong, and Schramm, 2007; Malanima, 1998). The space-time dynamics is relatively uncomplicated in that the main cities—Bologna, Firenze, Genova, Milano, Napoli, Padova, Palermo, Roma, Venezia, and Verona—were established by the fourteenth century and the period until Italian unification, when our analysis ends (in 1861), did not see dramatic growth or radical shifts in the ranking of the key centers. We have a manageable seven time instants—1300, 1400, 1500, 1600, 1700, 1800, and 1861—for which we have population size data on the 555 towns that existed during the period, and we can thus rank the towns by population size and examine any changes in rank rather easily. This core of towns has remained in the top fifteen by population size since the beginning of the Renaissance. In the subsequent analysis, we will only examine the rankings of the top 100 towns by size at each of the seven time instances, although over the temporal period, a total of 195 distinct towns enter the analysis. Of course some enter the top 100 and then leave, only to enter again by the end of the period, and our analysis cannot account for towns that move up and down the rankings during the intertemporal periods for which we do not have data. In the data set, 360 towns that exist in fact never enter the top 100 and are thus really villages, or at least settlements that never reach significance. We need to note that although we are dealing with the top 100 towns by population size in each of the seven time periods, in the second period (from $t = 1400$), the total number of towns falls to 95, while in all other periods, the number is slightly greater than 100 due to ties in the rank orders.

In terms of the relative stability of population during this period, total figures for all the 195 towns that enter the top 100 shows a population of 2.1 million (m) in 1300 declines to 1.71m by 1500 (probably due to the Black Death and war). It only recovers to 2.22m in 1700 and then rises to 3.97m by 1861, indicating the early beginnings of political unification and industrialization. As we will see, this comparative stability is complemented by highly stable rank-size relations. However, this is only true of the core towns as there is considerable volatility in the smallest towns in terms of their ranks, which show the essential struggle to compete in the face of the predominance of the core towns of the Italian city states.

This is a good case, for it lets us introduce a number of tools for analyzing and visualizing space-time dynamics using a particularly tractable and easy-to-understand example. We first graph the seven rank-size relations as logarithmic transformations of $P_i \sim r(P_i)^{1/1-\alpha}$, which give $\log P_i = \Phi - \vartheta \log r(P_i)$ where the slope ϑ is related to the scaling parameter $1/(1 - \alpha)$. Note that we drop the time subscript t or τ for the moment and relate the rank to size as $r(P_i)$. This is the same form as in equation (4.9) above. We show these distributions in figure 4.5(a), where their similarity is even clearer when they are collapsed onto one another. There are various ways of estimating the scaling parameter. The most traditional, which is in fact the most biased, is to estimate ϑ using ordinary least squares (*OLS*), from which we can then compute the scaling parameter as $\hat{\alpha}_{OLS} = 1 + \vartheta^{-1}$. A less biased way, however, is to use the maximum-likelihood method, which has been adapted to power functions by Newman (2005) and consists of solving the likelihood equation for each distribution as

$$\hat{\alpha}_{ml} = 1 \ + \ N\left[\sum_i \log \frac{P_i}{P_{\min}}\right]^{-1}, \tag{4.17}$$

where $\hat{\alpha}_{ml}$ is an estimate of the scaling parameter, N is the number of observations in the city data set (which can vary for each time period), and $P_{\min}$ is the minimum population defining the lower bound of each city-size distribution. Because the conventional argument is that such size distributions are more likely to be in their steady state in the fat or upper tail, then the lower tail should be truncated at some minimum and, as Clauset, Shalizi, and Newman (2009) suggest, this might be a matter for experimentation. We do indeed explore this in the software available for

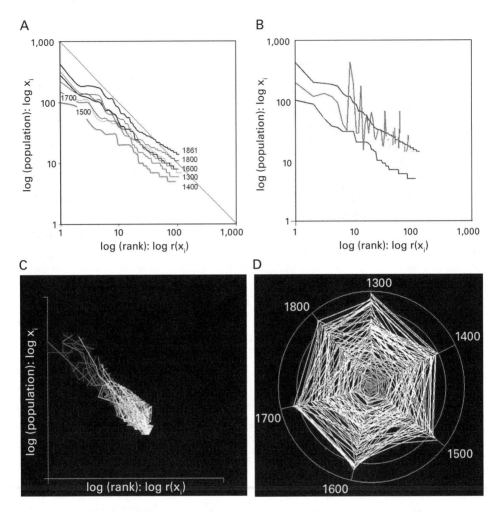

Figure 4.5
Visualizing space-time dynamics in terms of rank shift. (a) Zipf plots for the seven time instants. (b) Shifts in rank between 1400 and 1861. (c) Shifts in ranks of all towns throughout the seven time periods in rank space. (d) The rank clock. The colors from black through gray are chosen according to the chronology of appearance of towns from 1300 and according to their rank. The first town, the top rank at 1300, is colored red; the last town is the last rank to enter at the latest time period and is colored blue, with the red-blue spectrum mapped onto the equivalent gray tones.

Table 4.1
Estimates of rank size

Year	$\hat{\alpha}_{ml}$	$\bar{\alpha}$	$\hat{\alpha}_{OLS}$	r^2
1400	2.836	2.696	2.393	0.851
1500	2.606	2.522	2.283	0.858
1600	2.500	2.438	2.262	0.873
1700	2.631	2.506	2.277	0.871
1800	2.698	2.594	2.403	0.893
1861	2.582	2.562	2.365	0.899

these visualizations and having estimated $\hat{\alpha}_{ml}$ for M different minimum values P_k, we form the average as

$$\bar{\alpha} = \frac{1}{M} \sum_k \left\{ 1 + N \left[\sum_i \log \frac{P_i}{P_k} \right]^{-1} \right\}. \tag{4.18}$$

In table 4.1, we present these estimates, whose values are very close to one another, speculating that these values are close enough to suggest that there have been no major transitions in top-ranked cities during this 500-year period.

Visualizing Scaling: The Rank Clock

There are two obvious ways in which to further explore the space-time dynamics in these distributions. First, we can examine the shift in ranks between two periods. For any two different time instants, we can examine the rank shift, which can be visualized by plotting one of the size distributions using the ranks associated with the other distribution. In figure 4.5(b), we show this shift for the Italian city sizes, plotting the distributions in 1400 and 1861, where shift is based on plotting the year 1400 sizes using the 1861 ranks. The picture is one of greater volatility than we have seen so far. We can plot all the shifts for every time instant in rank space if we trace the rank and size of each city through their evolution, as we do in figure 4.5(c). Here the colors are set as follows: the largest and most ancient town is colored red throughout, and towns that are smaller, enter the top rank later, or both are colored according to the spectrum red-orange-yellow-green to blue, mapped onto gray scale. We also use this color map for the rank clocks that we use to visualize these and other distributions introduced later in this chapter. This visualization in rank space, however, confuses the picture even more. Whereas figure 4.5(b) does imply considerable shifts in rank, figure 4.5(c) tends to play these down, for the balance of color shows that the oldest and largest ranks from early in the evolution of the city system tend to remain in place throughout. What we need is something a little more visually intrusive to pull out any deviations that are significant, and to this end, we introduce the idea of the rank clock.

The rank clock focuses entirely on changes in the ranks, with the trajectory of each object—in this example, a town or city—representing its rank order in a circular space. At any time, the top-ranked object is always at the center, while the lowest rank is at the circumference, or the furthest point to the edge if the circumference has not been reached. Time is arrayed in a clockwise direction from the usual north-noon point of the clock. The time period over which the analysis takes place is marked by the circumference, whose length is 2π, in short, the complete clock. To compare two rank clocks, it must be assumed that the temporal circularity is directly comparable between examples, and it is thus only relevant to do this if one assumes that the temporal behaviors of two or more different systems are comparable within the sweep of the circumference. Moreover, systems with different numbers of objects are usually not directly comparable, although a suitable scaling of the clock might generate some comparability. As the examples introduced below will show, different clock configurations define very different kinds of scaling system, thus representing a new kind of morphology for analyzing and visualizing space-time dynamics.

We show the clock in figure 4.5(d). Two features are immediately clear. First, the towns that are top-ranked in the late Middle Ages and early Renaissance retain their dominance to the middle of the nineteenth century (and casual observation suggests that this is still the case in the early 21st century). This is clearly seen in the concentration of the color red around the pole of the clock. Second, most of the volatility, where colors diverge markedly from circularity, occurs toward the edge of the clock. Here we see little evidence of towns that are large and become small leaving the pole of the clock, spiraling out to the edge, or of towns that rise in rank dramatically spiraling into the clock. Unlike our later cases, there are only a handful of such examples in the Italian city system. Torino does not enter the picture (the top 100 ranks) until 1500, and then it rises swiftly to rank 4 by 1861. Siena, on the other hand, occupies rank 7 in 1300 and drops to rank 45 in 1861. It is hard to make these out in the clock in figure 4.5(d) but the software that is available enables this to be illustrated quite easily, as we will show for other examples (Batty, 2006a).

The clock also focuses us on other measures of change. Changes in rank from time period to time period are clearly seen as changes in the distance traveled around the clock, since this kind of morphology is rooted in a circular geometry. For the rank $r_i(t)$ of each object i at time t, we can define an individual distance—a first-order difference—that can be plotted on the clock. We do not show the distance clock here (Batty, 2006a), but distance is defined as

$$d_i(t) = |\, r_i(t) - r_i(t-1) \,|. \tag{4.19}$$

Aggregate distances can be defined over all cities for each time period as

$$d(t) = \sum_i |r_i(t) - r_i(t-1)|/N(t), \tag{4.20}$$

where $N(t)$ is the number of cities in the distribution at each time period. If cities are unique in terms of measures of their size, then this might be set as $N(t) = 100$, $\forall t$, although in the Italian example, because of ties in size, the number varies slightly at each time. This gives a measure of the average changes in rank that occur for all cities remaining in the set. In fact, second-, third-, fourth-, and greater-order changes can be computed if so required, as can related measures of cumulative change such as

$$d(\tau) = \sum_{i, t=1,\ldots,\tau} |r_i(t) - r_i(t-1)|/N(t). \tag{4.21}$$

The average switch in ranks over all cities and all time periods, in this example from 1300 to 1861, and over all 195 cities that appear in the top ranks, is

$$d = \sum_t d(t)/T, \tag{4.22}$$

where T is the total time over which the change takes place; in this example, it is 561 years. We show these measures for the Italian system in table 4.2, where it is clear that on average, a typical city shifts about 15 ranks (d) over 100 years, consistent with shifts ($d(t)$) taking place over each 100-year interval, which range from 12 to 17.

Our last measure of space-time dynamics involves the speed at which cities enter or leave the top-ranked set of cities at different times. It is easy enough to count the number of cities composing the top set of ranks at time t that are still in the new top set at some time later or earlier than t, say $t + \tau$ or $t - \tau$. We define this number as $L(t, t + \tau)$ for the cities that existed at time t in the top ranks and are still in the top ranks at $t + \tau$; and $L(t - \tau, t)$, the number of cities that were in the top ranks

Table 4.2
Distance measures: Changes in rank for all cities over all time periods

Time period	$d(t)$	$d(\tau)$
1300–1400	16.949	16.949
1400–1500	16.797	21.612
1500–1600	17.064	23.775
1600–1700	11.780	25.371
1700–1800	13.063	23.940
1800–1861	14.518	21.446
Average distance d	15.519	

at time $t - \tau$ and are still in the top ranks at time t. A little reflection suggests symmetry in these counts, that is $L(t, t + \tau) = L(t - \tau, t)$. We can work out the average number of cities that perpetuate with respect to a given time period t as

$$L(t) = \sum_{m \neq t} L(t, m)/(N - 1), \tag{4.23}$$

while the average number of cities that perpetuate from the top ranks at any time over all other time periods is

$$L = \sum_{\ell \neq t, m \neq \tau} L(\ell, m)/(N - 1)^2. \tag{4.24}$$

These measures assume that the number N of cities is fixed and needs to be modified if these vary by time period, as in the Italian example; proportionate or probability measures can also be computed if required.

The half-life has to be defined with respect to the entire time period T. We do not have explicit functions that describe the process by which cities persist in the top ranks. Thus we need to figure out these half-lives by inspection and interpolation, since the time intervals are usually not fine enough to compute these half-lives exactly. We can do this for the whole series, of course, or we can do it for each time instant where it will vary. Essentially, for each time t, we need to solve the equation $L(t, t)/2 = L(\tau, t) = L(t, \tau)$. Because no generic formal function for this persistence exists, we need to examine the matrix $L(t, t + \tau)$ but, assuming exponential decay, it is likely that the number of cities in the top ranks will decline through time from the point at which they are considered. If they do not, then this is evidence of extreme persistence and regularity in the system, and the degree to which the half-life approaches the maximum number of ranks is a measure of this stability.

Our Italian example is useful, for the number of time periods is small enough to enable a casual examination of the structure of the matrix elements $L(t, t + \tau)$, which we present in table 4.3. Although the maximum number of towns in any one distribution is 118, 195 different towns appear in the matrix from 1300 to 1861. There is, in fact, a remarkable persistence in the core set of towns, with an average of about 74 appearing to perpetuate throughout the period. In the remaining towns, there is considerable volatility in the rankings. So in this system, which at first sight seems simple and robust with little change among the largest towns, there is considerable movement in the smaller towns, which continually and aggressively compete for pride of place. From table 4.3, it is clear that the half-life in all cases is perhaps a little longer than the 561 time periods over which these distributions of towns are observed, implying very little spatial restructuring and rather low levels of growth. For the 118 towns in 1300, some 65 or 55% of these still exist in the top 109 in 1861. An average half-life of around 600 years would be a good guess, but this is much larger than the other examples of city systems we will now examine.

Table 4.3
Number of cities $L(\ell, m)$ that persist from times ℓ to m

Time ℓ, m	1300	1400	1500	1600	1700	1800	1861
1300	*118*	81	80	71	70	67	65
1400	81	*94*	81	65	62	64	59
1500	80	81	*106*	82	76	73	70
1600	71	65	82	*104*	85	80	76
1700	70	62	76	85	*103*	89	79
1800	67	64	73	80	89	*107*	86
1861	65	59	70	76	79	86	*109*
L(t)	72	69	77	77	77	77	73
L	74						

City Systems at Different Scales

We now have a small arsenal of tools to explore the space-time dynamics of scaling systems, and it is worth briefly summarizing what we have presented. First, a good measure of stability with respect to size is the Zipf plot, specifically the scaling parameter α or its equivalent ϑ, which is the slope of the plot. This gives the degree of competition between the elements with $\vartheta = 1$ and $\alpha = 2$, the pure Zipf case where $P_i \sim r(P_i)^{-1}$. When $\vartheta < 1$ or $\alpha > 2$, then the larger objects in the distribution are closer in size to the smaller, implying less competition and vice versa. Of course, the measure of fit of the power law to the Zipf plot and how these change through time are also measures of how close these plots are. The two methods we have suggested—maximum likelihood and OLS—are standard, notwithstanding various known biases (Newman, 2005; Clauset, Shalizi, and Newman, 2009). Second, the best graphic is the rank clock rather than any measure associated with the Zipf plot for individual objects can be traced in terms of their trajectories, and the whole set of objects can be visualized as a morphology. The clock is composed of distinct trajectories, five of which we visualize in figure 4.6, although our more general quest is to compile different examples of these morphologies, which will ultimately provide some sense of the dynamics of different kinds of scaling systems. We will attempt this now, although we will show different morphologies for cities later in this section and for firms and tall buildings in the next.

In the clock in figure 4.6, objects that remain at the same rank are marked by exact circles around the clock, which close on themselves. Objects that rise inexorably in their rank spiral into the clock, while objects that decline systematically spiral out. Objects can, of course, enter and leave the clock as many times as is possible, and there are obvious extreme cases: if an object were to enter at t, leave at $t + 1$, enter again at $t + 2$, and so on, then this pattern would simply appear as

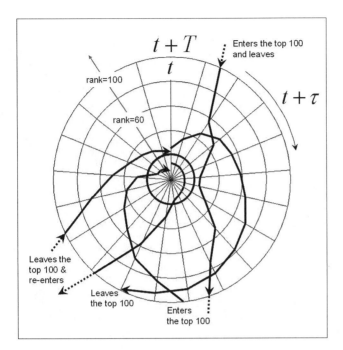

Figure 4.6
Possible trajectories defining the morphology of the rank clock.

a dot at every other time period at the rank where it entered. Note that we do not have the rank of the object before it enters the top ranks, so objects quite literally appear on the clock when they enter. If we did have these ranks, then we could compute their actual trajectories. One way of doing this would be to construct the clock for a much bigger system of objects and then reduce this to a lesser number of top ranks. Figure 4.6 also shows objects that display such oscillations, namely, one that enters and then leaves to re-enter again, and its opposite, one that leaves, re-enters, and leaves again. The pole of the clock is significant, for in many scaling systems, one of the objects often remains the biggest. In city systems, this is quite common: for example, for the last 200 years, New York City has been at rank number 1 in the US urban system, London at rank 1 for the United Kingdom, Paris at rank 1 for France, and so on.

The two measures of difference, the first based on distance and the second based on guesstimates of the half-life, are useful in measuring actual shifts in rank. The distance, of course, depends on the time periods in question and represents the shift that takes place in numbers of ranks over one standard time period. In the case of the Italian data, this is 100 years (with 61 years appropriately adjusted for the statistics in the seventh period). Examining table 4.2, we see that the average rank shift is about 15 over 100 years, and this suggests that over 500 years the shift would be about 75. In fact, the second statistic is the half-life, which we estimated to be about 600 years, meaning about 50 to 60 would shift out of the top ranks entirely during this period. As cities enter and leave the rankings, then simply taking the average distance d and scaling it by 5 or 6 will give an overestimate of the shift, but it suggests that judicious use of half-lives and distance measures provide a rich picture of these dynamics. The cumulative distance $d(\tau)$, in fact, is probably a better measure of shift, because this does take into account cities that enter and re-enter the rankings as the system evolves.

We are now in a position to select and track some very different city systems that exist at very different scales. We have analyzed city-size distributions for the top 50 cities in the world from 430 BCE to 2000 CE with variable time intervals using Chandler's (1987) database, which is the largest scale and longest time period that we have dealt with. At the next scale down, the continental, we have examined changes in rank size for the top 100 cities in the United States from 1790 to 2000 at 10-year intervals (from the US Census Bureau; see Gibson, 1998); at the country level, the top 100 in the United Kingdom from 1901 to 2001, also at 10-year intervals (ONS, 2010); and finally for a much smaller region, the top 172 towns in Israel from 1950 to 2005 with variable time intervals (Benguigui, Blumenfeld-Lieberthal, and Batty, 2009). Israel is more like a metropolitan region with respect to scale, although each of these examples have very specific geopolitical and cultural characteristics that add to the variety of this selection.

For each of these systems (including the Italian example), we need to choose either a fixed set of top ranks or simply let the number of cities in the wider database condition the number of cities being ranked at any time instant. For example, in the Israeli example, there are 172 towns that exist at some time over the 55-year time period from 1950 to 2005. In 1950, there are 34 towns defined, while by 2005 there are 164. However, 8 of the towns appear at some point in the rankings for the 13 time instants, entering and then leaving the space by 2005. In this sense, we do not constrain the Israeli example, whereas for the UK example, we have 458 towns that do not change throughout the entire 100-year period. This is complicated by the fact that we only examine the top 50 towns at any of the 10 time instants, and in this case, 70 of the 458 towns are considered, but at each point in the rankings, there are only 50 towns. To be able to make good comparisons, we need to be clear about these limits and how the statistics of rank shift are affected by changes in numbers of cities and times.

The World system of cities is the most volatile. There are no cities in the top 50 in 430 BCE that are still in the top 50 by 2000, and there are only 6 that still exist in the top 50 from the Fall of Constantinople in 1453. In fact, the half-life of the original set of cities is about 200 years, which reduces to about 100 years by the twentieth century. In short, the rise and fall of civilizations, particularly Greece and Rome, the coming of the Dark Ages, the parallel growth and decline and growth again of China, and the explosion of cities in the developing world can all be gleaned from the trajectories of cities in this database. The morphology of the rank clock is shown in figure 4.7(a), and here it is quite clear that there is no sense in which there is a group of persistent core cities, as in the Italian example. The longest-lived example of a city in the top 50 is Suzhou in China, which exists for 2,158 years of the 2,430 years covered, and even this city is no longer in the top 50 (although it is growing fast and could re-enter the list, unless it is absorbed into greater Shanghai). The half-life of cities in the World system is clearly reducing fast and is now no more than 75 years, falling at the rate of 20 years for every additional 25 years of time. The system nearest this rate of change is the US system, plotted in figure 4.7(b), which displays all the features of the examples in figure 4.6. New York City remains at the center of the clock, unchanging since 1790, while cities like Chicago, Los Angeles, and Houston spiral into the top ranks as the population has diffused to the Midwest, California, and the Southwest over the last 150 years. Cities such as Charleston, in the old colonial East, spiral out of the clock as the United States begins to industrialize from the mid-nineteenth century on.

Our two other examples are equally as varied. In the case of Great Britain (England, Scotland, and Wales), the half-life is about twice that of the United States and the current World system, at about 150 years, and it shows little sign of changing. This is because by 1901, all the key settlements were established—the core cities had developed during the nineteenth century, and suburbanization in the twentieth has not made much impression on the overall pattern of urban development. The clock is shown in figure 4.7(c), where the core stands out as in the Italian system. The Israeli system, however, is quite different in that it is almost impossible to guess the half-life as new settlements have been established and grown continually during the last half-century, when the country has developed and consolidated. In such a growth situation, the half-life is effectively an order of magnitude longer than the time during which development has taken place. A crude guess would be between 50 and 70 years, but in a growth situation where all 34 of the original settlements in 1950 are still in the top ranks in 2005, this is hard to assess, for the number of settlements dropping out of these ranks is very small. The half-life is really a forward-looking measure, notwithstanding the symmetry of the flow matrix $L(\ell, m)$. Better measures involve distances, which we will now discuss. The rank clock for the Israeli system is shown in figure 4.7(d).

The fit of the rank-size functions to the data for all city systems over all time periods is shown in table 4.4; the proportion of the variance explained ranges from 85% to 95%. The scaling parameters all tend to be greater than 2, suggesting that

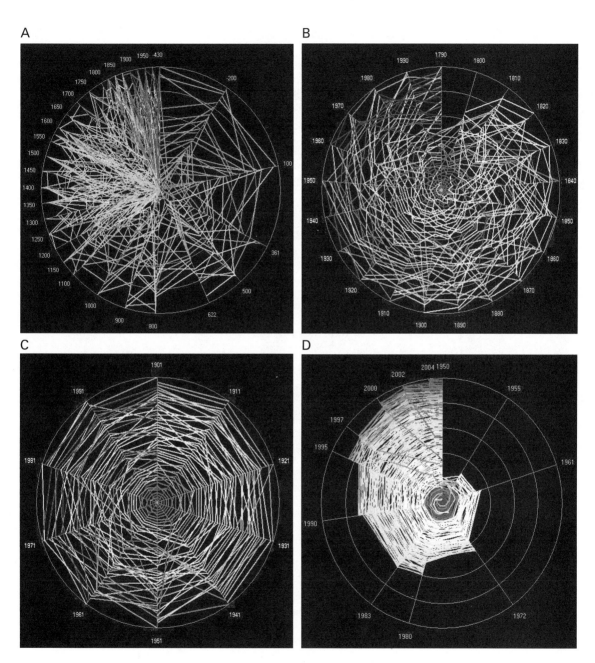

Figure 4.7
Rank clocks for (a) the World, (b) the United States, (c) Great Britain, and (d) Israel.

Table 4.4
Comparative measures of rank shift for five city systems

City systems	Number N	Time T yrs	$t \to t+1$ yrs	Min $\hat{\alpha}_{ml}$	Max $\hat{\alpha}_{ml}$	Min r^2	Max r^2	Min $d(t)$	Max $d(t)$	d	L
World cities	47–50 (390)	2430	80 (25–260)	2.127	3.236	0.906	0.945	4.914	10.973	7.785	14
US cities	24–100 (266)	210	10	1.952	2.674	0.911	0.944	2.242	7.559	4.667	25
UK cities	50 (458)	100	10	2.746	4.062	0.948	0.951	1.941	9.857	4.220	41
Israeli cities	34–164 (172)	55	4 (1–11)	1.748	2.035	0.782	0.850	1.578	5.917	4.032	83
Italian cities	94–118 (555)	561	80 (61–100)	2.493	2.836	0.851	0.899	11.780	17.064	15.519	74

the largest cities tend to be less dominant than in the pure Zipf case ($\alpha \sim 2$), which is widely regarded as being the situation when the system has adjusted to its ultimate steady state (Gabaix, 1999). The average distance shift in ranks must be interpreted in the light of the average time periods over which the size trajectories are observed. Interestingly, the city systems since industrialization (the United States, United Kingdom, and Israeli) show shifts of rank between 4 and 8 over 10-year periods, whereas the shift in rank for the World system is about 8 over periods of some 80 years. This distance measure is much less reliable since the time periods vary so massively, while in the case of the Italian system, the shift is about 16 over 80 years, which *pro rata* gives a shift of some 5 ranks for very 10 years, similar in fact to the other national city systems.

The last point we need to make is based on a visual comparison of the rank clocks. Clearly, the US and Israeli clocks begin with less than 100 top cities as these systems do not have 100 cities at their start. The US clock then reaches 100 in 1840, after which the top ranks remain stable. The Israeli clock illustrates the entire growth trajectory with no constraints on numbers. Both clocks show growth, with the Israeli example showing the persistence of core cities and massive growth of new ones. The US system has much less of a focus on the core cities, which is consistent with the rapid diffusion of large cities in the West and South. The British picture is one of more classic slow growth with many trajectories showing circularity, in which ranks remain similar over time and the core is maintained. The World system is by far the least stable in that cities of the ancient era are clearly quite different from those of the Middle Ages, the Chinese empire, and the modern world.

Disaggregate Populations: Firm Sizes and Building Heights

Our last two examples involve rather different human systems. The first, based on size distributions of US firms from 1955 to 1994, is taken from Fortune 500 data (CNN, 2009). The second, size distributions for tall buildings (skyscrapers) for the World and for New York City, is culled from the Emporis buildings database (Emporis, 2009). The firm-size distribution is in fact nonspatial (other than pertaining to the United States), while the distribution of skyscraper heights is at a level of spatial detail far below that of the city-size distributions for which we first developed these visualizations. In terms of firms, we have focused not on all 500, but on 100, since by now it must be clear that as the number of objects increases, the rank clock is more and more impressionistic and requires zoom capability to see individual trajectories (see O'Brien, 2012). Firms can be ranked, as can cities, by any measure of their size, and in this case, we have revenue and profit-earnings ratios, which give a measure of how well a firm is doing. We will not show this latter set of rankings here (but see Batty, 2007). However, in figure 4.8(a), we plot the revenue clock, and

A

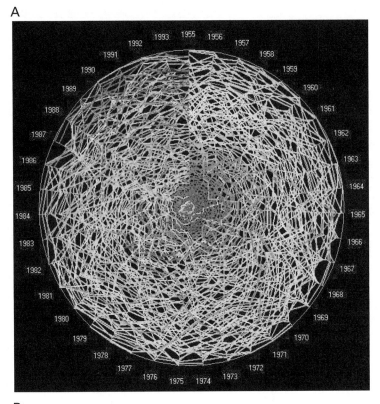

B

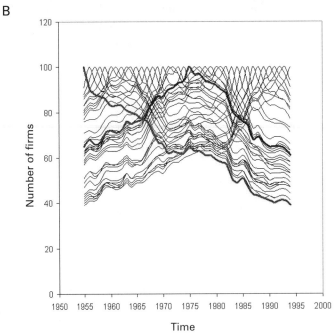

Figure 4.8
(a) The Fortune 100 rank clock and (b) the persistence-decline of firms by rank, 1955–1994.

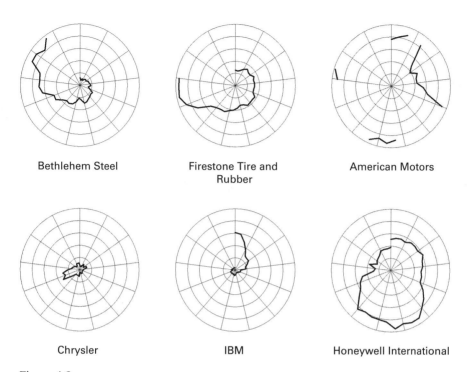

Figure 4.9
Individual rank trajectories for selected Fortune 500 firms, 1955–1994.

it is immediately clear that there is large but regular volatility in the rankings of firms.

Essentially, a core of firms does stand out, but there is considerable erosion in their ranks. The rate of this erosion is rather regular. This is best seen in figure 4.8(b), which is a visual plot of $L(\ell, m)$ showing the number of firms that stay in the top 100 rankings as time elapses. We can measure the half-life from this, which is about 25 years, and figure 4.8(a) marks out the firms that stay in the top rankings from the beginning year, 1955. The firms that enter or leave the rankings pivoted around mid-year of 1975. Figure 4.9 shows the trajectories of six of these firms; quite clearly, steel and heavy industry such as rubber spiral out of the rankings, while high-tech firms (e.g., IBM) spiral in. American car manufacturers like American Motors have mixed fortunes but the largest, like Chrysler, more or less maintain their rank, with General Motors (not shown) ranked as number 1 throughout the 40 years. The picture is very different now in 2012, and the analysis could be brought up to date so that the relevance of these statistics of change can be explored further. In table 4.5, the rank-size relations show extremely good fits, with r^2 near 99% for all distributions. The change in rankings from the distance measures are based on years and indicate quite rapid change compared to all the city-size distributions.

Table 4.5
Comparative measures of rank shift for firms and tall buildings

Scaling systems	Number N	Time T yrs	$t \to t + 1$ yrs	Min $\hat{\alpha}_{ml}$	Max $\hat{\alpha}_{ml}$	Min r^2	Max r^2	Min $d(t)$	Max $d(t)$	d	L
Fortune 100 firms	100 (343)	41	yearly	2.205	2.739	0.987	0.995	3.397	7.988	5.158	56
NY City skyscrapers	12–119 (516)	101	yearly	3.048	6.259	0.873	0.959	na	na	na	39
World skyscrapers	1–101 (500)	101	yearly	3.997	7.462	0.627	0.962	na	na	na	19

Our second example is dramatically different, as indicated earlier. So far, all the buildings taller than 12 stories or 40 meters in our data sets have *not* changed in height since their construction. Only in places of very rapid growth, like Hong Kong, are tall buildings now being systematically demolished and rebuilt, and it might be argued that the instant changes in rank such demolition and reconstruction occasion are tantamount to the construction of new buildings. This means that buildings do not rise in rank, they simply appear at a particular rank and never get any higher than their original rank. Buildings simply do not get bigger in this characterization of the data. If the system is in the steady state, buildings may stay at the same rank, but generally, once a building appears at particular rank, it will then decline in rank as taller buildings are constructed at later periods. We examine two data sets here: the first for New York City, which is the archetypal high-rise city, and the second for the 60,000 high buildings in the Emporis (2009) Global Database.

The first high building in each data set is taken as being constructed in 1909, although there were skyscrapers built before then. In New York, there is a rapid increase in the number of such buildings constructed in the 1920s and 1930s. Due to ties in height, the top 100 contain some 119 in 1916, which means that the particular rank clock shown in figure 4.10(a) reaches out to 120 on its radius rather

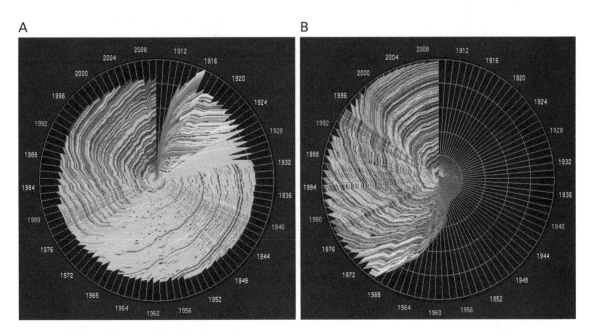

Figure 4.10
Rank clocks of the top 100 high buildings in (a) New York City and (b) the World from 1909 until 2010.

than 100 before falling back to 100 or 101. The clock is based on a complete spiraling out and downward of buildings from the time they enter the top ranks. The early buildings before the late 1920s are colored red, orange, and yellow, and these are blanked out by the flurry of construction shown in the green color in the early 1930s. Some of these buildings stay in the top 100 for the rest of the time period but in general, later buildings are greater in height. To produce a clearer picture, it is necessary to zoom into the clock and explore these trajectories. This is not possible on the printed page, so the interactive software is crucial to using these tools effectively (Batty, 2006a; O'Brien, 2012). Figure 4.10(b) shows the clock for the World data, where it is quite clear that globally, some high buildings constructed in earlier eras persist at high ranks into recent times. The picture of growth is considerably later than in the New York City case, since high building has diffused around the world as countries in the developing world have grown. If one takes a cross-section of these clocks at any time, it provides a picture of the time when a building was constructed and its rank, which need to be compared to the persistence matrix $L(\ell, m)$.

Numerical statistics are presented for these two examples in table 4.5. The rank-size functions for New York City fit with reasonably good approximations as the r^2 statistics show, with later distributions having better fits. The same is true of the World data, although the performance earlier in the time series is not as good. It ultimately becomes a lot better and is similar to the other distributions we have dealt with here. One issue that makes these distributions very different is the degree of competition, as reflected in the scaling parameter α. The values are much higher than 2, meaning that the slope of the rank-size curves are considerably flatter than the pure Zipf case where $\vartheta = 1$. This is of interest in that it suggests much less competition between the construction of high buildings in an intraurban context, but also that the growth dynamics here is quite different from that which characterizes cities and firms, which actually grow in the biological sense due to accretion. This entire dynamic requires considerably more research so we can explain the genesis of high buildings using models of copycat and fashion effects. In terms of the distance statistics, these are tricky to compute due to the fact that all the time periods are not distinct; in earlier periods, buildings persist but new ones are not constructed, and thus all ranks remain the same. Moreover, the distance measures all point in one direction—downward in terms of rank shift. This suggests, again, that we need some modifications to these visualizations to account for systems in which objects do not grow, per se, but do lose their position in the rankings due to the appearance (growth) of other objects. Half-lives are also tricky to estimate in this context. It appears that in the 1920s these are about 10 years, rising to 20 years in the 1970s, but they are impossible to estimate from this data for later times.

Extending Basic Ideas about City Size

Scaling systems involving human populations clearly display a form of regularity at the macro level, which masks dynamic volatility at the micro level. We measure this volatility by the extent to which objects change their position in relation to one another through their rank-size distributions. We do not yet have a detailed understanding of the way these dynamics play out, but we do know that competition between these various elements is intrinsic to the way they capture growth from one another. In fact, in city systems, traditional theories in which central places grow from the bottom up, gradually deriving their functions as they get larger and serve ever-larger populations, go a long way to explaining how cities scale at the macro level. Random proportionate growth plays a part in the way individual cities grow relative to one another, but as yet, we do not have good models of precisely how individual cities compete (but see Rozenfeld et al., 2008). When it comes to the growth of firms, the logic is more complicated by the process of mergers and acquisitions, while in the case of systems that are manufactured such as buildings, a very different substrata of dynamics is at work, dictated by the way developers define the need for high buildings and complicated by the construction process and by investment decisions.

Our visualization of these changes suggests that different scaling growth regimes display different morphologies, and we have made a start at classifying these as rank clocks. But we still need much better methods for making statistical comparisons between such systems, where we find it hard to control the number of objects or the number and length of time intervals over which such distributions are observed and measured. In particular, the concept of half-lives needs further refinement, as do the various measures of distance we have introduced here. The implicit notion that the size of the scaling parameter relates to the degree of competition between the elements comprising any particular distribution needs to be made much more definite, while links between distributions through their individual elements need to be tracked in a more synthetic manner. Trajectories on the rank clock help to focus this tracking, but our analysis does not yet link individual elements together.

In this sense, network interpretations of the way the elements in these distributions connect to one another are promising, and much of what we have learned about how different objects grow and change relative to one another can probably be unpacked into changes occasioned by the development and growth of new network links. We need to pursue at least three lines of inquiry: first, to develop a more robust set of space-time statistics for scaling systems, building on what we have introduced here; second, to generate network equivalents of such scaling systems that will add a richness to the analysis by, for example, expanding city-size distributions to the networks that support interactions, trade, and migration between cities; and, last but not least, we need more examples at different spatial scales that allow us to draw clear links between cities, buildings, and firms, which are all manifestations of how populations agglomerate in achieving economies of scale. In the next chapter, we will explore how these ideas might be extended to networks, returning to the connectionist ideas of chapter 3. There we will focus on how scaling and proportionate growth pertain to networks, illustrating how structure can be manifestly represented at different spatial scales in hierarchies.

5

Hierarchies and Networks

To a Platonic mind, everything in the world is connected to everything else—and perhaps it is. Everything is connected but some things are more connected then others. The world is a large matrix of interactions in which most of the entries are close to zero, and in which, by ordering those entries according to their orders of magnitude, a distinct hierarchic structure can be discerned.

—Herbert Simon, *Models of Discovery and Other Topics in the Methods of Science* (1977, p. 258)

Hierarchy is implicit in the very term "city." Cities grow from hamlets and villages into small towns, and thence into larger forms such as a "metropolis" or "megalopolis"; world cities are called a "gigalopolis." In one sense, all urban agglomerations are referred to generically as cities, but this sequence of city size from the smallest identifiable urban units to the largest contains an implicit hierarchy in which there are many more smaller cities than larger ones. This is enshrined as central place theory, which was first developed by Christaller (1933/1966), extended into a formal economic logic by Losch (1940/1954), widely applied in the 1950s and 1960s as summarized in Berry (1967), and is now the cornerstone of retail models based on spatial interaction such as those we will explore in chapter 9. Hierarchy is, of course, implicit in the rank-ordering of cities within city systems, as we illustrated in the last chapter on rank-size relations, but there we did not explicitly consider their embedding into Euclidean space. Here we will redress this by extending our simple models of proportionate growth, first to cities in a cellular landscape, and thence to the networks that tie cities together. In this chapter, our models will still remain parsimonious, stripped to their essence, but will be focused now on how cities relate to one another in space and time. We can extract hierarchy from networks of relations, as chapter 3 illustrated, and here we will present some simple examples to give substance to this argument. However, as throughout this book, we are more intent on introducing tools that might be applicable to many different aspects of the city system rather than providing an exhaustive catalog of possible models for every situation that might be envisaged. In this sense, our science is one of methods and exemplars that we hope others might apply in different contexts to generate the many insights that are required to understand the form and functioning of cities.

Implicit Hierarchies of City Size

Out of the Urban Soup: The Simplest Model

In the last chapter, we assumed that cities somehow emerged from hunter-gatherer societies as an implicit result of human beings' need to belong to a group for reproduction and defense, but also to satisfy needs for social interaction and organizing labor to increase material well-being, consistent with the economies of agglomeration. In terms of the origins of life itself, the conventional wisdom is now largely fashioned around the notion that, in the beginning, life began through some chance spark setting off a reaction in an undifferentiated chemical soup, leading to the formation of the various nucleotides that constitute the buildings blocks of life—RNA and DNA. In the same way, we can speculate that societies and cities began with household units randomly located across a landscape where the spacing of individuals was determined by food available from hunting or gathering. These units, of course, made contact in their quest for survival, and although the dominant mode was one in which households competed with one another for territory, which was synonymous with survival, there was a dawning realization that cooperation rather than competition could ensure greater prospects for survival. Hamlets and villages were formed initially to ensure strength in numbers for protection, but in time, the social contact that resulted reinforced a division of labor, leading to increased prosperity.

We will begin with the same model presented in the last chapter, in which population units—households, and clusters or groups of such households—grow randomly but in proportion to their size. As demonstrated previously, some units in this undifferentiated urban soup grow more than others simply due to the fact that they continually get ahead, while others fall behind and often disappear. Eventually, clusters that are differentiated by size, which we call cities, appear in this landscape, and it is these that give structure to the urban soup. Hierarchy is an intimate part of this structure, but before we show how such hierarchies emerge as a natural part of the growth process, we will take one step back and show how cities in this artificial world first organize themselves according to size. We define a hierarchy here as a natural ordering that is initially based on size, with size being measured in many different ways. In cities, size is typically based on the number of individuals, households, or workers—on populations—but it may also be based on the area over which such location occurs or energy is used, or on the field of influence over which individuals in the hierarchy have control. Let us begin with the simplest of possibilities:

places of the same size are randomly scattered over a uniform plane. In such a world, insofar as clusters exist, these are random occurrences. The big difference between this model and the one introduced in the last chapter, based on random growth through proportionate effect, is that our landscape of cities or settlements is now spatial and based on a grid. As we will see, this is essential to the way in which hierarchical structures defining the system of cities or central places evolve.

We will assume a process in which a place grows randomly but at a rate of growth that is applied proportionately to its size. So if a place i at time t has size $P_i(t)$ and the growth rate $\lambda_i(t + 1)$ is chosen randomly, the place grows (or declines) as

$$P_i(t + 1) = (1 + \lambda_i(t + 1))P_i(t). \tag{5.1} [4.3]$$

Equation (5.1) is the same as equation (4.3), and if we iterate on (5.1), we can examine the long time limit of the varying growth rates, which reflect a random but proportionate compounding effect for each place or grid square i. We will repeat what we said in chapter 4 about such a process, for its consequences are surprising at first until one pauses to reflect. In a system of many places, the distribution of growth rates will be uniform at any time t over a range of locations from small to large, which might also be from negative to positive. However, the chances of any particular place getting a series of very high growth rates allocated to it one after another, and thus growing very big, is increasingly small. Equally, the same is true for a place becoming increasingly small, and of course in this model if a place gets too small, it might be said to disappear, so there is some intrinsic asymmetry within the process. In fact, a decreasing number of places get infinitesimally small, but usually we establish a lower bound below which a place cannot exist. It is very easy to work out what happens if we apply this growth process to a small number of objects, with random growth rates chosen from a given range, and then apply these using equation (5.1) over and over again. An increasingly small number of the objects grow big, most remain small, quite a lot disappear, but the crucial issue is: does the resulting size distribution show any kind of order? We demonstrated this in the last chapter as a scaling law of population sizes where there are far fewer bigger objects than smaller, but let us take a worked example on our grid of places, which, although somewhat artificial, graphically demonstrates the point.

Our example is based on a grid of dimension 21×21, giving 441 objects or locations for which the initial populations are uniformly distributed, with $P_i(t) = 1$, $\forall i$, $t = 0$. The rates of growth $\lambda_i(t + 1)$ are chosen randomly in the range $-0.1 < \lambda_i(t + 1) < 0.1$. The proportionate growth model in equation (5.1) quickly sorts out the objects into a size distribution, and by time $t = 100$, the frequency distribution shows every sign of being lognormal. In fact, I have run the model for 1,000 iterations, which I refer to somewhat euphemistically as "years," and during this simulation there is much movement between those objects in terms of their relative size.

This, of course, is what we demonstrated empirically in chapter 4, but here we will explore the implications of this within such models. As before, the usual way of extracting the signature of these size distributions is by organizing their frequencies hierarchically in terms of size and then plotting them in this order, which is against their rank. The rank-size distribution or Zipf (1949) plot is in fact the counter-cumulative, which we derived in equations (4.5) to (4.9) in chapter 4. This rank-size distribution is plotted for $t = 1,000$ in figure 5.1, and it is immediately apparent that the signature is that of a lognormal distribution where the plot is visualized as a log transformation of population size against rank.

In fact, many have shown that proportionate random growth of the kind we have described leads to lognormal size distributions (Pumain, 2000). Gibrat (1931) produced the first comprehensive argument for cities and income distributions, but Yule (1925) knew of the model and its consequences a little earlier. If we were to continue the simulation beyond $t = 1,000$, then more and more populations would converge to zero and, ultimately, we hypothesize that in discrete systems of this kind, all activity would be attracted to a single cell. In fact, in city systems, such a simulation is bounded from below by indivisibilities, and it thus makes sense to modify our model by introducing a size threshold below which populations cannot fall. Then whenever a population cell i falls below this number, it is restored to that number, this mechanism acting as a safety net or subsidy of sorts. This can also be viewed as a way of killing off a city and introducing a new one at the same time, thus incorporating a perfectly balanced birth and death process. Then, formally,

$$if \quad P_i(t) < \Psi \quad then \quad P_i(t) = \Psi, \quad \forall i. \tag{5.2}$$

Examining the lognormal distribution in figure 5.1 reveals two regimes—the long tail, which is almost linear, and the short tail, which accounts for the order of the smaller settlements. It is tempting to think that the long tail could be approximated by a linear or scaling relation and, in effect, if we use the cut-off mechanism just postulated, then in a purely phenomenological sense, this effectively cuts off this short tail. Running the simulation again with equation (5.2) now operative and with the same parameters on growth, the "almost" straight line distribution in figure 5.1 is generated by time $t = 1,000$, and this has immediate similarities to the same process we introduced in the last chapter, which leads to figure 4.2(b). The cut-off in fact clearly works, and what we end up with is a distribution that is no longer lognormal. It is now scaling, since it can be approximated by a power law for which the population size $P_i(t)$ varies inversely with the rank $r_i(t)$ as $P_i(t) \propto r_i^{-\gamma} = K r_i^{-1/(\alpha-1)}$, with $\gamma = (\alpha - 1)^{-1}$ the so-called scaling parameter of the distribution. Note that this scaling is equivalent to the derivation in equations (4.5) to (4.9), where α is the parameter of the probability density function, proportional to the frequency of city sizes. As cities are moving up and down this hierarchy, it is tempting to think of the

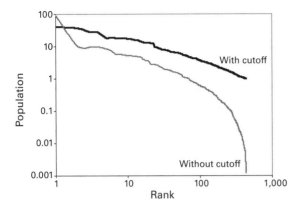

Figure 5.1
Generating a lognormal distribution using proportionate effect and power-law scaling from proportionate effect with a minimum size threshold.

fat (long) tail as a "steady state" to which cities are "attracted," and indeed, theorists such as Gabaix (1999) demonstrate that this is the case for the Gibrat process, which can converge to a pure scaling law—Zipf's Law—with the parameters $\alpha = 2$ and $\gamma = 1$. Note that our use of the terms long tail and short tail is an artifact of the way we present these distributions pictorially as rank-size plots in figure 5.1. In other contexts, the terminology is reversed (see Anderson, 2006).

Distributional Regularity and Hierarchical Volatility
Power-law scaling or something similar has been known as an empirical fact about cities and many other distributions, such as incomes, for over one hundred years. There is evidence that those writing about cities at the end of the nineteenth century, such as Weber (1899) and Auerbach (1913), accepted it as a consequence of urban growth, and of course Pareto (1896/1967) coined his famous law of income in this way, but in terms of frequencies rather than ranks. As we have noted in previous chapters, the most popular exposition of rank order that conforms to power-law scaling is provided by Zipf (1949) in his remarkable book, in which he examined many such distributions, from word frequencies to cities. Zipf argued that these distributions were not only scaling, conforming to the power law, but also that many such distributions—indeed, the implication that all such distributions—were such that the power law was a pure inverse, suggesting that the rank of the population in continents, countries, and counties, at any level or scale, would conform to $P_i(t) \propto r_i(t)^{-1} = 1/r_i(t)$. This is the strong form of Zipf's Law. It implies that city-size distributions are fractal, since if one examines the relationship at any scale, then the distribution is the same. This is self-similarity in its pure form and means that if the

distribution is rescaled, then this is simply a "scaling" up or down of the original distribution. Imagine that the rank-size is rescaled by s to another rank $sr_i(t)$. Then the rank size scales as $sP_i(t) \propto (sr_i(t))^{-1} = s^{-1}r_i(t)^{-1} \propto r_i(t)^{-1} \propto P_i(t)$, which implies that the scaling is the same over any order of magnitude (Batty and Shiode, 2003).

The law of proportionate effect with a lower bound is akin to a random walk with a reflecting barrier (Sornette, 2004; Saichev, Malevergne, and Sornette, 2010). The model is simplistic in the extreme, since it does not include any form of competition or interaction between the objects. This is extremely odd, since cities compete and interact and many models of their formation emphasize such interactions. Our model of proportionate effect with the lower bound clearly generates distributions that appear to be scaling and follow Zipf's Law, but in many ways this model is unstable. The time over which such distributions emerge and the volatility of the top-ranked cells or places is enough to suggest that the model does not have enough inertia to mirror real places, despite the evident volatility we examined in the previous chapter, where we saw cities rising and falling in size (and of course rank) over quite short time periods—centuries rather than epochs. The fact that it produces size distributions that concur with reality is not sufficient to indicate this is a good model. For example, as we move through the time periods, the distributions generate change not in their scaling but in their shape. We can see this in two ways. In figure 5.2, we show the pattern of distribution after 100, 1,000, and then 10,000 iterations ("years"), which reveals the convergence to extreme distributions as the simulation continues. In figure 5.3, the distributions for small numbers of iterations are much flatter and gentler than those for larger numbers.

Figure 5.3 however also shows the distributions for $t = 1,000$, $t = 2,000$, ... $t = 10,000$, which clearly get steeper—implying the parameter γ gets larger as the population grows. In fact, it appears that the parameter is converging on the pure Zipf case of unity, although from these results this is inconclusive. There is a variety of

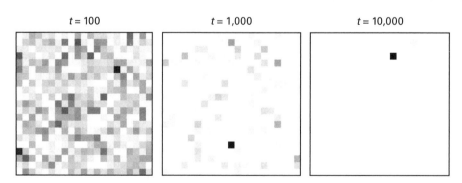

Figure 5.2
Emergence of the rank-size distribution using proportionate effect with cut-off.

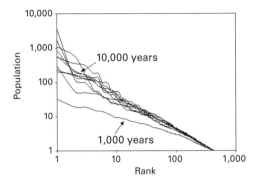

Figure 5.3
Power-law scaling as the population distribution emerges.

theoretical evidence that suggests this is the case for growth by random walk with a reflecting barrier, as Gabaix (1999) and Blank and Solomon (2000) show. In figure 5.3, the parameter γ rises from $t = 1,000$ to $t = 10,000$ as 0.668, 0.862, 0.907, 0.977, 1.008, 0.984, 0.980, 0.978, 1.053, to 0.962, which shows the final value hovering around 1 with the straight line logarithmic fits explaining more than 99% of the variance for each time slice. This is a fairly remarkable result. What we have shown is that an almost nihilistic model with no spatial competition can generate highly ordered simple hierarchies, which in fact mirror the empirical evidence that has been compiled for many cities in many places during the last hundred or more years. Figure 5.4 shows the rank size of incorporated places in the United States from 1970 (some 7,000 places) to the year 2000 (some 25,000). In figure 5.4(a), the entire distributions are shown, and these are clearly lognormal. When we cut off the short tails, the remaining long (fat) tails are quite straight, implying power laws, as in figure 5.4(b). In fact, figure 5.4 is the empirical equivalent of figure 5.1, and for the four time slices from 1970 to 2000, the γ parameter varies from 0.986 to 0.982 to 0.995 to 1.014 ,with the variance explained a little lower than the theoretical model as 0.98, 0.97, 0.97, and 0.97. The same kind of dynamic analysis has been done for France by Guerin-Pace (1995), and a thorough review is presented by Pumain (2006b).

Although the model produces aggregate distributions uncannily close to those we observe in most places, what is quite clear is that when we unpack the simulations, there are many inconsistencies that imply this model is not nearly as good as these results suggest. Of particular concern is the lack of apparent structural consistency as the simulation proceeds and the ranking of cells change. During the 10,000-"year" simulation, the number of different cells at the top of the rank order is 18. We have only sampled the rankings at every 50 time periods, and it is thus likely

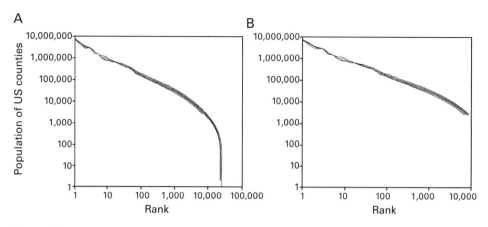

Figure 5.4
Lognormal and power-law scaling of the US population based on "incorporated places," 1970 to 2000.

that there are many more than 18 cells that appear at the top of the ranks during the simulation. To give an idea of the volatility of these ranks, we show how the top-ranked cells 1, 6, 12, and 18 from these top orders change over the 10,000-year history in figure 5.5. These cells appear at different times, as we indicate, but what is quite clear is that the length of time they occupy in the top position is small, thus implying that there is not enough inertia in the model. This is manifestly the case, and as the simulation time periods have no relationship to real times, then it is unclear what the 10,000 model "years" actually mean in terms of the evolution of actual urban systems, such as the US system shown in figure 5.4. As we implied in the last chapter, the volatility of these distributions through time is much greater than any real system, where change is slower and inertia greater.

The last thing to show, before we try to improve the model and generate hierarchies that imply spatial interaction and competition, is the effect of changing the geographic dimensions of the space within which the simulation takes place. We have changed the grid from 21 x 21 to 51 x 51 and then to 101 x 101 and run the simulation with the cut-off for 10,000 "years." We show the three rank-size distributions in figure 5.6, where it is quite clear that the slopes are similar, implying that the model does indeed hold up as we scale the system in geographic size. This might be expected, since there is no interaction between the parts, but what is of interest is the increased size of the population as the system scales. This is a bit of a mystery, but it probably occurs because there are more and more opportunities for extreme growth as the spatial system gets larger, yet it requires further investigation, as do many other aspects of these simulations. It does not, however, detract from the main result that the model scales spatially. It is surprising that so simple a model, which

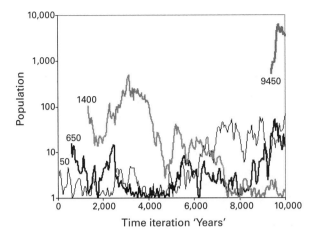

Figure 5.5
The first, sixth, twelfth, and eighteenth top-ranked population cells and their progress through the simulation.

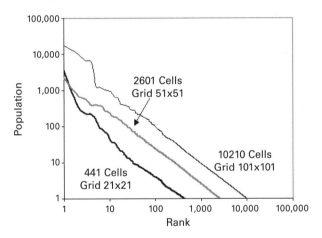

Figure 5.6
Consistent scaling behavior for different sizes of lattice.

has had so much effort devoted to it in terms of simulations and mathematical analysis, is still far from being thoroughly understood.

Generating Hierarchies through Competition, Interaction, and Spatial Diffusion

The hierarchy generated by the model of proportionate effect is the simplest possible—a simple rank-order or unidirectional hierarchy where the order of objects is simply one of size and where each object is independent of any other. This cannot be a good model for the growth of cities, because is does not admit competition or interaction of any kind. Cities are completely disconnected from one another. Simon's (1977) opening quote to this chapter is largely irrelevant to this definition of hierarchy, for nowhere in such a model are there clusters of connected activity that provide the kind of connectivity from which hierarchic structure can be derived. What we require is some measure of interaction—but not too much, to meet Simon's (1977) point—between cities or places, between the points on the lattice. To explore this, we will add some simple diffusion to adjacent grid cells at each stage of the model simulation. In short, at each time step, a fixed proportion θ of the population in each cell i diffuses to its nearest neighbors k in the von Neumann neighborhood comprising the cells that are north, south, east, and west of the cell in question. Thus for cell i, the population at time $t + 1$, $P_i(t + 1)$, is now computed as

$$P_i(t+1) = (1 + \lambda_i(t+1))P_i(t) + \theta \frac{\sum_{k \in \Omega_i} P_k(t)}{4}, \tag{5.3}$$

where the neighborhood for diffusion is defined as $\Omega_i = k_{north}, k_{south}, k_{east}, k_{west}$. In this model, minimal action-at-a-distance is admitted, and within a short time interval that is proportional to one dimension of the lattice, every cell influences every other cell. This kind of diffusion is still somewhat elementary in that it is only based on the notion that a proportion of people move to be with their neighbors, without specifying any particular reason for such movement, other than the implication that this is based on a social or economic rationale.

We have run the model in equation (5.3) retaining the cut-off in equation (5.2) for $t = 1,000, 2,000, \ldots , 10,000$ time periods, and this generates the rank-size distributions shown as Zipf plots in figure 5.7. The level of diffusion used involves setting the parameter $\theta = 0.3$, implying 30% of the population in each cell gets redistributed into adjacent cells in each time period. In fact, what this leads to are not scaling distributions, as the model without diffusion does, but lognormal distributions. The cut-off is in fact discounted by the diffusion. This is a little different from the similar model posed by Manrubia and Zanette (1998), which is based on the same process but without a cut-off and with only positive growth, which

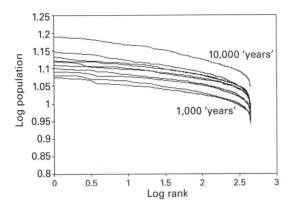

Figure 5.7
Lognormal distributions generated from proportionate effect with diffusion.

produces a scaling law. Yet, in a sense, whether these kinds of model produce power-law scaling or lognormal distributions is of less concern because at a phenomenological level, all multiplicative processes such as these variants belong to the same class of model (Sornette and Cont, 1997).

The diffusion in this model is so intensive in each time period that all cells are affected, and also so extensive, since the number of time periods is far greater than the size of the system—which in this example is based on the 21 x 21 lattice—that it is impossible to track all interactions that accumulate between all pairs of cells. Action-at-a-distance occurs through the medium of adjacent cells, and the number of combinations of diffusion paths is thus enormous. Effectively, this diffusion leads to densities that fall around the cells with the largest populations, just as a city core attracts and diffuses activity around it. We take an impressionistic view of the hierarchy formed, in which we simply plot the hierarchy by associating cells with their higher-order center (based on population size), deciding whether or not they are connected simply through adjacency. In figure 5.8, we first illustrate the patterns of growth for the model at $t = 100$, $t = 1,000$, and $t = 10,000$, and it is clear that the type of pattern produced occurs within 100 time periods and simply repeats itself—in different locations, of course—through time. In figure 5.8, we also show simplifications of these pictures by first identifying the top-ranked cell, the next 3 followed by the next 8, then the next 24, and finally the next 64 around these cores. This provides a crude picture of population density, which we can represent as a hierarchy.

We do this for the pattern at $t = 10,000$, where we simply associate each cell at each level with the cells above it if they are connected directly or indirectly to that level through cells of similar value. The hierarchy produced is plotted as a

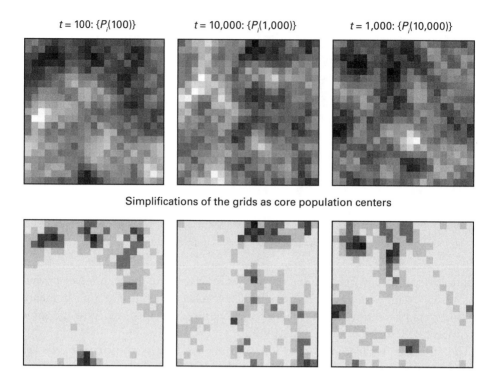

$t = 100: \{P_i(100)\}$ $t = 10,000: \{P_i(1,000)\}$ $t = 1,000: \{P_i(10,000)\}$

Simplifications of the grids as core population centers

Figure 5.8
Patterns of diffusion.

semi-lattice in figure 5.9 in a similar way to the structure introduced in chapter 1 in figure 1.1(c). It is not possible to uniquely associate every cell with a single cell at the next level of hierarchy because we do not have network links between cells that we can cut to define separate regions. In fact, this representation of hierarchy is much more realistic and supports the long-standing notion of overlapping fields of influence articulated rather well by Alexander in his article "A City is Not a Tree" (1965). There is considerable structure in this hierarchy introduced through the diffusion process, but this model appears just as volatile as the pure Gibrat process. Over the simulation period of 10,000 "years," of the 441 distinct cities or cells, all these cells occur at the top of the hierarchy at some point, while the pattern of these top-ranked cells would appear quite random. We show this pattern in figure 5.10, where it is quite clear that there are no particular clusters of cells or individual cells that predominate over any others. It is quite clear from these simulations that there is too little inertia in the system to mirror our experience of real city systems, for it is most unlikely that for the simulation times used here, all cells would at some point dominate. Other models are thus required.

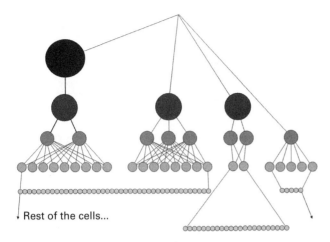

Rest of the cells...

Figure 5.9
A hierarchy for the pattern at $t = 10,000$.

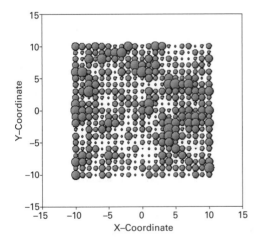

Figure 5.10
Top-ranked cells during the 10,000-"year" simulation. The bubbles range in size from 1 to 83 time periods, in which the relevant cells dominate with an average of 23 time periods.

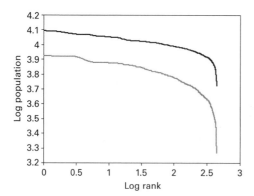

Figure 5.11
City-size distributions for the agglomeration model at $t = 1,000$ and $t = 10,000$

The distributions generated from the Gibrat process with diffusion are somewhat flat, and as the level of diffusion is increased, the hierarchical structure begins to disappear. We have attempted to inject more structure into the model by departing from the Gibrat process and introducing agglomeration economies into the model, adding a term reflecting current city size. Our model thus becomes

$$P_i(t+1) = (1 + \lambda_i(t+1))P_i(t) + \theta \frac{\sum_{k \in \Omega_i} P_k(t)}{4} + \phi P_{it}^{\eta}, \tag{5.4}$$

where ϕ and η reflect the proportionality and the scaling imposed by agglomeration economies. We have set $\phi = 0.2$ and $\eta = 1.08$. With these parameters, we do indeed succeed in sharpening the distribution of city sizes, but the lognormality of these distributions remains as we show in figure 5.11. There do not appear to be any real qualitative differences produced by this model. To introduce a different form of hierarchy into the urban soup, we require much more explicit networks of interaction, exploiting results from the burgeoning science of networks (Barabasi, 2002; Watts, 2002. These we will now present.

Network Hierarchies: The Gibrat Interaction Model

So far our models of hierarchy have focused on evolution, whereas Simon's (1977) definition tends to assume that such hierarchies are already developed. To detect them, we thus need to observe the interconnections between the system's parts in order to define the clusters of highly connected subsystems that form the whole. The model needs to be extended to make explicit these interconnections, and this requires us to generalize Gibrat's model to networks. We do this by adding both cells and their links randomly, one in each time period. The mechanics of the model are contained in the following equations. In each time period, for a node that is already established and linked to other nodes, we consider the random addition of a link volume as $\delta_{ij}(t + 1) = 1$, where $P_{ij}(t)$ is the total number of links from node i to j. The total links associated with i, which sum to the new population size of i, are thus $P_i(t + 1)$, and the equation for total links is thus

$$P_i(t + 1) = \sum_k P_{ik}(t) + \delta_{ij}(t + 1), \tag{5.5}$$

where the number of links is updated in each time period as

$$P_{ij}(t + 1) = P_{ij}(t) + \delta_{ij}(t + 1). \tag{5.6}$$

Whether a link is added depends on both the size of the node and its distance to other nodes, which is reflected in an exponentially weighted gravitational function of the form

$$\delta_{ij}(t + 1) = \begin{cases} 1 & if \quad rnd(\varsigma_{ij}(t + 1)) > KP_{ij}(t)\exp(-\varphi\, d_{ij}) \\ 0 \end{cases}. \tag{5.7}$$

The term $rnd(\varsigma_{ij}(t + 1))$ determines a random choice based on size of the potential interaction, where d_{ij} is the distance from node i to node j and the parameter φ reflects the frictional effect of this distance. Essentially, this process is one of preferential attachment in that links are added in proportion to the size of existing links and the population that the node has already attracted. It has been very widely exploited by Barabasi (2002) and his colleagues, who have shown that the model generates "scale-free" networks in which the number of in-degrees and out-degrees scale according to a power law (Barabasi and Albert, 1999).

This process does not show how nodes are established in the first place, and thus we must add a mechanism for the birth of new nodes, akin to that added by Simon (1955) in his classic model of the rank-size process. A new node at i is added if a random variable $rnd(\upsilon(t + 1))$ is greater than a predetermined threshold z, which is given as

$$\delta_i(t + 1) = \begin{cases} 1 & if \quad rnd(\upsilon(t + 1)) > z \\ 0 \end{cases}, \tag{5.8}$$

where the value of z is small compared to the probability for the addition of new links, as reflected in equations (5.5) to (5.7) above. For the 21×21 lattice, we choose the threshold for the addition of new nodes as $z = 0.1$, and this implies that at the beginning of the process in each time period, there is a 1 in 10 chance that a new node is added. Of course, as the process continues, this chance falls, for if a node is chosen that is already established, the node is abandoned. In terms of the generation of links to established nodes, then a node is first chosen randomly but in proportion to its size $P_i(t)$, and then a link to another node j from i is chosen in proportion to its inverse distance function as defined in equation (5.7). In this way, the network builds up through preferential attachment to existing nodes. We have assumed the overall dimension of the system to be 300×300 x-y coordinate units for each grid square, and thus we have set the deterrence parameter φ in equation (5.7) as 0.001, which implies an average distance between all 441 cells of around 1,000 units.

In figure 5.12, we illustrate the final distribution of population by node $\{P_i(1000)\}$, and alongside this, the distribution of link volumes between nodes for all links greater than 1, those greater than 2, and finally those greater than 4. A hierarchical pattern is revealed by these figures, and it would be possible to cut the link volumes at points where the cluster density falls below various thresholds, thus uniquely partitioning the space into different areas, which then orders the hierarchy. We have not done this, for our concern is not hierarchy per se, but ways of generating it. In figure 5.13, we plot the size of population per node against their rank as a Zipf plot, which is the logarithmic transform. We have not connected the points in this plot because of the comparatively low volumes generated in this example, but the plot is mildly lognormal. Fitting a straight line to this gives a scaling parameter of 1.05 with 90 percent of the variation in this plot explained. This is remarkably close to the pure Zipf scaling where $\gamma = 1$, and it is confirmation that this model of preferential attachment based on Gibrat (1931) does indeed generate the same profile, as in the simpler non-network cases which we discussed above.

Interlocking Spatial Hierarchies

Before introducing our final model, for which we will turn back to theory and generate the rank-size rule for hierarchies based on central place theory, we will show how hierarchies are generated by examining spatial activity at different thresholds. If we have a varying density of activity, we can portray the structure of this activity at different density thresholds, for which the resulting hierarchy is one based on intensity. In fact, in the models that have been used earlier to generate the intensity of activity at different nodes in the network, we portrayed these patterns at different levels of threshold, and in figure 5.8, for example, we extracted the hierarchy by varying the intensity threshold. There are many spatial systems that reveal such patterns. Our first example is for the hierarchy of retail activity in London, represented by an index of retail intensity that is a linear weighted sum of some 42 separate indicators, each suitably normalized, and tagged to the postcode geography, which at its finest scale represents retailing at an average resolution of some 50 meters (Thurstain-Goodwin and Batty, 2002). We have interpolated a surface from

Nodal population distribution
$\{P_i(1,000)\}$

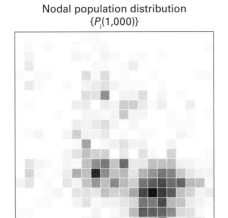

Network links $\{P_{ij}(1,000) > 0\}$

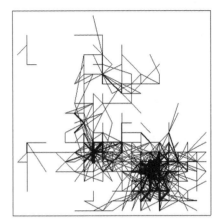

Network links $\{P_{ij}(1,000) > 1\}$

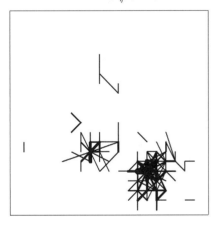

Network links $\{P_{ij}(1,000) > 4\}$

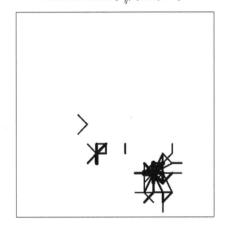

Figure 5.12
Patterns of network connectivity.

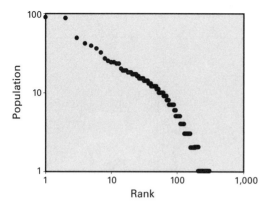

Figure 5.13
Rank-size distribution of the nodal network distribution.

this data and have then sliced it at some 5 different levels, which provide a picture of the retail hierarchy shown in figure 5.14. This is an implicit hierarchy, similar to those that can be derived from the population distributions illustrated earlier in figures 5.8 and 5.12. What we do not have from this analysis is the detailed interaction pattern that links consumers to the retailing activity through their movements to purchase retail goods at different points or centers on this surface. But the pattern is consistent with all that we have seen previously, and the distribution of retailing activity is rank size.

A second and more elaborate example that pertains to population distributions is based on using density thresholds to define boundaries around cities. As we relax the threshold, more population is encompassed in the growing urban cluster, and in this way we can grow cities from their smallest cores. In figure 5.15, we have taken Greater London and, starting with all population with a density greater or equal to 70 persons per hectare (pph; 7,000 persons per square kilometer), we relax the density threshold, first to 50 pph, then to 14, and finally to 5, which represents a fourfold hierarchy. At each stage, more population emerges that is part of the growing set of clusters, and as each new piece of cluster comes to be part of the mass, we link it to its nearest other cluster at the previous level of hierarchy. In this way a new area comes to be part of the growing hierarchy. If there are ties in the sense that the new area is as close (adjacent) to one cluster as another, we merge it with the largest of these. If there are several exact ties, which in this case there are not, then the merger is arbitrary. The growing cluster of London at these four levels is shown in figure 5.15(a). Alongside, in 5.15(b), we show the rather intricate hierarchy that results. Obviously, were we to take a continuous gradation of densities, each individual basic unit of population would be added into the growth, and

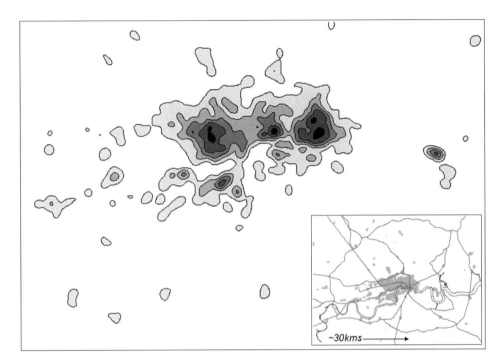

Figure 5.14
The implicit retail hierarchy in Central London.

starting from the zones with the greatest density (which in terms of this data is based on populations in wards), each ward would be added to its previous cluster, thus setting up as many levels of hierarchy as there are wards. In this region there are about 1,500 such units, which gives some idea of the intricate way city systems become structured across many levels.

Our third example takes flow data from a network of nodes that are stations on the tube network in Greater London, which we used to illustrate centrality in figure 3.11 in chapter 3. Figure 5.16(a) shows the directed graph, which is formed from the order of the betweenness centrality for each pair of stations, selecting the directed links in the following way. We form a hierarchy in the most simplistic way by first rank-ordering the stations or nodes from the highest to the lowest with respect to the centrality, and then we take the top station and link it to the stations that are adjacent to it with a lesser centrality value at the next level down. If there is no such station, we then select the next station with the second highest centrality and do the same. We thus proceed down the hierarchy in this fashion, but at each level of hierarchy, we look upward at all the higher centrality stations traversed so far and link the station to the node with the highest. This produces a tree structure

A

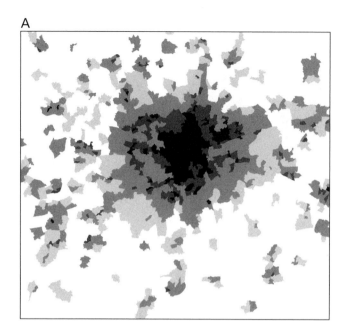

B

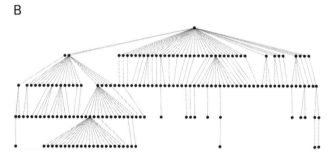

Figure 5.15
The explicit population hierarchy in Greater London. (a) The four-level gradation of density based on 70 (black), 50 (dark gray), 14 (mid gray) and 5 pph (light gray). (b) The complementary four-level hierarchy.

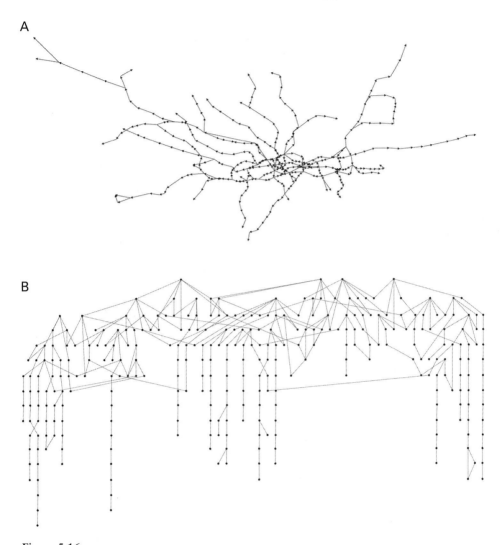

Figure 5.16
A strict hierarchy from a treelike graph of the London tube system. (a) The directed acyclic graph of
nodes (stations) and links (lines) in the tube (Underground) based on a strict ranking of betweenness
centrality. (b) The Sugiyama hierarchical graph derived from ordering the nodes by betweenness
centrality.

with no cycles, and we unravel the graph using the Sugiyama graph-drawing algorithm (Sugiyama, Tagawa, and Toda, 1981). Figure 5.16(b) shows the resultant hierarchy, which is a noncyclic graph.

Central Places: Rank Size from Geographical Dependence

Although we have illustrated models that produce quite distinct hierarchies, we have only introduced space as action-at-a-distance from distinct nodes. Insofar as competition has entered the argument, this has been either through intersecting and overlapping diffusion or through an implicit ordering where larger places get preferential treatment relative to smaller places, as in the network models of the earlier sections. One of the first expositions of how geographical areas based on spheres of influence around towns and cities are consistent with the rank-size rule was developed by Beckmann (1958). His argument is so clear that we will repeat it here, thus providing some sense of closure to our more general discussion of hierarchy through rank-size scaling. Beckmann (1958) defined two key elements in the way cities are organized with respect to their functional and spatial dependence. He first assumed that a city—or, rather, a small seed that sparked off the growth of a city—was proportional in size to the population on which it depended in its surrounding hinterland or sphere of influence. He then noted that each city had a "span of control," which related to the number of lower-order hinterlands that could be said to depend spatially and economically on the center city or seed at its core. This second kind of dependence leads directly to a series of hinterlands at different orders, increasing in number and decreasing in geographical area as they descend the hierarchy, and it is from this that the rank of any city can be established. This is entirely consistent with the theories of central place as expounded by Christaller (1933/1966) and Losch (1940/1954).

Formally, the initial dependence ξ of the city seed p_n on its wider population P_n for any order of city n is

$$p_n = \xi P_n, \tag{5.9}$$

where the order n is from the largest city, which we call order N, to the smallest, which is defined by the index order 1. The second spatial dependence involves the fact that the population of the higher-order level P_n is a sum of s populations at the next lower-order P_{n-1}, defined as

$$P_n = p_n + s P_{n-1}$$
$$= \frac{s}{1-\xi} P_{n-1} . \tag{5.10}$$

Recurrence on equation (5.10) leads to

$$P_n = \left(\frac{s}{1-\xi}\right)^m P_{n-m}, \tag{5.11}$$

and at the bottom of the hierarchy, where the population is at the lowest level $P = P_1$, then the power-law dependence is clear:

$$P_n = \left(\frac{s}{1-\xi}\right)^{n-1} P. \tag{5.12}$$

If we assume that the seed city is small or even zero, then $\xi = 0$, and equation (5.12) simplifies to $P_n = s^{n-1}P$.

Using the reverse order, which is from 1 to N, the total number of cities at each level is s^m and the total number up to m is given as

$$\sigma(m) = 1 + s + s^2 + s^3 + \ \ldots \ + s^m, \tag{5.13}$$

where this is a diverging geometric series whose sum up to level m is $(s^m - 1)/(s - 1)$. Thus the rank of the first city level m is $\{[(s^m - 1)/(s - 1)] + 1\}$, and the rank of the city that is midway through this order—the average rank—is

$$r(m) = \frac{s^m - 1}{s - 1} + \frac{s^m}{2}. \tag{5.14}$$

Examining components of this sum, we assume that $1/s - 1$ is small relative to other terms, and thus equation (5.14) can be simplified to

$$r(m) = s^m \left(\frac{1}{s-1} + \frac{1}{2}\right). \tag{5.15}$$

The rank-size relation is based on population size and rank, and if we multiply the relevant equation for population in equation (5.12) (which we convert from order n to m as $n = N - m + 1$) with the rank in equation (5.15), we get

$$P_{N-m+1}\, r(m) = \left(\frac{s}{1-\xi}\right)^{N-m} P\, s^m \left(\frac{1}{s-1} + \frac{1}{2}\right). \tag{5.16}$$
$$= \Phi\,(1-\xi)^m$$

This equation is a constant if $\xi = 0$, and thus the entire argument hinges on this. To an extent this is arbitrary, although it is easy to assume that the hinterland population dominates and the core or seed is near zero. If this holds, we can simplify equation (5.16) as $P_{N-m+1}\, r(m) = \Phi$. Writing this in a more familiar way, where we suppress the order indices and define population at a rank r as P_r, we get $P_r\, r = \Phi$ or

$$P_r \sim \frac{1}{r},\tag{5.17}$$

which is the pure Zipf case. Many assumptions have been made to get this far, and of course we have not tried to generalize Beckmann's (1958) discrete case to a continuous one. Nevertheless, it would appear that this kind of geographic—indeed geometric—reasoning, which assumes that space is nested hierarchically through its economic dependence, does lead to rank-size distributions of activity size such as population. At one level, of course, this is all too obvious, in that we have assumed hierarchical order and simply shown that geometric series describing such order can be manipulated to produce a rank size. This is, in fact, an indirect argument reflecting scaling. Surprisingly, the Beckmann model has not been widely exploited, and no one (as far as this author knows) has developed a stochastic version of it. Nevertheless, it does serve to remind us that there is a deep underlying rationale for the existence of rank-size distributions that is essentially a spatial or geometric ordering, or a hierarchical ordering in fact, in the geographical sense (Beckmann, 1968).

Hierarchy in the Design of Cities

The models we have used in this chapter to generate spatial hierarchies whose signature is the scaling of population are essentially stochastic and dynamic, although the Beckmann (1958) model in the previous section took a more deductive approach in which its dynamics were implicit, at best. Yet there are other ways of generating spatial hierarchies. It is possible, for example, to generate such distributions as the outcome of various optimization procedures, taking either a top-down static approach or even a quasi-dynamic one. By way of conclusion, and in our quest to square the circle and show how hierarchical systems in cities should feature in design, it is worth noting the long tradition in spatial interaction modeling, in which scaling distributions of population and trip-traffic distribution (such as those introduced in chapter 2 and exploited further in later chapters) can be derived using optimization theory. These models are based on maximizing utility-like or entropy and accessibility functions subject to constraints on the dispersion of such activities through their cost structures. Berry (1964) was one of the first to illustrate such an approach in his derivation of population distributions that conform to rank size using entropy-maximizing techniques; this approach was used by Wilson, Coelho, Macgill, and Williams (1981) in their quest to embed behavioral land use and transportation models into contexts in which behavior was considered optimizing, either at the individual or collective plan-making level.

In terms of these different approaches, the rank-size distribution provides a sharp illustration of the problems we face in explaining the evolution of complex systems.

Nothing could be more different than the generation of a distribution from a stochastic process, in which all the constituent elements are independent from one another and where the growth is from the bottom up—our first model—from the kinds of top-down optimization processes in which accessibility is maximized, subject to some constraints on cost or energy expended. Yet the outcomes in terms of the distribution of the elements being optimized are the same. In a way, all this shows is that how we approach city systems conditions the techniques we use to generate the outcomes we expect. This is the problem of equifinality that plagues all complex systems thinking—multiple, plausible but different models that shade into one another, generating identical outcomes—and is something that we must live with in our science of cities, indeed perhaps in many sciences. We will not say more here, but by now it must be familiar to our readers that our science can never be a watertight set of constructs to be tested unambiguously against an agreed-upon set of data. We will return to this dilemma many times in this book.

In city design—which, although it is linked to optimization, originates from very different intellectual mindsets and professional concerns—approaches using ideas from hierarchy theory are also well established. Although many of these use hierarchy in terms of the structure of the problem-solving process, in which problems are partitioned into a hierarchy of subproblems, the notion that we need equivalent simplifications to those we have sought here is instructive. We anticipated much of this in an earlier section in quoting Alexander (1965), who argued that the notion of strict hierarchy was far too simplistic an organizing concept for design. Alexander, among many others drawing on ideas from organically evolving systems that latterly have been exploited in neo-Darwinism by writers such as Dennett (1995) and Dawkins (1986), argued for a paradigm in which interaction rather than hierarchy was a required design construct. Overlapping hierarchies—semi-lattices, as we illustrated in figures 1.1 and 5.9—are much more appropriate vehicles for the organization of cities into spaces at different levels of geographical scale.

In essence, this argument suggests that strict hierarchical subdivision is too simplistic a concept for the design of neighborhoods and town spaces, although it has been widely used by architects operating in top-down fashion. Overlapping hierarchies, although simplifying interaction, capture the diversity of behavior and are much more suitable pictorial vehicles for progressing good urban design—something we will return to in the last part of this book, which turns our argument from science to design, or at least from science to design science. In chapter 10, we exploit the idea of hierarchy in design quite explicitly, building on Alexander's (1964) ideas and focusing the argument on how opinions about design are pooled, thus reflecting ways in which ideas that interrelate can be structured as overlapping hierarchies. In a sense, this argument has also been anticipated in urban systems science; Christaller's central place hierarchies were overlapping, while the whole point of spatial interaction modeling and its link to retail center definition has been to relax the notion of hierarchy, letting it remain implicit in the space of flows.

There are other, more direct reasons for thinking of cities as overlapping hierarchies or lattices, and this simply emerges from the fact that there are many such hierarchies. We have only examined the simplest here—those based on how the population in an aggregate sense arranges itself. But once one disaggregates populations into the multiplicity of categories that define them, and once one adds other kinds of activities that arrange themselves hierarchically in space, such as transport and other network systems, land uses, styles of buildings, social friendship nets, and so on, then the idea of overlap becomes the rule, not the exception. It is hard to escape the fact that good analysis should tackle this notion directly. The idea that most of our analysis tends to simplify the system beyond this obvious reality poses a dilemma. What we require are good, simple, and plausible models that show us how different kinds of hierarchies interlock and couple.

In terms of city-size distributions, then, the challenge is to build on the network characterizations of Gibrat's model, possibly interlocking the network model illustrated earlier in this chapter with some sort of dual but countervailing network based on friendship patterns rather than the economics of travel that were implicit in the model demonstrated. This opens the door again to bipartite graphs, and we will exploit these extensively for simple network systems—quite literally, urban street networks—in the next chapter. Interlocking networks that lead to coupled but consistent and simple scaling of aggregate activities would seem to be the goal, and these are best illustrated in terms of street systems that provide the essential skeleton of how cities function with respect to their delivery of energy, information, and materials. We know that most distributions we see in cities are scaling or near-scaling, and the goal must be to show how these might be unpacked and linked at the network level, where we are able to grapple with the diversity that characterizes cities. In this way, our understanding of cities is enriched, suggesting ways we might be able to design our patterns of interaction more effectively.

However, it is worth noting how far we still are from this goal. For this we make no apologies, since as we have argued, this new science of cities is but at the beginning, and we are only just acquiring the tools to help us unravel the mysteries of why simple models account for very strong regularities in hierarchies of size in location, flows, and networks. Krugman (1996), in a somewhat candid and prescient article on rank-size scaling entitled "Confronting the Mystery of Urban Hierarchy," summed up this dilemma admirably when he said: "The usual complaint about (economic) theory is that our models are oversimplified—that they offer excessively neat views of complex, messy reality. This … is an interim report on my efforts to understand why in one important case the reverse is true: we have complex, messy models, yet reality is startlingly neat and simple. In large part this is a report of failure—that is, while there must be a compelling explanation of the astonishing empirical regularity in question, I have not found it." In fact, to find it, our thesis here is that we need much more powerful tools and representations of networks and their processes, and that the key to this lies in extending graphs into their bipartite form. The next chapter will examine such graphs by way of an example of how we can enrich the study of cities. But beyond this, we need conceptions of systems that are tripartite, indeed *n*-partite. The following exposition will suggest directions in which we might go.

6

Urban Structure as Space Syntax

Structure in the technical sense of a self-regulating system of transformations is not coincident with form: even a stack of pebbles cannot be said to have *form* . . . , but a mere heap cannot become a *structure* unless we place it in the context of a sophisticated theory by inter-calculating the system of all its "virtual" movements. We are thus brought to physics.
—Jean Piaget, *Structuralism* (1971, p. 36)

In the approach to cities we adopt here, physical form is all-important. From form we infer processes that create the structures we see in cities, thus enabling us to build models of these processes, that in turn will simulate forms. Form and function are thus central to our concern, and in terms of our perspective here, cities conceived as locations, which in turn are functions of interactions, are the structures we explore throughout this book. Terms such as form, function, structure, process, and dynamics are not closely defined, but as Piaget's (1971) opening quote implies, we must be clear about how all these terms relate to one another before we proceed. In fact, the term "morphology" is often used to portray all these aspects of cities, and we will follow this usage here. Morphology is, of course, used in many fields, although its original use by Goethe (1790/2009) to describe the formation and transformation of natural entities generated organically is closest to the vision of cities espoused here. *Urban morphology* we define as the study of form and structure in cities that focuses on the dynamics of change and the rules underlying these dynamics. An alternative perspective, which is the one adopted in linguistics, switches this focus to the grammatical and syntactical rules by which language is constructed. In the development of theories of urban morphology, the focus on organic growth is represented in terms of how cities grow naturally from the bottom up, filling space and repeating key motifs at different levels of hierarchy as, for example, developed using ideas from fractal geometry (Batty and Longley, 1994), which we will introduce more explicitly in later chapters.

The focus on generative process following the linguistic model has been articulated in the development shape grammars and various kinds of procedural modeling.

There the focus is on rules generating new objects that grow buildings and cities at a somewhat finer scale than fractal growth models. We will explore fractal growth and ways of implementing the fractal model using cellular automata in chapter 8, but in this chapter, we will begin our exploration of urban morphologies in terms of network structures that underpin locations. We will focus on street and other route systems whose articulation is in terms of an implicit grammar of form, which has been called space syntax (Hillier, Leaman, Stansall, and Bedford, 1976), although the early generative focus of this work is still only implicit in the way network systems are put together. In fact, we begin with this approach, largely because it provides us with excellent examples of illustrating why we need to deepen our analysis of cities using more powerful tools than simple planar graphs and spatial networks. Space syntax enables us to show how networks can actually be predicted as a consequence of two sets of objects that are related—how streets are defined as explicit objects and how they are joined together by another set of objects, their junctions or intersections. In this context, we can begin to produce a much more powerful syntax using bipartite graph representations that build on the ideas of chapter 3. Thus we will illustrate a generic approach to thinking of urban structure as sets of relations that potentially dig much deeper into the immediate superficiality of urban form, which we perceive as multiple networks intersecting one another across multiple layers.

Traditional Representations

So far we have assumed that urban form is represented as a pattern of identifiable urban elements such as locations—points or areas—whose relationships to one another are often associated with linear transport routes, such as streets within cities. In some respects, these street systems represent the skeletal frame around which cities develop, and as such they provide a substrata on which much else is built. These elements can be thought of as forming nodes in a graph, the relations between the nodes being arcs that represent direct flows or associations between the elements. This was the conception of structure that lies at the basis of the various relational tools and methods outlined in chapters 2 and 3. These relations need not be physically rooted in the detailed geometry of buildings, for at coarser spatial scales, they are likely to be more abstract, such as migration flows between regions. At finer spatial scales and more local levels, however, they are usually taken to be linear features, such as streets or corridors. The focus of such analysis is either modeling the explicit flows between the nodes defining the route systems, or on implicit flow analysis using the relative proximity or "accessibility" associated with different locations. This can involve calculating distances between nodes in such graphs and associating these with densities and intensities of activity that occur at different locations and along the links between them. For example, clusters of work activity are usually associated with high levels of accessibility. Much planning and design is thus concerned with changing the efficiency of the way cities operate and with manipulating the equitable distribution of spatial resources as functions of these patterns of accessibility through the development of new or changed transport infrastructures.

There is a long tradition of research articulating urban form using graph-theoretic principles. Idealized systems of routes defining the structure of towns and their traffic go back to the late 1950s, as for example in the work of Smeed (1961), Holroyd (1966), and others (see Haggett and Chorley, 1969). Before that, we can trace notions about networks binding different locations together to the work of the German location theorists (see Isard, 1956) and back to Kohl (1841), whose ideas were briefly presented in chapter 2. Nystuen and Dacey (1961) developed such representations as measures of hierarchy in regional central place systems, while Kansky (1963) applied basic graph theory to the measurement of transportation networks. Graphs are implicit in the definition of gravitational potential based on the weighted sum of forces around a point first applied to population systems by Stewart (1947), and are illustrated for flow systems in chapter 2, while subsequent work on identifying accessibility as a key determinant of spatial interaction is based on an implicit graph-theoretic view of spatial systems (Hansen, 1959; Wilson, 1970). In a similar manner, graphs have been widely used to represent the connectivity between rooms in buildings (March and Steadman, 1971) and to classify different building types (Steadman, 1983). They have long been regarded as the basic structures for representing forms when topological relations are firmly embedded within Euclidean space. In the development of generative and procedural models of growing systems at the building or coarser scales, as in shape grammars, for example, relations between buildings and more aggregative spatial objects are implicit in the rules that are used to form and transform such structures as they develop through time (Stiny, 2006; Lipp, Scherzer, Wonka, and Wimmer, 2011).

In the most general representations introduced in chapters 2 and 3, we defined locations or points in Euclidean space as nodes or vertices $\{i, k\}$, and the links or arcs between them as $\{\ell_{ik}, i, k = 1, 2, ...\}$. The value of the link can be binary, one of presence or absence, or some actual physical distance d_{ik}. For systems at a fine scale, such as those we deal with here, where the focus is often on connectivities within neighborhoods and buildings, the linkage is usually binary, defined as

$$\ell_{ik} = \begin{cases} 1 & \textit{if a relation exists between i and k} \\ 0, & \textit{otherwise} \end{cases}. \tag{6.1}$$

In this context, such relations of presence or absence are usually symmetric, that is, $\ell_{ik} = \ell_{ki}$; direct or indirect links exist between any two nodes, thus implying that the underlying graph is strongly connected; and self-linkages ℓ_{ii} are not usually considered important and thus set equal to zero, $\ell_{ii} = 0$. We will adopt these assumptions here, although they do not in any way reduce the generality of our argument. Accessibility in such binary graphs is computed in terms of their connectivity, and the direct linkages of points or nodes (in-degrees and out-degrees) to one another are given as $\ell_i = \sum_j \ell_{ij} = \ell_k = \sum_j \ell_{jk}$, where $i = k$. Shortest route distances through the graph, given by d_{ik}, also provide access measures, and these need to be weighted inversely to provide an equivalent index of access as, for example, $V_i \propto \sum_k d_{ik}^{-1}$, where the same symmetry as for direct connectivities is implied. These measures are similar to some of those, such as closeness centrality, introduced in chapter 3.

In fine-scale analysis, the graph is usually planar or assumed to be approximated as such, in that the topological and Euclidean structures of the set of relations are identical; that is, the graph is the street or corridor network. It is assumed that there are no cross-overs of the arcs that do not imply nodes, and no one-way arcs (one-way streets). Figure 6.1(a) represents such a graph, focusing on accessibility of the nodes, which we refer to as the *primal problem*. There is, however, a related problem of relations defined on the same graph, which figure 6.1(b) illustrates. If we trace the relations between the arcs of the original graph, which in the street network problem is equivalent to finding relationships between each street segment, this provides another graph representation, called the *dual problem*. This is not the same as the more common but different specification of the dual, which relates to the

A   B

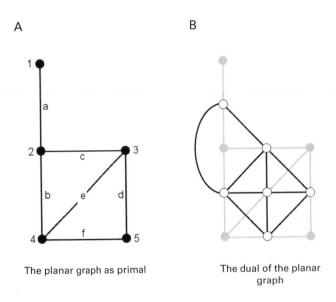

The planar graph as primal

The dual of the planar graph

Figure 6.1
Conventional graph-theoretic representation of the street network.

network of relations between the interstices formed by the areas bounded by the links in the original planar graph (March and Steadman, 1971). These relations are no longer embedded within physical space in quite the same way as the initial links, for they now represent abstract relations between streets. These are relations through the joining of streets at junctions, whereas the primal problem is posed as one of relations between junctions where the links are the streets themselves.

This dual problem has not been widely developed in applications of networks to cities, for the natural focus is on articulating the network of communications links as arcs and their intersection in specific locations as nodes. In earlier chapters, networks—insofar as they were embedded in Euclidean space—were treated as planar graphs, or as graphs whose nodes anchor the representation in such space. The focus on linear features rather than locations has rarely been developed, for the dual privileges lines or streets as the objects of interest, rather than locations or street junctions. Moreover, the dual breaks the clear link between Euclidean and topological space, which makes visual analysis of the dual more difficult. Nevertheless, there is a tradition in which this dual has been widely developed, and this is "space syntax," developed initially for spaces within buildings by Hillier and Hanson (1984). The theory has its roots in quite sophisticated speculation that the evolution of built form can be explained in analogy to the way biological forms unravel (Hillier, Leaman, Stansall, and Bedford, 1976). In its current and widely applied form, it is more a toolbox of simple techniques for measuring street accessibility in towns and associating this with unobstructed movement between spaces and lines of sight (Hillier, Penn, Hanson, Grajewski, and Xu, 1993). But the key distinguishing feature in space syntax, in contrast to similar methods that have been developed for network morphologies in cities (see Sevtsuk, 2010), is that precedence is given to linear features, such as streets, in contrast to fixed points that approximate locations (Hillier, 1996).

Figure 6.1 illustrates that there is a clear path between the primal and dual problems that has rarely been exploited, certainly not within space syntax. In this chapter, we will explore how the primal and dual problems represent different but complementary perspectives on network morphology, developing a unifying framework so that one can easily move between the two sorts of problem. In this way, we will show how space syntax can be translated into a more familiar locational analytic frame. We need to explain space syntax first, and we will do this in the next section, but then we will establish our framework showing how connectivity and distance in both the primal and dual problems can be more easily understood. We then illustrate how spatial averaging is involved in computing accessibility, introducing Hillier and Hanson's (1984) original example, the French village of Gassin, which we take as an exemplar. We show how the ability to move from one problem to its dual enables a much more satisfactory visual analysis, showing finally how we might add distance back into space syntax and, at the same time, indicating how more than one type of network might be handled using this extended formalism.

In the next chapter, which builds directly on this, we introduce a simple example for several street blocks in the West End of Central London, presenting all these results for both primal and dual problems, while we also suggest that these methods can be used to look at a variety of different forms of street system, planned and unplanned. There we will also progress this analysis to deal with distance, thus bringing this analysis back to street systems and suggesting ways in which different networks might be coupled together using this extended formalism. In both these chapters, our concern is to simplify space syntax and develop a consistent version, while pointing the way to further generalization of these kinds of problem and their relation to current developments in the evolutionary and statistical theory of networks (Dorogovtsev and Mendes, 2003).

Explaining Space Syntax

In space syntax, the focus is on lines, not points, streets, or corridors, nor the junctions that anchor them in two-dimensional space, as illustrated in figure 6.1. This is not particularly controversial, although it is often difficult to approximate a street by a centroid. However, where the analysis departs from the dual formulation in figure 6.1(b) is that the syntax map is no longer planar: street segments do not have to be anchored by nodes at two ends—a street can have any number of junctions or intersections greater than one. In this sense, junctions are not so important, for in space syntax, streets are related to one another if there are any junctions between them, not through particular junctions, unless there is only one. Streets are very definitely not locations in this interpretation, and thus the relations between any two streets can never be uniquely embedded in Euclidean space. This directs the analysis onto the topological relations between streets; it forces the representation of distance between two streets to be distance in the graph-theoretic rather than the Euclidean sense, thus removing the relational graph from the physical space in which it is defined in the first instance. In some contexts, this is a logical approach when the focus is on the importance of streets as objects in their own right rather than on the objects—junctions—that link them together.

Figure 6.2(a) shows how the simple graph from figure 6.1(a) can be relabeled to generate a different relational structure in which arcs have one or more nodes associated with them, which is the essence of space syntax representation. The new street graph, which is composed of lines a, b, c, d, and e, where a is now composed of the original segments between junctions 1 and 2 and 2 and 4, is not planar, and thus it is not appropriate to refer to this as a graph any longer. It is usually called an "axial map," and the lines that compose it are called "axial lines." There is some

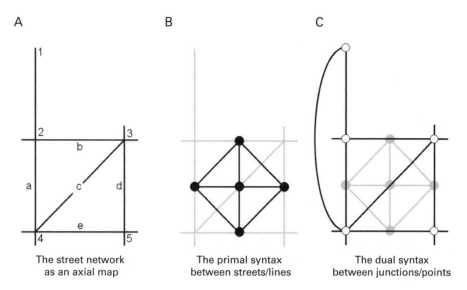

A   B   C

The street network
as an axial map

The primal syntax
between streets/lines

The dual syntax
between junctions/points

Figure 6.2
Space syntax representation.

controversy about how such lines are defined, but a general consensus seems to be that these are "lines of sight" or possibly lines of unobstructed movement, although this latter usage is more controversial. The former usage tends to limit space syntax to the building or urban design scales, where corridors or streets rather than generic transport routes are important and where detailed urban morphology and geometry are the focus. Figure 6.2(b) shows the space syntax graph, which is defined by associating any two streets if they have a junction in common. There is an immediate and clear difference from the planar graph: that is, a street increases in importance as the number of junctions associated with it gets greater. In terms of the traditional problem, as the importance of a junction increases, the greater the number of lines or streets associated with it, but the dual of this primal is different from the primal of the space syntax problem, as we also show below.

We should be clear at the onset about the primal and dual problems as we define them in space syntax. The primal problem is in fact a generalization of the dual of the traditional planar representation, with the focus on relationships between streets. The dual space syntax problem is then the problem of relating street junctions through streets. A visual representation of the graph for this is shown in figure 6.2(c). This dual is associated with the primal—the planar graph—in the original problem; the axial map is a subset of this graph, which has also been called a "visibility graph" (Turner, Doxa, O'Sullivan, and Penn, 2001). However, to make progress in understanding these problems and their implications for urban analysis, we need a much more powerful framework, which is outlined in the following section. This will enable us not only to move between one form of problem and the other, but also to relate accessibility measures between each problem. It will ultimately provide a much simplified form of space syntax.

A Unifying Framework: Representing Duals and Primals, Points and Lines, by Bipartite Graphs

The key to a more unified understanding involves an elemental representation in which it is recognized that morphological relations are essentially predicated between two distinct sets of objects, in this case locations and linear features represented as points and lines. These sets can be any features of urban morphology, such as streets and their junctions, building parcels and streets, and even one set of streets arrayed against another, or streets against railways, and so on. But whatever the two sets, they must be distinct and their relation to each other must be unambiguous. In space syntax, the first set, defined as $L = \{\ell \mid i,k = 1, 2, \ldots n\}$, are streets, while the second are street junctions, defined as $P = \{\rho \mid j,l = 1, 2, \ldots m\}$. If a street contains a junction or a junction a street, this is defined in the $n \times m$ matrix, whose elements are

$$a_{ij} = \begin{cases} 1, & if \ \ell_i \supset \rho_j \ \ or \ \ \rho_j \supset \ell_i \\ 0, & otherwise \end{cases}. \tag{6.2}$$

Equation (6.2) is visualized in figure 6.3(a) for the street network in figure 6.2(a). This is a bipartite graph of relations between lines and points, from which it is clear that the number of points associated with any given line i is

$$\ell_i = \sum_{j=1}^{m} a_{ij}, \tag{6.3}$$

and the number of lines associated with any point j is

$$\rho_j = \sum_{i=1}^{n} a_{ij}. \tag{6.4}$$

Equations (6.3) and (6.4) define the respective out-degrees and in-degrees of the associated bipartite graph. In the following sections, we will drop the full range of summation, since this will be the same for every such operation. The notation in this and the next chapter, although close to similar representations elsewhere in this book, does differ slightly in terms of the symbols used for variables and parameters. However, an attempt has been made to keep differences to a minimum so that comparisons can be made to similar graph and matrix notation in other chapters.

The primal-dual nature of this representation is already implied in the line-point asymmetry and the direct connectivity indices for lines and points in equations (6.3)

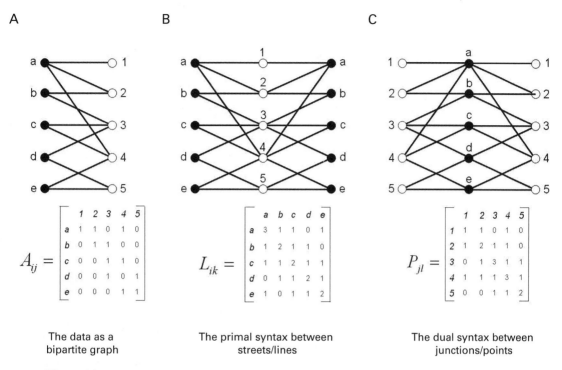

Figure 6.3
Space syntax as bipartite graphs.

and (6.4). As we shall see, lines are not privileged over points or vice versa. In fact, the planar graph and space syntax representations are particular cases within the framework, and these can be compared quite easily. Noting that for the planar graph case, the number of points for each line is always fixed at $\ell_i = 2$, $\forall i$ (as each street segment has a node at its beginning and end), then a measure of the deviation from planarity can be formed for any graph as $\Psi = \sum_i \ell_i / 2n$. For the village of Gassin, which we use later, $\Psi = 1.065$, which implies that there are an average of only 6.5% more nodes associated with street segments than in the planar case. The indices $\{\ell_i\}$ and $\{\rho_j\}$ are our first measures of direct access, and as we shall see, these will be central to our interpretation of accessibility in space syntax.

The measures simply count the number of points per line and lines per point, but the more usual approach is to examine the number of common points for any pair of lines or the number of common lines for any pair of points. These form the primal and dual characterizations of the problem. The number of points in common for any two lines is given by the matrix, whose elements ℓ_{ik} are defined from

$$\ell_{ik} = \sum_j a_{ij} a_{kj}.$$ (6.5)

The best way to visualize this is to connect the reverse bipartite graph to the original one, as shown in figure 6.3(b), where the number of common paths between any line i and any line k is given by counting the number of paths from i to k. This way of representing common points between lines immediately shows that $\mathbf{L} = [\ell_{ik}]$ is symmetric, which is also reflected in the in-degrees and out-degrees of the $\mathbf{L}$ matrix that form our next measures of line accessibility. These are calculated as

$$\tilde{\ell}_i = \sum_k \ell_{ik}, \quad \tilde{\ell}_k = \sum_i \ell_{ik}, \quad and \quad \tilde{\ell}_i = \tilde{\ell}_k, where\ i = k. \tag{6.6}$$

In essence, ℓ_{ik} is the space syntax graph but the practice has been to slice this graph, losing the count information that is associated with any relation between a pair of lines, thus making the matrix binary. Thus

$$Z_{ik} = \begin{cases} 1, & if\ \ell_{ik} > 0, i \neq k \\ 0, & otherwise \end{cases}. \tag{6.7}$$

Note that the slicing in equation (6.7) also loses information about the strength of the self-loops. In fact, this type of slicing is unnecessary, for valuable information about the strength of relations is lost; instead, we suggest that applications of space syntax should henceforth be based on ℓ_{ik} rather than Z_{ik}. However, this is a detail that does not make a substantial difference to the ensuing analysis.

The dual problem follows directly and can be stated in an analogous manner. First, the number of lines common to any two points can be calculated as

$$\rho_{jl} = \sum_i a_{ij} a_{il}, \tag{6.8}$$

and the measures of direct access or connectivity in the graph based on in-degrees and out-degrees are given as

$$\tilde{\rho}_j = \sum_l \rho_{jl}, \quad \tilde{\rho}_l = \sum_j \rho_{jl}, \quad and \quad \tilde{\rho}_j = \tilde{\rho}_l, where\ j = l. \tag{6.9}$$

The equivalent bivariate graph representation is illustrated in figure 6.3(c), where it is clear that the matrix $\mathbf{P} = [\rho_{jl}]$ is symmetric and counts the number of paths between any pair of points in terms of their common lines.

The primal and dual problems interlock with one another in an intriguing way, which has direct practical implications for how point accessibility can be translated into line accessibility and vice versa. To demonstrate this, we need to shift to matrix notation, which provides a much more parsimonious form for laying bare the nature of this interlocking. As in chapter 3, we will define all matrices and vectors in bold uppercase and lowercase type, respectively, starting from the basic $n \times m$ matrix of relations $\mathbf{A} = [a_{ij}]$. We will transpose this matrix as $\mathbf{A}^T$, but where we need to sum the elements of such matrices using the unit vector $\mathbf{1}$, we will not make any distinctions in terms of the transpose operation; use will be clear from context. We can now state the primal (space syntax) problem from equations (6.3), (6.5), and (6.6) as

$$\ell = \mathbf{A}\,\mathbf{1}, \quad \mathbf{L} = \mathbf{A}\,\mathbf{A}^T, \quad \text{and} \quad \tilde{\ell} = \mathbf{L}\,\mathbf{1}, \tag{6.10}$$

from which it is clear that $\mathbf{L}$ (and $\tilde{\ell}$) is symmetric: $\mathbf{L} = \mathbf{L}^T = (\mathbf{A}\,\mathbf{A}^T)^T = \mathbf{A}\,\mathbf{A}^T$, and $\tilde{\ell}^T = \mathbf{1}\,\mathbf{L}^T = (\mathbf{1}\,\mathbf{A}\,\mathbf{A}^T)^T = \mathbf{L}\,\mathbf{1}$. The dual problem has a similar structure,

$$\rho = \mathbf{1}\,\mathbf{A}, \quad \mathbf{P} = \mathbf{A}^T\mathbf{A}, \quad \text{and} \quad \tilde{\rho} = \mathbf{1}\,\mathbf{P}, \tag{6.11}$$

with analogous symmetries. The relation between the two problems is easy to illustrate. In equation (6.11), if we post-multiply $\rho = \mathbf{1}\,\mathbf{A}$ by $\mathbf{A}^T$, we derive

$$\tilde{\ell} = \rho\,\mathbf{A}^T, \tag{6.12}$$

and if we pre-multiply $\ell = \mathbf{A}\,\mathbf{1}$ by $\mathbf{A}^T$, we get

$$\tilde{\rho}^T = \mathbf{A}^T\ell. \tag{6.13}$$

The meaning of these relations is slightly tortuous; the number of common points for each pair of lines in equation (6.12) can be seen as a convolution of the number of lines for each point with respect to the existence of any line at each point. The number of common lines for each pair of points has an analogous interpretation. In fact, in later chapters, particularly in part III where we develop ideas about social interactions, networks, and exchange relevant to processes of conflict resolution, we exploit these relations on the basic bipartite graph much more extensively, but it illustrates a key theme we consider important to our science; that is, that it is not enough to simply begin with matrices of interactions, but instead we need to consider how such interaction matrices are formed. In this, the bipartite structure is a key element, providing a level of richness to the analysis that has been largely absent from this field hitherto.

In fact, ℓ, $\tilde{\ell}$, ρ, and $\tilde{\rho}$ will be the key indices of direct accessibility and connectivity, which we will use and compare in the following section, but before we broach the whole subject of distance in the graphs of these primal and dual problems, we must note the origins of the approach. In chapter 3, we noted theories of social power and sociometry dating back to the mid-twentieth century and before in which relational structures defining links between individuals and various attributes were described by bipartite graphs. In social network theory, these came to be referred to as two-mode graphs, in comparison with unipartite sets of relations called one-mode graphs (Borgatti and Everett, 1997). However, the idea of interpreting relations between two sets in the field of urban analysis is from Atkin (1974), who pioneered "Q-analysis." This analysis begins with relations arrayed in the form of the matrix $\mathbf{A}$ with dual and primal characterizations similar to those here, but being represented in a geometry called a simplicial complex (the primal) and its conjugate (the dual). Q-analysis was never widely exploited, perhaps because of its rather arcane presentation, and it was rarely linked to the theory of graphs.

From a rather different perspective, this kind of primal-dual framework was exploited by Coleman (1973) in his interpretation of social exchange. It was also generalized and linked to graph theory by Batty and Tinkler (1979), and related to social power in design-decision-making by Batty (1981). Until quite recently, the framework has only occasionally been exploited, but it has been rediscovered in the great wave of recent interest in networks, their evolution, and their statistics. We will build on these ideas in part III of this book, when our focus shifts to interaction patterns between those involved in conflict resolution and consensus formation. It is currently being widely exploited in the analysis of social networks using small worlds by Watts (2003) and Newman (2003). There have been some attempts at examining alternative graph-theoretic relations in space syntax itself (see Kruger, 1989), and Jiang and Claramunt (2000) have suggested that the visibility graph, which is in essence the graph of the dual, be the subject of analysis, shifting the focus to points rather than lines, as we suggest in this chapter. Porta, Crucitti, and Latora (2006a, 2006b) have proposed an explicit primal-dual characterization which, like the approach here, seeks to broaden the space syntax method of measuring accessibilities in street systems.

Patterns in the Syntax

Accessibilities Based on Connectivity and Distance

The connectivity measures introduced above are measures of direct access to lines and points from the same elements that are immediately adjacent to them; that is, link with them directly. More appropriate measures of distance, although taking account of such adjacency, are based on indirect links between the system elements. The usual form is to calculate shortest routes between the elements, thence computing the associated in-degrees and out-degrees, which provide measures of potential or accessibility. In this section, we will introduce the standard measure and then propose another that has more desirable features, but in each case, these distances will be based on the interaction matrices $\mathbf{L}$ for lines and $\mathbf{P}$ for points. We will first illustrate our standard measure for the primal problem. We start with the matrix $\mathbf{L}$, which gives the number of points that are common between any pair of lines. What we require is a computation of the number of common points between all paths in the graph that exist between any two lines that are at different steps removed from one another. The elements of the basic matrix ℓ_{ik} are one step removed from each other and are direct links, while the number of two-step paths is given by

$$\ell_{ik}^2 = \sum_z \ell_{iz}^1 \ell_{zk}, \tag{6.14}$$

where ℓ_{ik}^1 is the basic matrix element ℓ_{ik}. Successive numbers of paths of length s are thence computed as

$$\ell_{ik}^{s+1} = \sum_z \ell_{iz}^s \ell_{zk}. \tag{6.15}$$

We compute a measure of distance, however, not in terms of the number of points associated with these path lengths but in terms of the actual path length, which minimizes the distance between any two lines i and k. Thus, formally,

$$if \quad \ell_{ik}^{s+1} > 0 \quad and \quad \ell_{ik}^s = 0, \quad then \quad d(\ell)_{ik} = s, \tag{6.16}$$

where s is the length of the path. In a strongly connected graph (which all graphs here are by definition), $d(\ell)_{ik} > 0$ when the path length s reaches n, if not before. This is a standard result of elementary matrix algebra, and equations (6.15) and (6.16) thus provide the algorithm that enables shortest paths in these kinds of graph to be computed.

We noted earlier that in space syntax, the matrix that is used is not $[\ell_{ik}]$ but its binary form $[Z_{ik}]$, defined in equation (6.7). However, this gives a distance matrix very close to $[d(\ell)_{ik}]$. The weighting produced by raising $\mathbf{L}$ to successive powers, which is what the algorithm in essence is doing, is of no relevance. In fact, even though the matrix $[Z_{ik}]$ has its self-elements $Z_{ik} = 0$, the two-step paths become positive, and the resulting distance matrix is highly correlated with that produced by the process in equations (6.15) and (6.16). It is, however, easier to present these operations using matrix notation. Thus, for the primal problem, successive powers of $\mathbf{L}$ are given by $\mathbf{L}^{s+1} = \mathbf{L}^s \mathbf{L}$. The distance matrix, which we now write as $\mathbf{D}(\ell)$, becomes stable when $s \leq n$. An exactly analogous process is used to generate the dual distance matrix, where the point to point matrix $\mathbf{P}$, which gives the number of common lines between any pair of points, is raised to successive powers $\mathbf{P}^{s+1} = \mathbf{P}^s \mathbf{P}$, with the distance matrix computed as $\mathbf{D}(\rho)$.

We compute the in-degrees and out-degrees, (but these are the same because of symmetry) of successive powers of the appropriate matrices as

$$\tilde{\ell}^s = \mathbf{L}^s \, \mathbf{1} \quad and \quad \tilde{\rho}^s = \mathbf{1} \, \mathbf{P}^s, \tag{6.17}$$

and there are multiple ways of showing how these degree vectors for the primal and dual problems interlock with one another. We state without further explanation the nature of this interlocking for each problem as

$$\left.\begin{aligned} \tilde{\ell}^s &= \mathbf{L} \, \tilde{\ell}^{s-1} = \mathbf{L}^{s-1} \, \tilde{\ell} = \mathbf{A} \, \mathbf{P}^{s-1} \, \rho^T \\ \tilde{\rho}^s &= \rho^{s-1}\mathbf{P} = \tilde{\rho} \, \mathbf{P}^{s-1} = \ell^T \, \mathbf{L}^{s-1} \, \mathbf{A} \end{aligned}\right\}. \tag{6.18}$$

The relationships in equations (6.17) and (6.18) provide a wealth of alternate interpretations for the meaning of path length in graphs of this nature. Further analysis along these lines, however, takes us away from the focus here, but we will pick this up below and also exploit the kinds of relations in equation (6.18) in depth in various chapters in part III of this book.

The aggregate distances from a line to all others in the primal problem and from a point to all others in the dual are computed in the usual way by summing the relevant distance matrices as in-degrees or out-degrees; that is,

$$\mathbf{D}(\ell) = \mathbf{D}(\ell)\,\mathbf{1} \quad and \quad \mathbf{d}(\rho) = \mathbf{1}\,\mathbf{D}(\rho). \tag{6.19}$$

In fact, these distances are measures of inaccessibility rather than accessibility and need to be inverted in some way to provide appropriate measures. In space syntax, $\mathbf{d}(\ell)$ is referred to as depth and is usually averaged with respect to the number of lines in the system n. This is necessary if lines (and zones of lines) within a certain distance or depth from a given line are to be identified, but it makes no difference to the relative distribution. The measure of access used in space syntax simply takes the mean values of distance for the primal problem and inverts each, providing an index that is called "integration." Variations in these indices exist (Teklenberg, Timmermans, and van Wagenberg, 1993), but for the primal and the dual, integration (or accessibility) for each element is usually defined as

$$\ell(d)_i = \frac{1}{[d(\ell)_i/n]} \quad and \quad \rho(d)_j = \frac{1}{[d(\rho)_j/m]}. \tag{6.20}$$

The main problem with these measures is that they ignore both the relative importance and the strengths of paths through the graph. First, information is lost through the fact that connectivity strengths are transformed to simple step-length distances, as in equation (6.16). Second, each step is given equal weight, whereas it might be assumed that as the step length gets greater, the relative importance of the step gets smaller. Third, the number of steps in the graph depends upon the size of the graph, and thus systems of different sizes cannot be compared. Some normalization has to take place to ensure comparison. Some of these issues have been tackled, but they are best resolved with a new measure of distance that relies on the basic path connectivity matrices $\mathbf{L}$ and $\mathbf{P}$ and on some notion that larger step lengths act like distances in Euclidean space, becoming increasingly less important.

Weighted Accessibilities

There are many possibilities, and we simply introduce one of these here. For the line to line interaction matrix $[\ell_{ik}]$, we weight each matrix power s by ω^s and form the linear combination

$$\tilde{D}(\ell)_{ik} = \sum_s \omega^s \ell_{ik}^s, \tag{6.21}$$

where ω^s declines with increasing path length s. If we set this weight as λ^s where s is now a power (as well as an index) and $0 < \lambda < 1$, then as $s \to \infty$, $\lambda^s \to 0$. The aggregate distance for line i can be computed as

$$\tilde{d}(\ell)_i = \sum_k \tilde{D}(\ell)_{ik} = \sum_s \lambda^s \sum_k \ell_{ik}^s = \sum_s \lambda^s \tilde{\ell}_i^s. \qquad (6.22)$$

We can fix the range of the summation over s to a value determined by the size of λ^s. When s is the size of the matrix, all step lengths are guaranteed to be positive and $\lambda^n \ll 1$, but usually the range can be fixed at the point s where all the step lengths become positive. The equivalent measure for the dual problem is defined as

$$\tilde{d}(\rho)_j = \sum_s \lambda^s \tilde{\rho}_j^s. \qquad (6.23)$$

This definition illustrates that each path length makes a specific contribution to the overall definition of distance, and this can be tuned by fixing the value of λ. In the Gassin example, we fix $\lambda = 0.05$. If we then measure the contribution of each path length s for the primal (line) problem as $\Phi = \sum_{ik} \ell_{ik}^s / \sum_{iks} \ell_{ik}^s$, we generate the following proportions: $s = 1$, $\Phi = 0.717$; $s = 2$, $\Phi = 0.178$; $s = 3$, $\Phi = 0.062$; $s = 4$, $\Phi = 0.025$; and $s = 5$, $\Phi = 0.019$, where the maximum path length (or depth) between streets in the Gassin axial map is 5.

We now have four measures of accessibility for each of the two problems: two based on direct or adjacent distances and two based on all distances. For the primal problem these are the vectors ℓ, $\tilde{\ell}$, $\ell(\mathbf{d})$ and $\tilde{d}(\ell)$; for the dual, ρ, $\tilde{\rho}$, $\rho(\mathbf{d})$ and $\tilde{d}(\rho)$. What we suspect in space syntax graphs where the average depth or step length is small—in Gassin it is 3.239—is that these measures are highly correlated with one another. To test this hypothesis, we have examined 1,000 randomly constructed systems of points and lines where the number of lines varies from 30 to 60 and number of points varies from 40 to 80. We also vary the density of relations between lines and points measured using the ratio of the total number of relations to the potential number, $\Theta = [1 - \sum_{ij} a_{ij}/(nm)]$, from 0.75 to 0.99. Note that in Gassin, the number of lines is 41, the number of points 63, and the ratio $\Theta = 0.948$, so these randomized runs are comparable with our real case. Because we have ruled out all disconnected systems from these random runs, the average density is $\Theta = 0.825$, the average number of lines is 45, and the average number of points is 59. The systems generated are somewhat dense axial maps with an average step length of around 2.6. As such, these provide a coarse first attempt at comparing various types of distance measures, but more work is required to support the tentative conclusions we draw here. We define an index of similarity between each pair of distances, which we show for the example of ℓ and $\tilde{\ell}$ in equation (6.24) as follows:

Table 6.1
Average similarities Ξ between the four distance measures. The comparisons are symmetric, and the statistics in brackets below the diagonal are standard deviations of the relevant similarity measure above the diagonal.

(a) Line distances	ℓ	$\tilde{\ell}$	$\ell(\mathbf{d})$	$\tilde{\mathbf{d}}(\ell)$
ℓ	•	0.927	0.775	0.914
$\tilde{\ell}$	(0.030)	•	0.767	0.972
$\ell(\mathbf{d})$	(0.069)	(0.082)	•	0.769
$\tilde{\mathbf{d}}(\ell)$	(0.041)	(0.020)	(0.049)	•

(b) Point distances	ρ	$\tilde{\rho}$	$\rho(\mathbf{d})$	$\tilde{\mathbf{d}}(\rho)$
ρ	•	0.898	0.638	0.880
$\tilde{\rho}$	(0.048)	•	0.626	0.959
$\rho(\mathbf{d})$	(0.163)	(0.178)	•	0.687
$\tilde{\mathbf{d}}(\rho)$	(0.064)	(0.031)	(0.117)	•

$$\Xi(\ell : \tilde{\ell}) = 1 - \sum_i \frac{\left[\left(\ell_i \Big/ \sum_k \ell_k\right) - \left(\tilde{\ell}_i \Big/ \sum_k \tilde{\ell}_k\right)\right]}{\left(\ell_i \Big/ \sum_k \ell_k\right)}. \tag{6.24}$$

This measure is chi-square-like and varies between 1—complete similarity—and 0—complete dissimilarity. The other measures are computed accordingly for both the primal and dual problems.

Comparisons of these distance measures are shown in tables 6.1(a) and (b) for the primal and dual problems, respectively. Three of the distance measures based on the in-degrees and out-degrees of the original data matrix $\mathbf{A}$, the basic interaction matrices $\mathbf{L}$ and $\mathbf{P}$, and the weighted distance matrices $\tilde{\mathbf{D}}(\ell)$ and $\tilde{\mathbf{D}}(\rho)$ are more than 80% similar to one another. The step-length distance matrix $\mathbf{D}(\ell)$ has around 70% similarity with these other three measures while the matrix $\mathbf{D}(\rho)$ has only 60% similarity. Nevertheless, this suggests that the direct measures of access, which ignore all indirect links, are quite good measures of the importance of lines or points in the primal or dual problems where the axial map is quite densely connected, which these 1,000 runs imply. As we shall see, these results are similar to the Gassin example reported below, although there is considerable volatility in the similarities between $\tilde{\ell}$ and $\mathbf{d}(\ell)$, and $\tilde{\rho}$ and $\mathbf{d}(\rho)$, which are revealed in figure 6.4. This suggests that where we have more points than lines, as in many space syntax problems, then it is among the points that the greatest discrimination with respect to accessibility occurs. This might seem counterintuitive, since space syntax privileges lines over

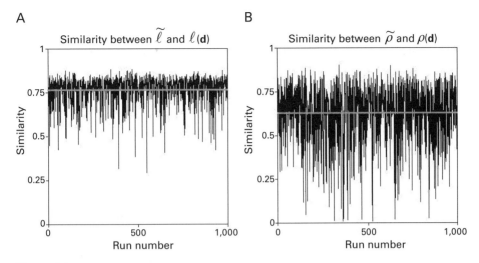

Figure 6.4
Variations in similarity between direct distance and indirect step-distance.

points and streets over their junctions, yet there is a sense in any problem where one set is numerically greater in its mass than another that this set will have greater significance. We will return to this in our analysis of Gassin below, but before we do so, we need to introduce one last idea about the meaning of distance.

The Algebra of Syntax

Averaging Lines into Points and Points into Lines
It makes sense to explore whether there are distance vectors associated with the relative accessibility between lines, which can be derived by consistently weighting the points and vice versa. This would amount to a perfect interlocking of the primal and dual problems, but it would also provide a form of natural averaging between lines and points. In short, what we require are vectors $[\overline{\ell}_i]$ and $[\overline{\rho}_j]$ such that

$$\overline{\rho}_j = \sum_i \overline{\ell}_i X_{ij}, \tag{6.25}$$

and

$$\overline{\ell}_i = \sum_j \overline{\rho}_j Y_{ij}, \tag{6.26}$$

where the matrices $[X_{ij}]$ and $[Y_{ij}]$ give the respective weights of each point in each line and each line as part of each point. Equations (6.25) and (6.26) can thus be regarded as types of steady-state equation.

The problem must be grounded, of course, in data that somehow relates to the structural matrix $[A_{ij}]$, with an obvious definition of these weights given as follows. We first express the relative importance of each point j to a given line i as

$$X_{ij} = \frac{A_{ij}}{\sum_l A_{il}}, \quad \sum_j X_{ij} = 1, \tag{6.27}$$

and the relative importance of each line i to a given point j as

$$Y_{ij} = \frac{A_{ij}}{\sum_k A_{kj}}, \quad \sum_i Y_{ij} = 1. \tag{6.28}$$

The problem is now well-defined. We seek vectors $\bar{\rho}$ and $\bar{\ell}$, which are solutions to equations (6.25) and (6.26), respectively, which in matrix terms are $\bar{\rho} = \bar{\ell}\,X$ and $\bar{\ell} = \bar{\rho}\,Y$.

There are two ways of proceeding. First, simple substitution of equations (6.25) into (6.26) and (6.26) into (6.25) leads to

$$\bar{\rho}_j = \sum_i \sum_l \bar{\rho}_l Y_{il} X_{ij}, \tag{6.29}$$

and

$$\bar{\ell}_i = \sum_j \sum_k \bar{\ell}_k X_{kj} Y_{ij}. \tag{6.30}$$

The matrix weightings in equations (6.29) and (6.30) can be defined as

$$\Omega_{jl} = \sum_k Y_{kj} X_{kl}, \quad \sum_l \Omega_{jl} = 1, \tag{6.31}$$

and

$$\Lambda_{ik} = \sum_j X_{ij} Y_{kj}, \quad \sum_k \Lambda_{ik} = 1, \tag{6.32}$$

where Ω and Λ are clearly Markov transition matrices. These can be interpreted as measuring the relative importance (probability or proportion) of a point (or line) being related to another point (or line). In chapter 3, we constructed the same matrices as those in equations (6.31) and (6.32) in equations (3.9) and (3.10), and then developed an equivalent Markovian interpretation in terms of defining averaging processes on one of these matrices.

We now write equation (6.29) as

$$\bar{\rho} = \bar{\rho}\,\Omega = \bar{\rho}\,Y^T X, \tag{6.33}$$

where the vector $\bar{\rho}$ gives the relative importance of each point, and equation (6.30) as

$$\bar{\ell} = \bar{\ell}\,\Lambda = \bar{\ell}\,XY^{T},\tag{6.34}$$

where $\bar{\ell}$ is the vector giving the relative importance of each line. As Ω and Λ are Markov matrices (and strongly connected), these provide steady-state equations, which can be solved:

$$\left.\begin{array}{l} \bar{\ell} = \bar{\ell}\,\Lambda = \bar{\rho}\,Y^{T} \\[4pt] \bar{\rho} = \bar{\rho}\,\Omega = \bar{\ell}\,X \end{array}\right\}.\tag{6.35}$$

This is a natural weighting that enables us to average the importance of lines into points and vice versa, so that if one solves the primal problem, there is a direct interpretation of the dual consisting solely of averaging the dimensionality of the primal into the dimensionality of the dual. Moreover, it also provides a justification for averaging one dimension into another using the relative importance of points and lines contained within the initial data, and thus might be applied, as we show below, to measures of distance other than those computed from the steady state.

Uniqueness of the Steady-State Accessibilities

The second way of showing the uniqueness of the steady state involves us in choosing any arbitrary distance vector, for lines, say, and then generating better and better approximations to the steady state through successive averaging. For example, from a given vector $[\bar{\ell}_i^1]$, we can compute a better approximation $[\bar{\ell}_i^2]$ by averaging or weighting the vector according to the sequence $\bar{\ell}_k^2 = \sum_i \bar{\ell}_i^1 \Lambda_{ik}$, $\bar{\ell}_k^3 = \sum_i \bar{\ell}_i^2 \Lambda_{ik}$, and so on. Using this relation, we can write the recurrence for any iteration s as

$$\bar{\ell}^{s+1} = \bar{\ell}^{s}\Lambda = \bar{\ell}^{1}\Lambda^{s}.\tag{6.36}$$

As Λ is a Markov matrix (and by definition strongly connected), the recurrence in equation (6.36) converges to a limit, that is

$$\lim_{s\to\infty}\bar{\ell}^{s} = \bar{\ell}^{s}\Lambda,\tag{6.37}$$

which is equation (6.34). The analogous process for the dual is based on the same form of recurrence, $\bar{\rho}^{s+1} = \bar{\rho}^{s}\Omega = \bar{\rho}^{1}\Omega^{s}$. In fact, equation (6.36) provides a straightforward solution to the steady state rather than simultaneously solving some combination of equations (6.33) to (6.35).

There is, however, a somewhat unusual simplification that occurs with the definitions used here, and to anticipate this, we suggest that the steady state is in fact implicit within the initial data. To show this, we must revert to the initial data by expressing the relative data matrices $\mathbf{X}$ and $\mathbf{Y}$ in terms of $\mathbf{A}$. Then, noting again that $\ell = \mathbf{A}\ \mathbf{1}$ and $\rho = \mathbf{1}\ \mathbf{A}$, and defining diagonal matrices of dimension $n\ x\ n\ \ell(\delta)$ and $m\ x\ m\ \rho(\delta)$ from the reciprocals $[1/\ell_i]$ and $[1/\rho_j]$, we can write $\mathbf{X} = \ell(\delta)\ \mathbf{A}$ and $\mathbf{Y}^T = [\mathbf{A}\ \rho(\delta)]^T = \rho(\delta)\ \mathbf{A}^T$. The steady-state relations in equations (6.34) now become

$$\left.\begin{aligned} \overline{\ell} &= \overline{\ell}\ \Lambda = \overline{\ell}\ \mathbf{X}\ \mathbf{Y}^T = \overline{\ell}\ \ell(\delta)\ \mathbf{A}\ \rho(\delta)\ \mathbf{A}^T \\ \overline{\rho} &= \overline{\rho}\ \Omega = \overline{\rho}\ \mathbf{Y}^T\mathbf{X} = \overline{\rho}\ \rho(\delta)\ \mathbf{A}^T\ell(\delta)\ \mathbf{A} \end{aligned}\right\}. \tag{6.38}$$

Let us assume that the steady-state vector for lines is the same as the raw data vector for lines, that is, $\overline{\ell} = \ell$. Then, using this in equation (6.38), it is clear that

$$\overline{\ell} = \ell\ \ell(\delta)\ \mathbf{A}\ \rho(\delta)\ \mathbf{A}^T = \mathbf{1}\ \mathbf{A}\ \rho(\delta)\ \mathbf{A}^T = \rho\ \rho(\delta)\ \mathbf{A}^T = \mathbf{1}\ \mathbf{A}^T = \ell. \tag{6.39}$$

In exactly analogous fashion for the dual, we can show that

$$\overline{\rho} = \rho\ \rho(\delta)\ \mathbf{A}^T\ell(\delta)\ \mathbf{A} = \mathbf{1}\ \mathbf{A}^T\ell(\delta)\ \mathbf{A} = \ell\ \ell(\delta)\ \mathbf{A} = \mathbf{1}\ \mathbf{A} = \rho. \tag{6.40}$$

In short, $\overline{\ell} = \ell$ and $\overline{\rho} = \rho$, which is a somewhat surprising result in that the steady state is in fact composed of the in-degrees and out-degrees associated with the original data. This suggests that a simple count of the in-degrees and out-degrees in the original bipartite graph based on $\mathbf{A}$ provides intelligible and meaningful measures of the importance of lines and points and streets and their junctions. These measures, of course, do not need digital computation and can be readily derived by simply inspecting the axial map.

However, what is of interest is the process of averaging. If we have a distance measure for lines, let us say any of the distance measures defined previously as $\tilde{\ell}$, $\ell(\mathbf{d})$, or $\tilde{\mathbf{d}}(\ell)$, we can derive averaged point estimates as $\rho' = \tilde{\ell}\ \mathbf{X}$, $\rho'' = \ell(\mathbf{d})\ \mathbf{X}$, or $\rho''' = \tilde{\mathbf{d}}(\ell)\ \mathbf{X}$. These would not be stable, since if we then reweighted these average point estimates by lines—that is, generated $\ell' = \rho'\ \mathbf{Y}^T$, or $\ell'' = \rho''\ \mathbf{Y}^T$, or $\ell''' = \rho'''\ \mathbf{Y}^T$ —these would not be the same as the original distances used because the unique vectors for these steady-state relations are ℓ and ρ. Nevertheless, we can compute a measure of difference from the steady state $\rho' - \rho$ for the case of $\rho' = \tilde{\ell}\ \mathbf{X}$, say (and all other distances for lines or points follow in the same way). This provides some index of how far the actual weighted measures deviate from the steady state, which we have shown to be a measure of direct access in the system. In a way, we demonstrated this earlier when we computed the distance differences from equation (6.24), illustrated in table 6.1 and figure 6.4.

Demonstrating the New Syntax: Accessibility in the Street Patterns

The Baseline Exemplar: Gassin

We have already introduced a little data pertaining to the village of Gassin implying that the axial map, like most, is sparse in comparison to such sets of relations for non-Euclidean systems. The map is shown in figure 6.5, along with the in-degree and out-degree distributions $[\ell_i]$ and $[\rho_j]$, which are computed from the raw data matrix **A**. The density of links is only 5.1% of the total possible for a system in which every line would be linked to every point, and vice versa. The average number of points per line—junctions per street $\sum_i \ell_i / n$—is 3.385, while the average number of lines per point—streets per junction $\sum_j \rho_j / m$—is 2.129, which is very close to planarity. We noted this fact earlier in that $\Psi = 1.065$, meaning that only just above 6% of the points are differently configured from an equivalent planar graph. In fact, of the 63 points, only 6 are associated with more than 2 lines, and these involve only 3 lines each. This is a worrying feature of space syntax in that the systems in question do not pick up the kind of variation that characterizes other measures of accessibility, such as those in spatial interaction theory. Of even more concern is the fact that as the relationships between lines—the key emphasis in space syntax—is based on the number of common points, and if most points have only two such lines, the distribution of topological distances between lines is likely to be rather narrow, as in fact we note in many applications where the depth or distance in graphs is seldom more than 6 or 7 step lengths. This means that information pertaining to distances from numbers of points and lines in common should not be thrown away, as it is in current practice for computing distance, thence integration, in space syntax.

We first examine the similarities between various distance measures for the primal and dual problem, just as we did for the randomized runs presented earlier. We have taken the four distance measures used in table 6.1, which are ℓ, $\tilde{\ell}$, $\ell(\mathbf{d})$ and $\tilde{\mathbf{d}}(\ell)$ for the primal, and ρ, $\tilde{\rho}$, $\rho(\mathbf{d})$ and $\tilde{\mathbf{d}}(\rho)$ for the dual problems, and added the weighted distance measures $\ell_i'' = \sum_j \rho(d)_i Y_{ij}$ and $\rho_j'' = \sum_i \ell(d)_i X_{ij}$, which appear to be more discriminating with respect to accessibility than any others. Table 6.2(a) shows these similarity measures for the primal problem involving lines, and table 6.2(b) for the dual problem involving points. For lines, there are strong similarities between the group of measures based on the in-degrees of the raw data, the basic distance, and the weighted distance matrices ℓ, $\tilde{\ell}$ and $\tilde{\mathbf{d}}(\ell)$; and within the group comparing the non-weighted distance measure $\ell(\mathbf{d})$ and its weighted variant from the dual problem ℓ''. Measures in these groups have similarities of around 0.9, while similarities in measures between the two groups are around 0.7. The similarity structure for the dual problem is more complex since the out-degrees ρ from the basic matrix **A** is hardly a distribution at all, but is more like a step function. In consequence, the direct and weighted distance measures $\tilde{\rho}$ and $\tilde{\mathbf{d}}(\rho)$ have less similarity, and thus it would appear that these measures are much more effective in picking up the structure in the syntax than any of the measures associated with the lines.

A

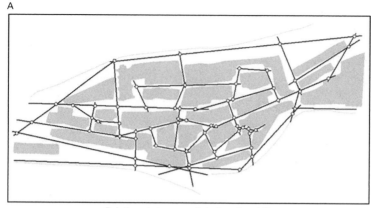

Relations between points (junctions) &
lines (streets)(Hillier & Hanson, 1984)

B

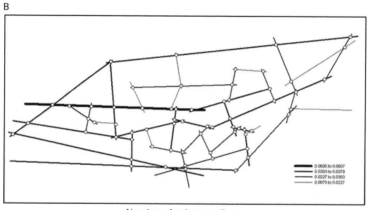

Number of points per line:
in-degrees ℓ

C

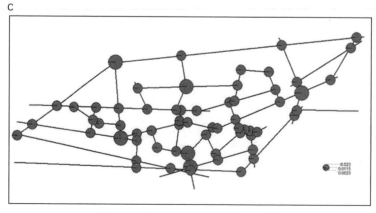

Number of lines per point: out-degrees ρ

Figure 6.5
The basic data for Gassin: Points and lines reflected in the matrix **A**.

Table 6.2
Similarities between five distance measures for Gassin

(a) Line distances	ℓ	$\tilde{\ell}$	$\ell(\mathbf{d})$	$\tilde{\mathbf{d}}(\ell)$	ℓ''
ℓ	•	0.922	0.781	0.888	0.748
$\tilde{\ell}$		•	0.768	0.916	0.735
$\ell(\mathbf{d})$			•	0.672	0.926
$\tilde{\mathbf{d}}(\ell)$				•	0.962
ℓ''					•
(b) Point distances	ρ	$\tilde{\rho}$	$\rho(\mathbf{d})$	$\tilde{\mathbf{d}}(\rho)$	ρ''
ρ	•	0.687	0.813	0.570	0.821
$\tilde{\rho}$		•	0.745	0.865	0.735
$\rho(\mathbf{d})$			•	0.540	0.875
$\tilde{\mathbf{d}}(\rho)$				•	0.970
ρ''					•

A better way of showing this structure and these similarities is in scatter graphs for the relationships between the in-degrees ℓ and their four related distance measures $\tilde{\ell}$, $\ell(\mathbf{d})$, $\tilde{\mathbf{d}}(\ell)$, and ℓ'', and the out-degrees ρ and their measures $\tilde{\rho}$, $\rho(\mathbf{d})$, $\tilde{\mathbf{d}}(\rho)$, and ρ''. These are plotted in figure 6.6, where it is clear how the lack of variation in the numbers of points on lines confounds the entire problem. This is an issue that requires much further investigation, for its importance clearly varies with the size of such applications. It does, however, pose very practical problems. Many applications reveal scatter plots like those shown in the second column of figure 6.6, which are the rule rather than the exception. This suggests that integration measures for these applications do not vary enough for them to be associated with volumes of movement, particularly pedestrian traffic, which they are often used for. In short, statistical correlations in many such applications are suspect because there is simply not enough variation in the basic data; hence our decision to use a measure of similarity, not correlation.

Visualizing Syntax as Nodes, Links, and Surfaces
We illustrate these four key distance measures for the primal and dual distributions in figure 6.7, which provides us with an ability to visually classify the configurational properties of the syntax. In fact there is nothing in space syntax that actually provides synoptic measures of morphology, since the only way to examine the overall pattern is to map the measures—that is, to translate the topological measures back into Euclidean space and to search for pattern visually. For the lines, we use conventional space syntax coloring, dividing the range in eight equal classes from

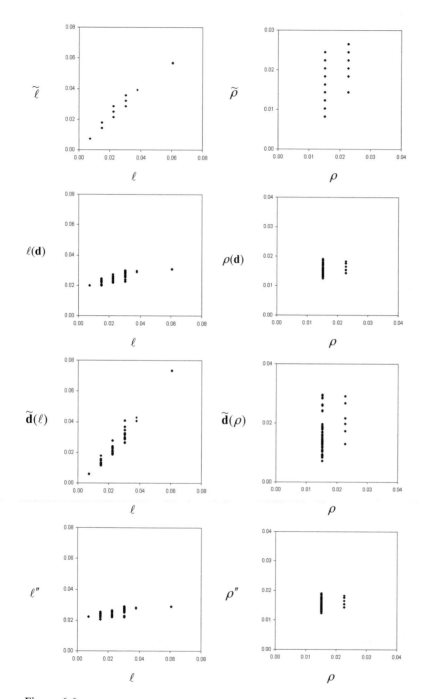

Figure 6.6
Scatter plots of access measures from the data ℓ and ρ against the direct and indirect distance measures.

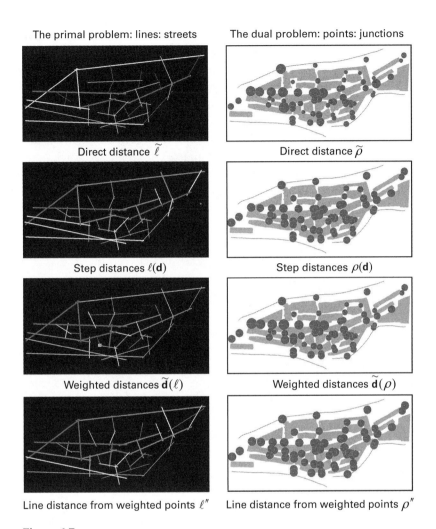

Figure 6.7
Comparison of distance measures for the primal and dual problems.

highest (black) to lowest (light gray), but we also vary the thickness of the lines to emphasize the intensity of the largest values, with the thickest lines being black and the thinnest light gray. These four line graphs are quite similar. The central spine through the village and the increased accessibility in the west is a common feature of each distance, while the lowest values are within the interior, where it is hardest to penetrate, and in the southeast of the built-up area. There is some sense in which the northern axis line exerts a significant influence on accessibility, although the fact that this is on the edge of the village reduces this impact. The strength of each point or junction for each of these measures is shown using proportional pie charts, where it is again clear that the junctions on the central spine dominate. In both the primal and dual problems, there seems to be slightly more discrimination with respect to $\tilde{\mathbf{d}}(\ell)$, whereas $\ell(\mathbf{d})$ and its derivative ℓ'' from the dual problem emphasize the importance of the northern axis, as confirmed by an examination of the related point distributions. As one might expect, there is a clear tie-up between the primal and dual problems in that the distance measures from one reinforce those from the other.

One of the biggest difficulties in space syntax is in providing a clear interpretation of the map pattern from classifying lines; our brain does not process such linear data nearly as well as aerial data when we wish to interpret place-related information. One of the advantages of moving from the primal to the dual, or from lines to points, is that points are place-related and it is easy to generate spheres of influence around them. Indeed, the mapping of accessibility is largely accomplished using surfaces and contours, which imply hinterlands of influence around fixed-point locations. This is easy enough to accomplish for points, but the spheres of influence around lines are trickier, although not impossible to generate. To illustrate how we might generalize this problem and provide a means whereby we can compare lines with points and vice versa in a way that is more consistent than the two representations in figure 6.7, we have used a surface interpolation technique to generate a field around these points. These are sometimes called "heat maps"; heat in this instance is high accessibility (dark gray), and cold low (light gray). This enables us to fix a subfield of influence around each point or line being mapped and to control the averaging of adjacent points with respect to the usual inverse distance weighting associated with such interpolation. We have chosen values such that the influence is as sharp as possible but not so sharp as to destroy the aerial pattern in the data.

In figure 6.8, we show the heat maps associated with the distance measures $\ell(\mathbf{d})$ and $\tilde{\mathbf{d}}(\ell)$ for the primal problem. It is quite clear these surfaces are highly correlated; they reinforce the conclusions drawn above about the importance of the central spine and the relative increase in accessibility as one travels west within the village. As before, $\ell(\mathbf{d})$ tends to emphasize the northern axis, but this is the only major configurational difference between the two maps. We generate the same two interpolations for points in the dual problems, which we show in figure 6.9, arraying

A

The line surface from $\ell(\mathbf{d})$

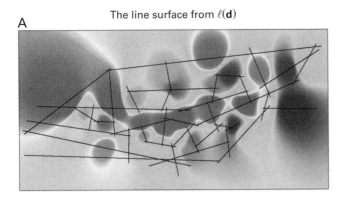

B

The line surface from $\tilde{\mathbf{d}}(\ell)$

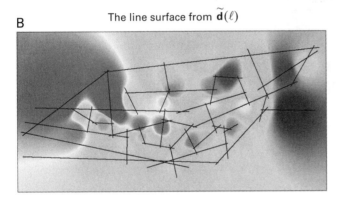

Figure 6.8
Surface interpolation from the line distances.

the points rather than the lines across the two surfaces. There is a sense in which these point surfaces reinforce the line surfaces, although the influence of each point is more distinct with slightly less of a ridge line character to these maps. The objection to such interpolation is that it ignores the influence of buildings and edges, although what it does do is reinforce the trends in the accessibility surface and give an immediate sense of overall variation. It is possible to clip such surfaces to building features, but what we have done here by way of showing how we might move forward is to simply impose the building extent onto these surfaces, leaving the reader to judge for him or herself the usefulness of the mapping.

In figure 6.10, we have interpolated between the weighted line ℓ'' and point ρ'' accessibilities and then intersected these surfaces with the buildings and boundary edges to the village, thus providing a sense of aerial accessibility within the street system. This is quite an effective technique: what it gives to interpretation that is missing in the conventional line diagrams in figure 6.7 is some sense of trends within

A   The point surface from $\rho(\mathbf{d})$

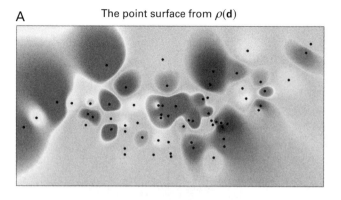

B   The point surface from $\widetilde{\mathbf{d}}(\rho)$

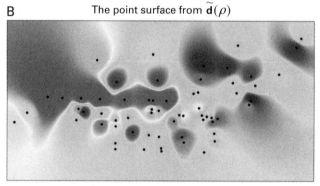

Figure 6.9
Surface interpolation from the point distances.

the whole system. There is much more work to do on adapting these visualization techniques to problems of urban morphology and its syntax, but the fact that we are now able to move from the primal problem to the dual gives some meaning to interpretations that begin with lines and move to points and back again.

Next Steps: Simplifying Space Syntax

The essential message of this chapter is that the techniques and practice of space syntax, which we consider a special case of accessibility within graphs, are but one way of looking at the problem of tracing relationships between the relative importance of streets that make up the urban fabric. The conventional formulation is the primal problem, but as we have shown, there is a dual problem that has equal significance and consists of measuring the relative importance of the points, junctions, or intersections that define the location of streets in question. We consider

A   The weighted line surface from ℓ''

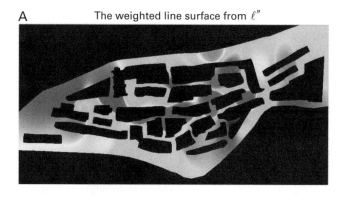

B   The weighted point surface from ρ''

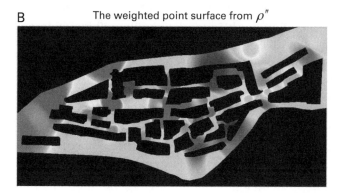

Figure 6.10
A new mapping for space syntax: Adapting surface interpolations to the building and street patterns.

that there are equally good reasons for considering the dual problem, perhaps more so because it is easier to map the accessibility of points rather than the accessibility of streets. We leave the reader to judge whether the problem should be approached through the primal or the dual, but in one sense this is of no importance: for every primal there is a dual and vice versa, and whether one measures accessibility in the primal (or the dual), it is possible to translate quickly and consistently from one to the other.

The more important issue in practical terms is how easy it is to interpret the primal or the dual; much of this chapter has been about such interpretations with respect to different distance measures. Our general conclusion is that it is much easier to map and interpret the dual, and that connecting space syntax to the wide arsenal of spatial analytic techniques, among which surface interpolation to produce heat maps is now routine, is much more meaningful with respect to the dual than the primal. So in terms of mapping accessibility, the various techniques introduced at the end of the previous section would seem to hold enormous promise in progressing practical applications. None of this necessarily involves simplifying space syntax. Indeed, readers unfamiliar with matrix algebra might think this new theory obfuscates rather than simplifies, although the algebra used is elementary and standard. Our point is that to see alternate ways of developing space syntax, we must take several steps forward to move one step back to a more simplified form.

In fact, the simplifications we now pose involve the various measures of distance that are computed for either the primal or the dual problem. We would argue that all these measures are so highly correlated in problems, which in the first instance are intrinsically embedded in Euclidean space, that their topological structure is quite simple. This is reflected in distance measures that take account of all step lengths in the syntax graph. Thus simply counting in-degrees and out-degrees ℓ and ρ provides quite a good measure of access for lines and points, and this of course can be done manually. Going one step further to compute the measures $\tilde{\ell}$ and $\tilde{\rho}$ from the interaction matrices $\mathbf{L}$ and $\mathbf{P}$ is easy to do, and again provides a good direct measure of access. Although digital computation might be needed for these measures, they could be produced manually for modest problems, and the act of doing so highlights the importance of what these measures mean in terms of relations between lines through their common points and points through their common lines. What, however, all this suggests is that the starting point for space syntax is not the axial map, per se, but the matrix of relations $\mathbf{A}$ between lines and points. For each problem, specifying this matrix formally provides a much more neutral statement of the problem, while at the same time producing an initial examination of its structure.

There are many directions implied here. The next chapter will continue extending this style of network morphology, linking the analysis here back to measures of Euclidean space, planarity, and physical distance. But it is worth mapping out the longer-term agenda for, as is the case for most of the ideas in this book, these are early days in building a science of cities and this is unfinished business. First, the notion that space syntax is a relation between any two sets of morphological elements, streets, and their junctions in the current kinds of application is in itself limited. We need to consider other elements, such as streets and land parcels, different types of streets, different types of land use, and so on. Second, we can establish chains of relations, such as streets and their intersections, then intersections and their relation to building plots, then building plots and their relations to land uses, and so on. Such frameworks need to be formally explored, for therein contain the ways in which space syntax can be linked to other elements of the urban system. Third, there is still more work to do on distance and accessibility, as well as on how we might consistently embed the physical distance in the street system into space syntax, thus making use of this information. In a sense, this chapter has not addressed this issue, yet there are promising extensions to the algebra developed here that might show how such connections can be made. We will consider these in the next chapter.

Fourth, we need to explore how space syntax and related networks relate to small worlds, the burgeoning statistical theory of graphs, scaling, the growth of networks, neural net conceptions, and so forth, which form a cornucopia of potential research directions already well-established. Some of these we have hinted at already, but we will provide links to other kinds of connections in later chapters. All of this constitutes a massive research program for space syntax, but only as one corner of a much wider research program in urban morphology, for which new theories of networks and graphs as well as new techniques of visualization and mapping will provide the momentum.

7

Distance in Complex Networks

Everything is related to everything else, but near things are more related than distant things.
—Waldo Tobler, "A Computer Movie Simulating Urban Growth in the Detroit Region," *Economic Geography*, (1970, vol. 46, p. 236)

In the development of network science, Euclidean distance and spatial embedding have been somewhat neglected, largely because the properties of networks are not as clear-cut when constraints imposed by physical space need to be observed. Moreover, in social networks, geometric distance is rather too simple a characterization of the main functional relationships such networks seek to represent. Thus, together with the difficulties of meeting the criterion of planarity, which is a major constraint, many social relations are not easily understandable or even observable with respect to distance, spatial extent, and physical influence. Yet as Tobler (1970) argues in the quote above, referring to what he calls the "first law of geography," distance is a primary, if not the central, construct in sorting and transforming spatial structures as they evolve over time. Distance and its variants such as travel time and travel cost are thus central to the way behaviors through movements and flows are related to networks.

The early evolution of social physics treated distance as some incontrovertible organizing principle, even though there were many spatial configurations that were clearly influenced by factors other than distance alone. Yet it is still surprising that one of the dominant approaches to urban networks, space syntax, which we presented and extended in the last chapter, tends to throw away the most obvious feature of spatial systems that we can measure, indeed the measure itself: distance. This is probably because this style of network analysis began in architecture, where lines of sight, convexity of spaces and enclosure, and other such concepts, are central to theories of good city form and the way people use buildings. Coming from the other direction, geography, insofar as it has embraced distance in networks, did not go far beyond the planar graph, and there the matter rested until the rise of network science, which largely came from physics, sociology, and political science.

The concept of network structure has somewhat ironically never been central to transportation modeling, where the focus has been on simulating flows and the network is largely regarded as something that is fixed, or at least predetermined in some way, before simulation begins. This approach is being slowly redressed as ideas from network science are being gradually adapted to transportation as, for example, by Xie and Levinson (2011).

In this chapter, we will bring distance back into networks, starting from the unified theory developed in the previous chapter. The theory embraces a range of network types, from strict planarity to space syntax to topology, but with the added requirement that networks are built from more elemental relations—from lines and their intersections as junctions in many cases, but also from a wider mix of physical and nonphysical objects. Our basic tool, used throughout this book, is the bipartite graph, which has little or nothing to do with transport networks, per se, though they are also graphs of various sorts. In the last chapter, we developed these ideas for manifestly spatial networks consisting of line segments and points, focusing on movement and flow that might be loaded on such networks. But we only skirted around the question of distance, and now we must bring it up front, for as we will see, to move to more complex network structures that mix different kinds of physical movement, distance is an essential organizing concept.

Representing Networks

Chapter 6 cast the generic problem of urban morphology in terms of the relations between any two sets rooted in Euclidean space. In terms of this characterization, we can assume that these are junctions and their streets—points and lines—but this representation can involve any two sets of objects that define the elements composing a network, and they need not be restricted to Euclidean space. These relations can be represented by an $n \times m$ matrix $[a_{ij}]$ where

$$a_{ij} = \begin{cases} 1 & \text{if} \quad i \Leftrightarrow j \\ 0, & \text{otherwise} \end{cases}, \tag{7.1}$$

and $\{i, 1, 2, \dots, n\}$ are now streets, and $\{j, 1, 2, \dots, m\}$ junctions or intersections; the sign $\Leftrightarrow$ means that a street is associated with a junction and vice versa. This is an entirely general representation that can be extended to any form of urban morphology specifying relations between two sets. The matrix $[a_{ij}]$ also forms a graph, but in this case, it is bipartite; if the two sets of elements can be rooted in Euclidean space, then they can also be represented as a network in such space as we show below.

What this representation enables is a basic form that does not privilege any one set over the other. In this sense, we can study how the set $\{i\}$ relates to the set $\{j\}$ or vice versa. If we look at the problem in terms of how streets {*i*} relate to each other, which is through {*j*}, we have the traditional space syntax approach, in which streets become the main focus. If we look at the problem in terms of how street junctions {*j*} relate to one another, then this is the traditional geographical graph problem. We call the first problem the *primal problem* and the second the *dual problem*. The last chapter introduced this framework in detail, but we need briefly to restate it here so that the general context is clear. The reader will notice that some of the equations are the same or similar to those in chapter 6, but we will not repeat equation numbers because our summary is self-contained. Mathematically neither the primal or the dual problem is more important than the other, although in practice, there may be very good reasons for preferring one form over the other. The primal problem thus examines relations between the streets in terms of the junctions, a key measure of which are the out-degrees of the bipartite graph, the number of junctions associated with each street, defined as

$$\ell_i = \sum_j a_{ij}, \tag{7.2}$$

while the dual problem examines the relations between the junctions in terms of streets whose key measure is the number of streets associated with each junction (the in-degrees of the bipartite graph), given as

$$\rho_j = \sum_i a_{ij}. \tag{7.3}$$

However, one of the main issues is that the traditional problem is not the dual of the space syntax problem, for the matrix [a_{ij}] has a rather different structure for each. The syntax problem is less restrictive in that $\ell_i \geq 2$ and $\rho_j \geq 1$, while the geographical graph problem always constrains the number of junctions associated with a street to 2; that is, $\ell_i = 2$ and $\rho_j \geq 1$. Thus, in essence, the space syntax problem is different from the traditional geographical network problem, and each have different primals and duals. These are related but not in a straightforward way, as is clear from chapter 6 and will be demonstrably so from the analysis that follows.

The meaning of these differences is illustrated in figure 7.1(a), where for the traditional problem, we show a simple cross-shaped network as a set of five street nodes and four street segments. This is indeed a planar graph, which we originally called the geographical graph problem. We also show the basic matrix [a_{ij}] for this graph alongside, from which it is clear that the number of junctions for each street is exactly 2, that is $\ell_i = 2$. In figure 7.1(b), we have aggregated the 2 street segments *a* and *b* to form a single street line *a'*, and it is now clear that there are three junctions associated with this line. The [a_{ij}] matrix shown alongside now has only 3 lines but still 5 junctions: in raw physical terms, there is no difference to the underlying street network but the space syntax problem produces an abstraction, still coincident

A   Traditional street network as a planar graph

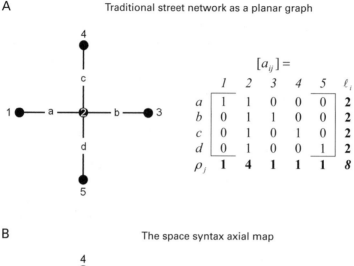

B   The space syntax axial map

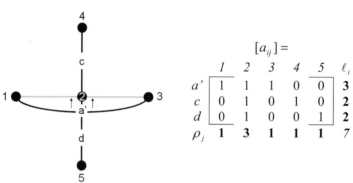

Figure 7.1
A traditional planar graph-street network and a space syntax representation.

with the street map at its basic level, but one in which the number of "streets" differs from the traditional planar graph. This abstraction is called an *axial map* and its street components are *axial lines*.

It is easy to guess the relative accessibilities in street maps such as those in figure 7.1. For the traditional map and graph in 7.1(a), it is quite clear that the central junction, or node 2, is the most accessible, and that since each street line has the same relationship to any other, the street line accessibility is the same for each. However, accessibility is much more difficult to guess in the space syntax problem. As street *a′* has 3 junctions and the other streets only 2 each, *a′* is the most accessible in that it relates directly to both streets. Since the junction at the center of the map is the same, it seems likely that this node 2 is still the most accessible. We will, however, compute these relationships exactly after we have examined the primal and dual problems and stated the various topological and Euclidean distance measures we will work with here.

In the next section, we will briefly introduce the unifying framework once again and then derive and restate the various topological distance measures associated with the primal and dual problem forms. We will take the analysis of figure 7.1 further and illustrate a simple space syntax problem for an area of the West End of Central London so that readers are clear about how our extended network analysis pertains to other applications. We then derive distance measures between the points and between the lines for any form of the generic matrix $[a_{ij}]$ where $\ell_i \geq 2$ and $\rho_j \geq 1$. In fact, we can generalize the problem a little further to systems where lines have only one junction associated with them. But at some point in computing distance on a line, we need beginning and end points, and although space syntax does deal with lines that have only one junction, other junctions must always be implicit.

We will then examine a pure syntax problem in which the streets are lines of sight and the importance of a place clearly depends on how far one can see. We use the basic example developed by Hillier and Hanson (1984) and reworked by Peponis et al. (1997), Batty and Rana (2004), and Carvalho and Batty (2004) for the French village of Gassin (used previously in chapter 6) to show that the topological and Euclidean measures of distance and accessibility produce entirely different patterns. We then discuss what we call a "mixed syntax problem" that involves not only line of sight measures as axial lines, but also lines of movement that do not have sight associated with them. This is the case when the technologies involved to move people are usually enclosed: trams, buses, and trains. We illustrate the problem for Central Melbourne, where the grid of streets lies on top of a heavy rail loop which is underground. This provides us with another perspective on accessibility, but it also shows how we can extend space syntax to deal with systems in which many kinds of route and transportation mode define the morphology of the city.

Primals and Duals: Networks as Lines through Points, Points through Lines

The representation in the matrix $[a_{ij}]$ allows us to look at the problem in two distinct ways: across each row or line in terms of a count of the objects or points associated with each line, and down each column, where each object or point is associated with a number of lines. These are the primal and dual problems, respectively. The number of common points between any two lines forms a network of relations, a weighted graph, whose basic form is computed as

$$\ell_{ik} = \sum_j a_{ij} a_{kj}, \tag{7.4}$$

where $[\ell_{ik}]$ is the number of points in common for any two lines. In space syntax and other analytical representations, this matrix is usually sliced to provide a binary form such that

$$Z_{ik} = \begin{cases} 1 & if \quad \ell_{ik} > 0, i \neq k \\ 0, & otherwise \end{cases}. \tag{7.5}$$

This means that no weighting is given to the actual number of points that any two lines have in common: association or not thus depends on having at least one point in common. The total number of points in common with respect to all direct associations between one line {i} and all other lines is calculated as

$$\tilde{\ell}_i = \sum_k \ell_{ik}, \tag{7.6}$$

which is the out-degree interpretable as a measure of direct distance, the opposite of nearness, with respect to the line in question. As [ℓ_{ik}] is symmetric, then the in-degree $\tilde{\ell}_k = \tilde{\ell}_i$, $k = i$.

This primal form is the classic space syntax problem with the measure of distance $\tilde{\ell}_i$, the number of streets that lie one depth away from the street in question. The dual problem repeats all this logic on points rather than lines. The relationship matrix [ρ_{jl}] is computed from [a_{ij}] as

$$\rho_{jl} = \sum_i a_{ij} a_{il}, \tag{7.7}$$

where ρ_{jl} is the number of lines that points j and l have in common. The associated measure of direct distance based on the out-degrees is given as

$$\tilde{\rho}_j = \sum_l \rho_{jl}, \tag{7.8}$$

and the same symmetry conditions on the in-degrees hold. Here [$\tilde{\ell}_i$] and [$\tilde{\rho}_j$] are two initial measures of distance, although these are really counts of volume, direct nearness, or adjacency, namely the number of points for each line, and the number of lines for each point—the number of points a line has in common with all other lines, and the number of lines a point has in common with all other points. We now need to develop more refined measures of distance based on any pair of lines and any pair of points computed from the matrices [ℓ_{ik}] and [ρ_{jl}], respectively. In fact, this takes us back to chapter 2, where we computed various measures of accessibility from trip volume matrices and from spatial interaction models as summations or potentials of all flows emanating from an origin or entering a destination.

The distance measures that take account of all relationships in the graph are computed by deriving all the numbers of lines or points in common for successive path lengths through the two graphs. These graphs are always strongly connected by definition, and thus successive path lengths—called step lengths—need only be computed up to the number of lines or points in the system, no more. The shortest routes will usually be found well before this size is reached. We will demonstrate this computational mechanism for the primal problem involving the line matrix only, for the dual follows directly. The number of points in common for two steps through the graph from line i to line k is calculated as

$$\ell_{ik}^2 = \sum_z \ell_{iz}^1 \ell_{zk}, \tag{7.9}$$

where $\ell_{ik}^1 = \ell_{ik}$, and the recursion on equation (7.9) for any step length $s + 1$ thus becomes

$$\ell_{ik}^{s+1} = \sum_z \ell_{iz}^s \ell_{zk}. \tag{7.10}$$

At some point where $s \leq n$, where n is the number of rows in the matrix, this recursion will converge when all paths through the graph become positive, that is, when $\ell_{ik}^s > 0$.

The first measure of distance $d(\ell)_{ik}$ is based on step length, and this is computed at each iteration of equation (10) as

$$d(\ell)_{ik} = s \quad if \quad \ell_{ik}^s > 0 \quad and \quad \ell_{ik}^{s-1} = 0. \tag{7.11}$$

This computation eventually converges. Two measures of overall accessibility or proximity for each line i can be computed from

$$\ell(d)_i = \sum_k d(\ell)_{ik}^{-1}, \tag{7.12}$$

and

$$\hat{\ell}(d)_i = \frac{1}{\sum_k d(\ell)_{ik}}. \tag{7.13}$$

These measures are likely to produce similar results, for the inverse weightings differ only marginally. In the case where inverse distances are computed for each link ik, as in equation (7.12), the inverse power (−1) could vary, with the theory of spatial interaction invoked to explain the meaning of this scaling.

There are many variants on these measures, with different types of normalization often being applied. One that works directly with the number of paths through the graph and the number of points in common for each pair of lines is based on a linear combination of the different sequential path length matrices $[\ell_{ik}^s]$, and weights these in such a way that successively longer step lengths get successively lesser weighting. This measure is defined as

$$\tilde{d}(\ell)_{ik} = \sum_s \lambda^s \ell_{ik}^s, \tag{7.14}$$

where if we set $0 < \lambda < 1$, each successive term in the sum in equation (7.14) assumes a lesser importance. In fact, λ must be tuned so that each term in the series reduces in value, where a typical value for λ is of the order of 0.05, the value used in chapter 6. The matrix $[\tilde{d}(\ell)_{ik}]$ is symmetric, and thus the in-degrees or out-degrees serve as equivalent measures. As we weight the measures in terms of a decreasing contribution of the number of points on sequential paths, then the measure is already in accessibility form. An appropriate aggregate is thus

$$\tilde{d}(\ell)_i = \sum_k \tilde{d}(\ell)_{ik}, \qquad (7.15)$$

which we called the weighted accessibility.

We now have five measures of accessibility, ℓ_i, $\tilde{\ell}_i$, $\ell(d)_i$, $\hat{\ell}(d)_i$, and $\tilde{d}(\ell)_i$, which are repeated for the dual as ρ_j, $\tilde{\rho}_j$, $\rho(d)_j$, $\hat{\rho}(d)_j$, and $\tilde{d}(\rho)_j$. These were broadly the measures introduced in the last chapter, where we built up the theory of the primal and dual space syntax problems. We now intend to explore how these topological measures can be augmented with measures based on Euclidean distance. The measures introduced below are not meant as substitutes for the topological measures, but as complements. In fact, Euclidean distance is a somewhat different concept from the topology of relations between lines of sight, which is in essence what space syntax is all about. However, as the topological measures are close to geographical space, since the starting point is the axial map based on its lines and points, which are firmly embedded within this space, it makes sense to ask what implications a topological measure of distance between lines and points embedded in a geographical network has for more traditional measures based on Euclidean space. Moreover, we also need to explore the best ways of visualizing syntax relationships in terms of the primal and the dual. To this end, an examination of the properties of axial maps in terms of Euclidean distances is warranted. But before we develop these distance measures, we need to explore the relations between the primal and dual problem for the traditional and then the space syntax problems in a somewhat deeper way. We also need to demonstrate the actual meaning of accessibility in networks for these two kinds of problem. To do so, we will return to the simplest graphs in figure 7.1 and then complement an analysis of these problems with a straightforward example of the use of space syntax for examining lines of sight along streets in a dense set of street blocks in a large city, which is typical of many applications.

Exploring Related Geographical and Space Syntax Problems

The Simplest Syntax Problem
The planar graph in figure 7.1(a) and its aggregation to the space syntax map in figure 7.1(b) imply primal and dual representations—networks—which we illustrate

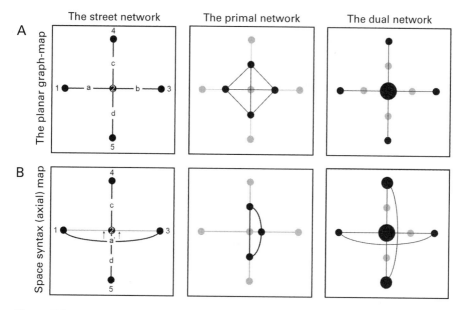

Figure 7.2
Dual and primal networks for the two map problems.

in figures 7.2(a) and 7.2(b), respectively. In the case of the planar map in figure 7.1(a) and 7.2(a), which consists of four street segments anchored by junctions at each segment and arranged as two crossed lines, for the primal problem, which consists of relating street segments to one another through counting the link between two segments when they have a junction or point in common, the network is quite simple, as shown in the second panel of figure 7.2(a). Here we have four line segments, with the solid black nodes showing their centroid locations and links from each to every other—6 links between 4 street nodes, a completely connected network. The dual problem links junctions to one another through their common lines. The junctions are shown in the third panel of figure 7.2(a); the size of each node reflects the in-degrees and out-degrees of the graph. The central junction has the largest degrees, and the edge nodes the lowest. The graph is not completely connected because to access the junctions at the end of each cross from any other apart from the central junction, one has to pass through the central junction in two steps. These networks only give one-step links but it is easy to see that in this dual problem, the distance matrix would be connected once the two-step distances are computed. In short, the traditional problem is straightforward and symmetric. Note that in the primal network the junctions are colored gray and the street location nodes are black, while in the dual problem, the street nodes are gray and the junctions black, showing that the focus in each case is on links between the nodes colored black.

The space syntax map in figures 7.1(b) and 7.2(b) is an aggregation of the planar map in which the two horizontal cross segments a and b now form one line, a'. The problem is thus reduced from four to three lines, but the original five junctions remain in place. The primal problem, the second panel in figure 7.2(b), now has all three lines connected, noting that we displace the node for the central cross line a' to the right of the central junction so we make sure we can still see it on the map. The dual problem, whose network is the third panel in figure 7.2(b), is considerably more complex. The impact of aggregating the two original horizontal segments a and b into a' essentially means that the two north-south end nodes are better connected than before, as the size of the nodes that mirror the in- and out-degrees of these junctions show. In one sense, this is quite obvious, because if we aggregate lines, then we reduce the diversity in the system; *ceteris paribus*, it would appear easier to navigate between places on the graph of the axial map than on a graph where all the street segments were associated with only two junctions at most.

For these very simple systems, what this analysis shows is that there are quite considerable differences produced by changing the focus from simple segments to more complex aggregates, the axial lines. We do not show the distance matrices associated with each of these primal and dual networks in figure 7.2, but they are easy to infer because each becomes completely connected at the next step distance, as application of equation (7.11) would show. In space syntax, the axial lines, which are essentially embedded in Euclidean space, are usually demarcated (often by color) with respect to their accessibility—the number of in-degrees or out-degrees taken from the distance matrix $[d(\ell)_{ik}]$. This is a depth value, computed as

$$d(\ell)_i = \sum_k d(\ell)_{ik} = \sum_i d(\ell)_{ki} = d(\ell)_k , i = k, \qquad (7.16)$$

which is called integration by Hillier and Hanson (1984). Accessibility is the inverse of depth, and the values in question, which are usually associated with each axial line, are those given earlier in equations (7.12) to (7.15). In fact, many such values can be used, including all those introduced in the last chapter and those introduced below.

A Typical Street Syntax in Central London

To illustrate the range of measures that are generated from treating the network as a space syntax problem where the criterion is to define streets as lines of sight, these being composed of one or more segments between junctions, we will present this for a mile-square block of Central London centered on the Soho-Regent Street area of the West End. We define the axial map in figure 7.3, where we see immediately that the definition of lines of sight is not completely unambiguous, particularly on curved streets. Nevertheless, despite these obvious difficulties as well as problems

Figure 7.3
Axial lines and junctions in the Regent Street area of Central London.

posed by the arbitrary boundary of the region, the axial map comprises some 41 lines and 30 junctions, and this suffices to illustrate the method. In figure 7.4(a), we show the axial map with lines drawn in proportion to the number of in-degrees (or out-degrees due to symmetry) $d(\ell)_i$; and in figure 7.4 (c), the same degrees for the dual problem, where the junctions are drawn in proportion to their in- (and out-) degrees $d(\rho)_j$. The distance matrices between lines and then between junctions are computed up to their maximal spanning distance, but because of their complexity, we only show those links that are one-step direct connections or two-step connections between axial lines (figure 7.4[b]) and junctions (figure 7.4[d]). Like all networks, once we get more than about 20 or 30 nodes, their visual form says little, hence our reason for simplifying these only up to two step lengths between the significant objects of the primal—axial lines—and the dual street junctions.

Several other visualizations are used in this kind of analysis: in particular, lines or junctions colored according to their distances from other lines; depth map analysis from any one line or junction to all others; visualizations of the radius, within which a certain set of lines or junctions lie; and various other linkage diagrams that can illustrate how any one object or set of objects (lines or junctions) relates to any other object or set of objects. We should also note that the implications of depth or reachability of one street or junction to another have not been exhaustively explored. If we have symmetric structures, then this symmetry is reflected in the various graphs and accessibility measures that are generated. In figure 7.5, we show an example of a Manhattan grid handled by space syntax in which each line in the grid is an axial

A

B

C

D

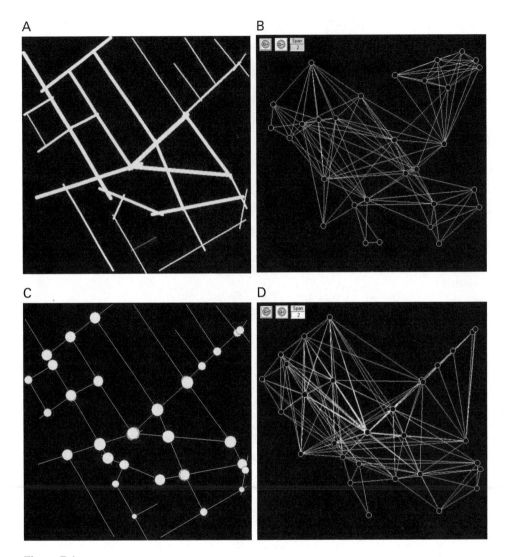

Figure 7.4
(a), (b) Primal networks embedded in and across Euclidean space. (c), (d) Dual networks embedded in and across Euclidean space.

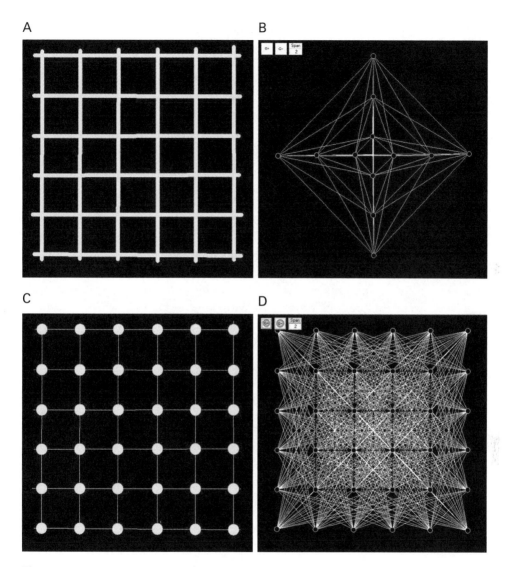

Figure 7.5
Space syntax on the Manhattan grid: Uniform accessibilities. The primal problem on (a) lines and (b) network; the dual problem on (c) junctions and (d) network.

line; that is, a 6×6 grid of line segments defined by 6 vertical lines and 6 horizontal. This problem has 12 lines and 36 junctions, which means that the relative distance networks are quite different in dimension. We see there is a possible total of 144 links between lines—132 if we assume as we usually do that self-links are not relevant, and as these are symmetric there are only 66. For the dual problem there are a total of 1260 without self-links and 630 symmetric links. This is clear from the network graphs in figures 7.5(b) and 7.5(d).

The second example we present is a regular fractal. Figure 7.6 shows the line and junction maps and their distance matrix networks, which are similarly symmetric. Moreover, a graph that is a tree has only one node less than the number of lines. In this case, there are 26 lines and 25 nodes. Essentially, the central node is really for two lines—the major axes of the tree—and thus we speculate and demonstrate that for a tree, the primal is equivalent to the dual with respect to the network of relationships. Figure 7.6(a) and 7.6(c), the line and node maps, show a symmetry and equivalence just as the networks do. This, of course, goes for all such symmetric structures in space syntax, extending Hillier's (1996) results in his book *Space is the Machine*. These explorations provide us with some sense that when we unpack networks in this way and build them as primal or dual problems from sets of bipartite relations, we generate a richness of structure that provides considerable insights into the properties of physical and network space. But to really embed these back into Euclidean space, we need to take a new look at distance. We will begin by introducing some ideas about how we measure such distance, and we will then rework our simplest example in figure 7.1 before we show what this means to the standard space syntax example in the village of Gassin. We will then be in a position to deal with mixed systems where lines of sight and lines of movement interact.

Euclidean Distance in Space Syntax Graphs

Weighted Distance Measures

As we have assumed in this book so far, we are dealing with spatial networks that are composed of straight lines—axial lines or straight line segments between nodes that have coordinates, from which we can compute straight line distances. In fact our treatment easily extends to "curved lines," which can be approximated at some level of resolution by finer straight line segments, but we will not invoke such generalizations here (Figueiredo and Amorim, 2005). We will begin with the dual problem, which is straightforward and for which we have coordinate pairs $\{x_j, y_j\}$ and $\{x_l, y_l\}$ for the points j and l defining a relevant line segment. Noting that the direct distance elements $[\rho_{jl}]$, which define the dual graph, can only be equal to 1 or 0 for any line i between j and l where $j \neq l$, since there can be no more than 1 line between any two points, then the direct Euclidean distance d_{jl} is

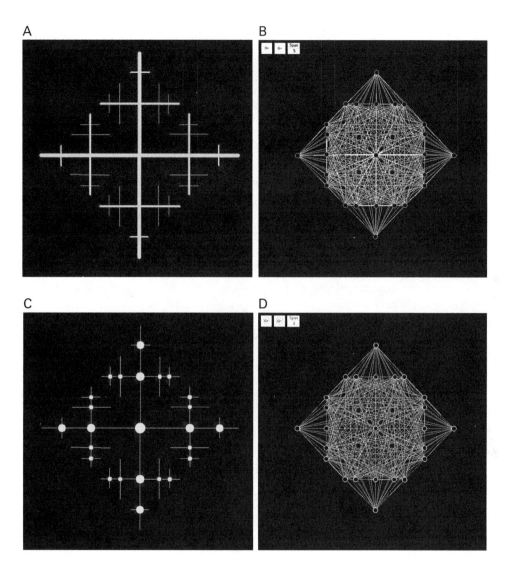

Figure 7.6
Space syntax on a fractal tree: The primal is the dual. The primal problem on (a) lines and (b) network; the dual problem on (c) junctions and (d) network.

$$d_{jl} = \rho_{jl}\,[(x_j - x_l)^2 + (y_j - y_l)^2]^{1/2}, \tag{7.17}$$

where the self-distances d_{jl} are clearly zero. Shortest routes between any points j and l can now be defined using the standard Dijkstra algorithm, which in the form used here, based on step lengths, is formulated as follows:

$$if \quad \rho_{jl}^{s} > 0 \quad and \quad \rho_{jl}^{s-1} = 0 \quad then \quad d_{jl} = \min_{z}\{\,d_{jz} + d_{zl}\,\}. \tag{7.18}$$

Recursion on equation (7.18) occurs until all step lengths in the graph become positive or until the number of iterations s approaches the number of points m in the graph.

In general, the matrix $[\rho_{jl}]$ is different from that for a planar graph, which we can write as $[p_{jl}]$. As a line can be associated with more than 2 nodes, which is not the case in a planar graph, in the general case some elements of $[\rho_{jl}]$ are positive and equal to 1, in contrast to $[p_{jl}]$, which is a more parsimonious structure, as is clear from figures 7.1(a) and 7.1(b). We defined a measure of the deviation from planarity in terms of the number of points associated with each line in the last chapter as $\Psi(\ell) = \Sigma_i \ell_i / 2n$, but other measures based on the graph distances might be

$$\Psi(\rho) = \frac{\sum_{jl}|\rho_{jl} - p_{jl}|}{\sum_{jl} p_{jl}} \quad and \quad \Psi(d) = \frac{\sum_{jl}|d_{jl} - d(p)_{jl}|}{\sum_{jl} d(p)_{jl}}, \tag{7.19}$$

where $d(p)_{jl}$ is the measure of distance computed for the planar graph associated with the dual syntax graph (which can be easily pruned from $[\rho_{jl}]$). It is also possible to generate trip lengths as in spatial interaction theory from these distances. If we have a loading of trips or movement volumes $\{T_{jl}\}$ on each link, then the standard mean trip length for the system can be computed as

$$T(\rho) = \frac{\sum_{jl} T_{jl}\,d_{jl}}{\sum_{jl} T_{jl}} \quad or \quad \bar{T}(\rho) = \frac{\sum_{jl} d_{jl}}{m^2}, \tag{7.20}$$

where the second equation in (7.20) represents the case when the movements on each link are absent or the same; hence they are set to unity.

Trip lengths can of course be computed for each point or node simply by summing equations (7.20) over l not j or vice versa, but the more appropriate measures are the inverse forms for the out-degrees of the distance matrix: the sum of the inverse distances, and the inverse of the sum of the distances with respect to each point. These are true Euclidean distance potentials, defined as

$$e(\rho)_j = \sum_l d_{jl}^{-1}, \tag{7.21}$$

and

$$\hat{e}(\rho)_j = \frac{1}{\sum_l d_{jl}},$$ (7.22)

which are measures of locational access. They could be weighted by the mass of the points as in the traditional social physics and spatial interaction theory shown in chapter 2 (see Wilson, 1970), but in this context we will not confuse the problem. Note that the inverse sum in equation (7.22) is proportional to the inverse unweighted mean trip length associated with the location j.

Euclidean distance measures for the primal problem are trickier in that we need to compute centroids associated with each axial line. In essence, an axial line can relate to more than 2 points, and thus there is a centroid for every such line of sight associated with the line. For example, in figure 7.1(b), the axial line a' is associated with three lines of sight from 1 to 2, from 2 to 3, and from 1 to 3. Thus it is logical to compute a centroid from these centroids using a simple averaging, although again variable weighting might be considered. We first compute a centroid for each line of sight associated with the axial line i as

$$\left.\begin{array}{l} \bar{x}_{ijl} = a_{ij}a_{il}(x_j + x_l)/2 \\ \bar{y}_{ijl} = a_{ij}a_{il}(y_j + y_l)/2 \end{array}\right\}.$$ (7.23)

These centroids need to be averaged, and this is accomplished by

$$\left.\begin{array}{l} \bar{x}_i = \sum_{\substack{j \neq l \\ l > j}} \dfrac{x_{ijl}}{m(m-1)/2} \\[2em] \bar{y}_i = \sum_{\substack{j \neq l \\ l > j}} \dfrac{y_{ijl}}{m(m-1)/2} \end{array}\right\},$$ (7.24)

where the summations are over all pairs of points associated with the line in question, which is equivalent to averaging the coordinates of all relevant points.

This simple averaging could be augmented with differential weights if views along an axial line to different points reflect differing degrees of importance, but here we will stick with the nonweighted form. We can now compute distance between any two axial lines as i and k by taking the distance from the centroid of line i, say, to the point j, which is common to the line k to which it is being linked. This distance is

$$d_{ik} = \ell_{ik}\{[(\bar{x}_i - x_j)^2 + (\bar{y}_i - y_j)^2]^{1/2} + [(\bar{x}_k - x_j)^2 + (\bar{y}_k - y_j)^2]\},$$ (7.25)

since axial lines are straight. This operation could easily be generalized to non-straight lines by replacing them with finer scale straight lines and operating recursively on equation (7.25). As each line has a mass—that is, it has finite length—then it is possible to compute an intra-line distance, a self-distance that must be specified as

$$d_{ii} = \sum_{\substack{j \neq l \\ l > j}} a_{ij} a_{il} \frac{[(x_j - x_l)^2 + (y_j - y_l)^2]^{1/2}}{m(m-1)/2}.$$

(7.26)

In this paper, we will set $d_{ii} = 0$ since we follow the tradition of space syntax, but other arguments can clearly be made for keeping this self-distance as a positive deterrent to mobility.

We are now in a position to compute the shortest routes between lines. We do this in exactly the same manner as for distances between points illustrated in equation (7.18); that is

$$\text{if} \quad \ell_{ik}^s > 0 \quad \text{and} \quad \ell_{ik}^{s-1} = 0 \quad \text{then} \quad d_{ik} = \min_z \{ d_{iz} + d_{zk} \},$$

(7.27)

where convergence is guaranteed by the time $s = n$. We can now compute the line accessibilities from $[d_{ik}]$ in the form of the sum of the inverse or the inverse of the sum. Then, in analogy to equations (7.21) and (7.22),

$$e(\ell)_i = \sum_k d_{ik}^{-1},$$

(7.28)

and

$$\hat{e}(\ell)_i = \frac{1}{\sum_k d_{ik}}.$$

(7.29)

All the other measures involving trip lengths that we noted in equation (7.20) apply, and the lines can be weighted with trip volumes if required. However, as in space syntax, we assume that this is not necessary for problems involving lines of sight, and all we require is an overall measure of line distance, which we define as

$$\bar{T}(\ell) = \frac{\sum_{ik} d_{ik}}{n^2}.$$

(7.30)

Applications to the Simplest Syntax

We now have two more measures of distance to add to our arsenal. To illustrate the subtle differences in meaning of these measures, we have computed all of them for the dual and primal problems that emanate from the planar street network graph in figure 7.1(a) and the axial street map in figure 7.1(b). For the street network in figure 7.1(a), which is a planar graph in terms of the dual problem formulation, where each line or street has exactly two points or junctions associated with it, the accessibilities of any point to any other are obvious by inspection. Point or node 2 is clearly the most central and in a commonsense way has the highest accessibility, while the four others have lower but equal accessibility. What we have done is smoothly interpolated between these point accessibilities, producing the surface (or heat map) shown in figure 7.7(a) in the time-honored way. There are clear edge effects in such surfaces, which are hard to control for, but in general, the pattern of point accessibility varies inversely with distance from the central point for all measures and step-length measures, as well as those based on Euclidean distances. For the primal problem involving accessibility from any line to all others, each line would appear to have the same accessibility due to the nature of the symmetry. This indeed is the case as we show in figure 7.7(b), where the pattern of accessibility is uniform over the entire space. Once again, this applies to all measures. In the planar street network case, it would appear from this simple example and from our intuition that all the measures co-vary with one another, notwithstanding differences in distribution.

In terms of the syntax problem in figure 7.1(b), two of the lines in the planar network case are collapsed into one, a' being formed from a and b. In this case intuition suggests that the point accessibility involving the nodes is much the same as before, but since the two north-south lines c and d involve more step lengths to reach line a', then the two north-south nodes 4 and 5 are slightly less accessible in terms of step lengths $\rho(d)_i, \hat{\rho}(d)_i$, as figure 7.7(c) shows. For the direct step lengths, weighted distance, and Euclidean distance accessibilities $\rho_i, \tilde{\rho}_i, \tilde{d}(\rho)_i, e(\rho)_i, \hat{e}(\rho)_i$, the nodes have the same pattern of accessibility as the planar street network, as figure 7.7(e) shows. In terms of lines, the binary step length and weighted distance measures $\tilde{\ell}_i, \ell(d)_i, \hat{\ell}(d)_i, \tilde{d}(\ell)_i$ show each line with equal accessibility (figure 7.7[d]), in contrast to the other measures $\ell_i, e(\ell)_i, \hat{e}(\ell)_i$ where the merged line a' is more accessible than the other lines c and d (figure 7.7[f]). This indicates that the number of lines of sight associated with a line like a' does reinforce the importance of the line, especially as this line is central to the morphology. However, as we will see, this is not a straightforward issue, since in systems where there are many short lines with some long ones dominating, the pattern of Euclidean accessibility can be quite different from the step-length accessibilities.

Before we move to a realistic problem, it is worth noting how surface distributions are produced from the point and line estimates. A technique of spatial averaging using a kernel, which is centered over each cell in question, diffuses the value of that cell over a given search radius and does this until all the cells are filled in one pass. For a system of five nodes, say, each value at each node is generalized in this way, with the diffusion being based on an inverse distance allocation from the points in question. The technique is stopped at the boundary of the system in this case where there is an abrupt cut-off, hence the skew introduced into the diffusion at the edge. While this could be controlled by putting the boundary further out,

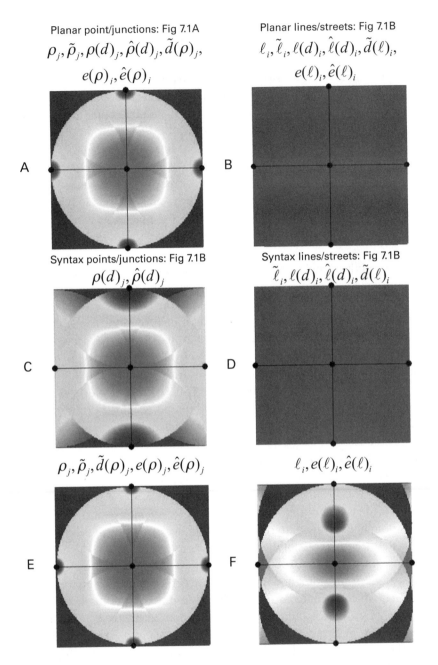

Planar point/junctions: Fig 7.1A

$$\rho_j, \tilde{\rho}_j, \rho(d)_j, \hat{\rho}(d)_j, \tilde{d}(\rho)_j,$$
$$e(\rho)_i, \hat{e}(\rho)_i$$

Planar lines/streets: Fig 7.1B

$$\ell_i, \tilde{\ell}_i, \ell(d)_i, \hat{\ell}(d)_i, \tilde{d}(\ell)_i,$$
$$e(\ell)_i, \hat{e}(\ell)_i$$

Syntax points/junctions: Fig 7.1B

$$\rho(d)_j, \hat{\rho}(d)_j$$

Syntax lines/streets: Fig 7.1B

$$\tilde{\ell}_i, \ell(d)_i, \hat{\ell}(d)_i, \tilde{d}(\ell)_i$$

$$\rho_j, \tilde{\rho}_j, \tilde{d}(\rho)_j, e(\rho)_j, \hat{e}(\rho)_j$$

$$\ell_i, e(\ell)_i, \hat{e}(\ell)_i$$

A   B   C   D   E   F

Figure 7.7
Accessibility surfaces for the primal and dual problems from the simple planar and axial maps shown in figures 7.1(a) and 7.1(b).

making the diffusion smooth, we prefer to visualize this elementary problem in its most basic form.

The Pure Syntax Problem: Applications to the Village of Gassin

We are not going to repeat the data for Gassin, since it was introduced in the last chapter and has been reproduced in a number of papers as a benchmark example of space syntax. It was originally presented by Hillier and Hanson (1984) and then reworked by Peponis et al. (1997), Batty and Rana (2004), and Carvalho and Batty (2004). Nor will we generate all forms of syntax map for the primal and dual problems, thus avoiding representing weighted axial lines across the standard color range for line accessibilities, or pie charts for the measures of accessibility at each point. We will stick to the heat map representations, since these provide good impressionistic pictures of variations in accessibility over the street map. For the primal maps, we will overlay the street line, and for the dual, the street junctions or points.

We refer to Gassin as a "pure space syntax" problem because there was never any intention in the original application of measuring accessibility in terms of Euclidean distance. Axial lines are lines of sight, and the longer the line of sight, the more likely the line is to intersect with other lines of sight at junctions. However, the longer the line of sight, the longer the distance associated with that line. Although we will not set a measure of Euclidean distance for the relation of each line to itself (as identified above in equation [7.26]), a long axial line is given a centroid that reflects its length. All other things being equal, it would be more distant, hence less accessible to other lines in the system. In contrast, in terms of measuring nearness to other lines without taking distance into account, its length would not be a factor per se. However, it would intersect with more lines, all other things being equal, and hence its accessibility to all lines would be greater. In figure 7.1(b), this was not the case because the symmetry of the system and the equal distance of the elemental line segments ensured that the axial line was best connected in both step-length and Euclidean distance terms.

In Gassin, which is shown in terms of its axial lines in figure 7.8, we have several long lines of sight, but in general these are not in the core area of the village where there is the greatest concentration of streets. Thus in terms of distances, the cluster of lines that mark the village core are very close to one another in distance terms, and in general we might expect accessibility computed between these lines to be much higher than that between the longer lines. As we will show, this is indeed the case, and in Gassin, the space syntax interpretation is qualitatively different from that based on Euclidean distance. The same difference is reflected in the clustering of the street intersections, which again reflects the clustering of the lines. As noted

The primal lines/streets problem | The dual points/junctions problem

Step-length distance $\ell(d)_i$ | Step-length distance $\rho(d)_j$

Weighted distance $\tilde{d}(\ell)_i$ | Weighted distance $\tilde{d}(\rho)_j$

Euclidean distance $e(\ell)_i$ | Euclidean distance $e(\rho)_i$

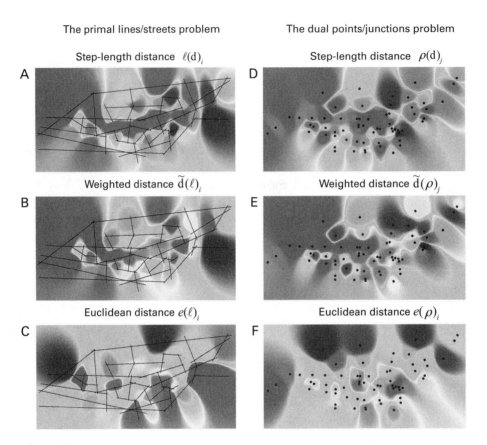

Figure 7.8
Key accessibility measures for the primal and dual pure syntax analysis of Gassin.

in the previous chapter, Gassin is close to planarity with $\psi = \Sigma_i \ell_i / 2n = 1.065$, hence the close association of street line accessibility with street junction accessibility.

We have computed the seven distance measures for both the primal and dual problems and show the correlations between them in table 7.1. Those for the primal problem in table 7.1(a), which deal with accessibility between the lines, partition very cleanly into two sets (shown in **bold** and *italic* type). The measures based on step length that do not have any implication for Euclidean distances are highly correlated. The two Euclidean measures are highly correlated, but the correlations between these two sets are low and negative. This relates to a qualitative difference in what is being measured; lines of sight in Gassin do not correlate very well with the clusters of junctions that provide the most accessible central areas in terms of Euclidean distance. In terms of the dual problem, shown in table 7.1(b), the points

Table 7.1
Correlations between the seven distance measures for Gassin

(a) Line access	ℓ_i	$\tilde{\ell}_i$	$\ell(d)_i$	$\hat{\ell}(d)_i$	$\tilde{d}(\ell)_i$	$e(\ell)_i$	$\hat{e}(\ell)_i$
ℓ_i	•	0.971	0.770	0.839	0.967	−0.088	−0.099
$\tilde{\ell}_i$		•	0.807	0.897	0.984	−0.094	−0.098
$\ell(d)_i$			•	0.973	0.810	−0.275	−0.272
$\hat{\ell}(d)_i$				•	0.889	−0.223	−0.222
$\tilde{d}(\ell)_i$					•	−0.122	−0.112
$e(\ell)_i$						•	_0.970_
$\hat{e}(\ell)_i$							•

(b) Point access	ρ_j	$\tilde{\rho}_j$	$\rho(d)_j$	$\hat{\rho}(d)_j$	$\tilde{d}(\rho)_j$	$e(\rho)_j$	$\hat{e}(\rho)_j$
ρ_j	•	0.393	0.156	0.215	0.297	−0.092	−0.087
$\tilde{\rho}_j$		•	0.792	0.900	0.984	−0.006	−0.085
$\rho(d)_j$			•	0.973	0.806	−0.122	−0.169
$\hat{\rho}(d)_j$				•	0.912	−0.093	−0.151
$\tilde{d}(\rho)_j$					•	−0.060	−0.141
$e(\rho)_j$						•	_0.949_
$\hat{e}(\rho)_j$							•

are more highly correlated, but the same distinction exists between non-Euclidean and Euclidean measures. In fact, for the raw out-degrees data, the correlations with all other measures are very low. This simply implies that the distribution of the out-degrees for the points—that is, the number of lines associated with each point—is either 2 or 3, implying a step-like function.

We would expect all these differences to be reflected in the surfaces associated with the spatial distribution of these measures. In figure 7.8, we show these for three of the accessibility measures we consider are the best reflectors of the difference between the distributions, namely the step lengths $\ell(d)_i$ and $\rho(d)_j$, which are the basic space syntax measures; the weighted distance measures $\tilde{d}(\ell)_i$ and $\tilde{d}(\rho)_j$, which are alternative measures of the syntax; and the Euclidean measures $e(\ell)_i$ and $e(\rho)_j$, which measure accessibility over the physical network. The patterns shown in figure 7.8 bear out the differences suggested at the beginning of this section. The step length and weighted measures generate the same surfaces as those illustrated in the previous chapter, with any slight differences a result of using a narrower range of colors and a smaller exponent of spatial averaging here. For both the primal and dual problems, we derive surfaces for which the central axis of the village is the area where streets and their junctions are most accessible to each other, with the west of the village more accessible than the north, south, or east (figures 7.8[a] and [d], and

[b] and [e]). The Euclidean distance measures in figures 7.8 (c) and (f) produce a quite different picture. The most accessible streets and their junctions are in the areas where the streets are shortest and the lines densest. These do not correlate with the longest lines of sight, and thus the picture is one where the clusters of high accessibility are broken up along the central axis but with a tendency toward highest accessibility in the southeast of the village. This is about all we can say for Gassin: that space syntax is very different from the street distance accessibility, and that this in itself is the basis of informed speculation as to how the visual quality of the town, the location of its key land uses, and the movement patterns therein all relate to these different measures of accessibility.

The "Mixed Syntax" Problem: Coupled Networks with Overground and Underground Routes

What we have in these two types of distance measure is a mix of accessibility indices based on lines of sight and physical travel distance. There are, however, many systems where travel is based not only on streets down which one can see but also on routes where one cannot see, as in underground railways, or even on routes where sight is of much lesser importance, such as on buses or trams. To conclude, we will apply these ideas to systems where we can easily define such differences, which in turn reflect a mixture of axial lines based on lines of sight or unobstructed movement and route segments that reflect planarity. Our application is to Central Melbourne, which is laid out on a grid but around which there is a heavy rail loop, buried underground. We show the axial map/planar route network in figure 7.9, where we distinguish between the two types of route. There is a much denser morphology of routes in the central business district (CBD) than we show in figure 7.9, and the central area is criss-crossed by surface-level trams. But the really distinctive

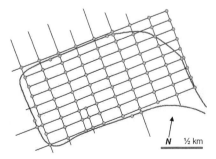

Figure 7.9
The street grid for Central Melbourne with the Underground Rail Loop.

Table 7.2
Correlations between the seven distance measures for Central Melbourne

(a) Line access	ℓ_i	$\tilde{\ell}_i$	$\ell(d)_i$	$\hat{\ell}(d)_i$	$\tilde{d}(\ell)_i$	$e(\ell)_i$	$\hat{e}(\ell)_i$
ℓ_i	•	0.993	0.899	0.938	0.994	−0.055	0.124
$\tilde{\ell}_i$		•	0.934	0.967	0.999	−0.098	0.100
$\ell(d)_i$			•	0.991	0.928	−0.170	0.039
$\hat{\ell}(d)_i$				•	0.962	−0.143	0.070
$\tilde{d}(\ell)_i$					•	−0.089	0.103
$e(\ell)_i$						•	*0.849*
$\hat{e}(\ell)_i$							•

(b) Point access	ρ_j	$\tilde{\rho}_j$	$\rho(d)_j$	$\hat{\rho}(d)_j$	$\tilde{d}(\rho)_j$	$e(\rho)_j$	$\hat{e}(\rho)_j$
ρ_j	•	0.169	0.075	0.079	0.145	−0.012	−0.207
$\tilde{\rho}_j$		•	0.916	0.944	0.999	−0.062	−0.193
$\rho(d)_j$			•	0.996	0.922	0.057	−0.110
$\hat{\rho}(d)_j$				•	0.949	0.028	−0.129
$\tilde{d}(\rho)_j$					•	−0.059	−0.191
$e(\rho)_j$						•	*0.823*
$\hat{e}(\rho)_j$							•

structure is the underground railway, which connects to the street level at some five key stations. If one wishes to loop around the CBD, the fastest and shortest way to do this is via this railway, so it will certainly have an impact on accessibility if Euclidean distance is taken into account. If you want to see places within the CBD, then the long straight streets provide perfect axiality. Thus the contrast between getting to a place fast and seeing the same place, immediately, could not be greater.

In table 7.2, we measure the correlations between the seven accessibility measures we have computed for both the primal and dual problems. The structure of these bears a remarkable similarity to those in table 7.1 for Gassin with the out-degree, step-length, and weighted measures being highly correlated with one another, in contrast to the Euclidean measures, which in turn are highly correlated but not with the first set of measures. As the distribution of points in each line has greater variability than the distribution of lines over each point (the planarity measure is $\Psi = 1.033$, which shows that the map is very nearly planar in these terms), these raw out-degree measures form the first set of dense correlations. However, in the dual problem that involves the points, correlations between these out-degrees and the other two sets of measures are low. In short, what we have here, even before we begin to explore the spatial distribution of these measures, is consistency between the primal and dual in terms of a major difference between the step-length and the Euclidean-type measures. Step-length measures, which pick up syntax as nearness in terms of the way lines of sight are close or far from one another, is dramatically different from the way lines and points are near to each other in terms of their physical distance. This is also clear from figure 7.9 for the railway has few points of contact with the street system, but its relative accessibility to the streets it touches is much higher than the more even distribution of street junctions and the lines that link these.

Figure 7.10 examines the surfaces for our three key measures $\ell(d)_i$, $\tilde{d}(\ell)_i$, and $e(\ell)_i$ and shows the basic out-degrees for the lines ℓ_i, which are illustrated in terms of line thickness. Note how the railway has very few points for each of its lines or tracks. The patterns for step-length and weighted distance accessibility in figures 7.10(b) and 7.10(c) generate the highest accessibility in terms of the nearness to different lines of sight, broadly in the center of the CBD. The area to the southwest

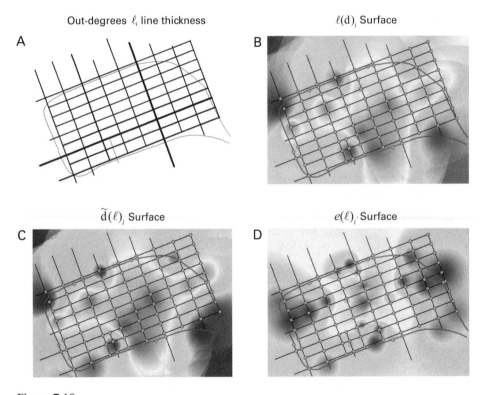

Figure 7.10
Line accessibility surfaces based on (a) out-degrees, (b) step-distances, (c) weighted distances, and (d) Euclidean distances.

of the physical center of the CBD map reveals a pocket of low accessibility—lines of sight with few common points, where the major grid is permeated by a couple of local narrower streets. However, when we look at the Euclidean distance in figure 7.10(d), the stations along the rail routes are pockets of lower accessibility because again at those points, there are a much smaller number of lines of sight that you can reach.

The dual problem that involves accessibility between points or street junctions is even clearer in its distinctions between the step-length distance and Euclidean measures. Figure 7.11(a) shows the distribution of lines for each point as simple pie charts, revealing that out of 93 junctions, there are only 5 that have more than 2 lines associated with them, and in those 5 cases, there are only 3 such lines. This indicates how close the network is in terms of planarity. The step-length measures in figure 7.11(b) and the weighted measure in figure 7.11(c) are highly correlated, and both show that it is the station areas that have the lowest accessibility.

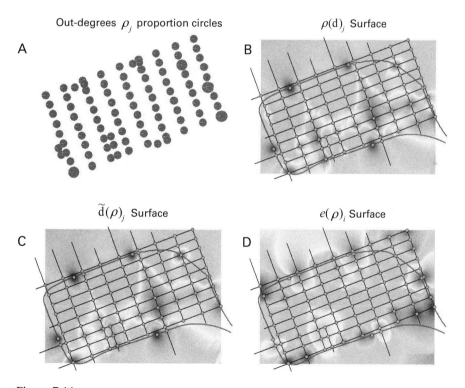

Figure 7.11
Point accessibility surfaces based on (a) out-degrees, (b) step-distances, (c) weighted distances, and (d) Euclidean distances.

This is because there are fewer points from which to see long vistas. In contrast, figure 7.11(d) shows exactly the opposite: the stations are the high points of accessibility and form the heartland of the CBD, where many roads intersect near to stations. In all cases, however, the areas on the very edge of the map have lowest accessibility, as one might expect from the imposition of arbitrary boundaries on the problem.

A comparison of figures 7.8 and 7.11 for Gassin and Melbourne is instructive, for there are many interesting comparisons to explore further. The wealth of interpretations that come from these two types of distance measure suggest that the way forward involves many syntaxes, rather than one, with the consequent challenge that the diversity of indices and surfaces associated with such multiple syntaxes needs to be integrated.

Proximity: Extending the Measures of Step-Length Distance

The critical difference between space syntax and geographical graph representations involves the nature and meaning of distance in the two types of problem. In syntax, the starting point is a topological representation of relationships between streets as lines, while in geographical problems the relationships constitute physical measures of distance between nodes. The fact that the two types of problem are rooted in the same underlying geometry of the street system generates a confusion of purpose that has plagued all critical debate about the meaning of space syntax since its inception. Although we have clarified this difference here, the fact remains that space syntax does not concern itself with geometric or physical distance, but simply with whether there is a relationship between two points or lines in space. Thus proximity, as it is measured in terms of adjacency in the bipartite graph, in the dual and primal topological graphs that are generated from the bipartite representation, or in the step-length distances that are generated from these graphs, forms the core analytical tool for dissecting the syntactical structure of urban space.

There are, however, other measures of proximity that appear to have important advantages over the traditional step-length distances in graphs. In spatial systems, Bera and Claramunt (2002) propose a subtle manipulation of the concept of adjacency, which is based on the weighted sum of a direct measure—whether a line (or point) is linked to another, and the commonality between the set from which the link originates, and the set associated with the adjacent destination. We can use several measures to express direct adjacency such as that used earlier in equation (7.5) as $Z_{ik} = 1$ *if* $\ell_{ik} > 0$; otherwise $Z_{ik} = 0$. We now need to extend this definition, noting it with respect to lines and points, and thus equation (7.5) becomes

$$Z(\ell)_{ik} = \begin{cases} 1 & \textit{if} \quad \ell_{ik} > 0, i \neq k \\ 0, & \textit{otherwise} \end{cases}, \quad Z(\rho)_{jl} = \begin{cases} 1 & \textit{if} \quad \rho_{jl} > 0, j \neq l \\ 0, & \textit{otherwise} \end{cases}. \tag{7.31}$$

We need to define the out-degrees (and in-degrees) of these measures, for these define the size of the set associated with lines and points. From equations (7.31), then

$$Z(\ell)_i = \sum_k Z(\ell)_{ik} \text{, and } Z(\rho)_j = \sum_l Z(\rho)_{jl}, \tag{7.32}$$

where it is clear from our previous definitions that the in-degrees and out-degrees are symmetric, since $Z(\ell)_{ik} = Z(\ell)_{ki}$, $i = k$ and $Z(\rho)_{jl} = Z(\rho)_{jl}$, $j = l$.

The new measure can be defined for lines (points follow by analogy) as

$$R(\ell)_{ik} = \alpha \, Z(\ell)_{ik} + (1 - \alpha) \sum_{z \in \Omega_i} \frac{R(\ell)_{zk}}{Z(\ell)_i}, \tag{7.33}$$

where the two components on the right-hand side of equation (7.33) are weighted by the parameter $0 \le \alpha \le 1$. The first component is simply the adjacency index as defined in equation (7.31), which gives a unit link from i to k if a link from one element to an adjacent one exists. The second is a relative measure that compares the number of elements adjacent to the origin set Ω_i to those linked to the destination set associated with k. This is a little like a first-order clustering coefficient, similar to that used to define clustering in small-world graphs by Watts (1999). The weighted sum essentially compares the direct link between i and k to the number of common intermediate links between i and k through elements z that are common to i. If there are few in common between i and k, then the measure will reduce in its impact where, for example, we have equal weighting. The weights themselves of course control the strength of the direct and indirect adjacencies.

This system of linear equations (7.33) has a unique solution for certain minimal constraints on the adjacency matrix (such as strong connectivity, for example). We have solved this system using iteration, and for the size of problems involved here— Gassin with 41 lines and 63 points, and Central Melbourne with 25 and 119—the procedure is fast, taking no more than 20 iterations for the dual or the primal in each application. A key feature of the solution is that the resultant matrix of relative adjacencies $[R(\ell)_{ik}]$ and $[R(\rho)_{jl}]$ is not symmetric, with the measure picking up the fact that the set of streets or junctions accessible from i to k is not the same as that from k to i. This however does not involve any unidirectional links, for essentially the basic adjacency graphs are not directed, implying that we need to examine both the out-degrees and in-degrees of the relevant matrices. For lines (and points follow directly), these are defined as

$$R(\ell)_i = \sum_k R(\ell)_{ik}, \, R(\ell)_k = \sum_i R(\ell)_{ki}, \text{ and } R(\ell)_i \ne R(\ell)_k, \, i = k. \tag{7.34}$$

In fact, we might expect these measures to be quite close to one another because the adjacent sets considered in the formula are only one step removed. This suggests that other measures incorporating higher-order adjacencies at larger and larger step lengths might be constructed. Although Bera and Claramunt (2002) do not extend their measures in this way, they do show how the measure can be weighted by variables that reflect geometric properties such as distance, perimeter, and area, thus suggesting, as we do here, how Euclidean distance information might be handled.

We have reworked the Gassin and Central Melbourne examples with these proximity measures and show the correlations between these and the seven measures used in tables 7.1 and 7.2 in tables 7.3(a) and 7.3(b), respectively. The same structure displayed in tables 7.1 and 7.2 is revealed by this comparison, in that the proximity measures have high correlations with all the traditional measures and low correlations with the Euclidean measures for the two examples, with respect to both the

Table 7.3
Correlations between the nine measures for Gassin and Central Melbourne

Table 7.3(a) Gassin

Line access	$R(\ell)_i$	$R(\ell)_k$	Point access	$R(\rho)_j$	$R(\rho)_l$
ℓ_i	0.868	0.887	ρ_j	0.165	0.222
$\tilde{\ell}_i$	0.945	0.966	$\tilde{\rho}_j$	0.953	0.982
$\ell(d)_i$	0.862	0.852	$\rho(d)_j$	0.831	0.818
$\hat{\ell}(d)_i$	0.949	0.941	$\hat{\rho}(d)_j$	0.928	0.918
$\tilde{d}(\ell)_i$	0.938	0.952	$\tilde{d}(\rho)_j$	0.985	0.989
$e(\ell)_i$	−0.175	−0.135	$e(\rho)_j$	−0.040	−0.005
$\hat{e}(\ell)_i$	−0.165	−0.123	$\hat{e}(\rho)_j$	−0.120	−0.091
$R(\ell)_i$	•	_0.983_	$R(\rho)_j$	•	_0.982_
$R(\ell)_k$	_0.983_	•	$R(\rho)_l$	_0.982_	•

Table 7.3(b) Central Melbourne

Line access	$R(\ell)_i$	$R(\ell)_k$	Point access	$R(\rho)_j$	$R(\rho)_l$
ℓ_i	0.957	0.975	ρ_j	0.067	0.089
$\tilde{\ell}_i$	0.975	0.993	$\tilde{\rho}_j$	0.985	0.996
$\ell(d)_i$	0.946	0.953	$\rho(d)_j$	0.960	0.921
$\hat{\ell}(d)_i$	0.977	0.982	$\hat{\rho}(d)_j$	0.980	0.948
$\tilde{d}(\ell)_i$	0.976	0.992	$\tilde{d}(\rho)_j$	0.989	0.998
$e(\ell)_i$	−0.087	−0.127	$e(\rho)_j$	−0.034	−0.050
$\hat{e}(\ell)_i$	0.112	0.085	$\hat{e}(\rho)_j$	−0.167	−0.173
$R(\ell)_i$	•	_0.987_	$R(\rho)_j$	•	_0.989_
$R(\ell)_k$	_0.987_	•	$R(\rho)_l$	_0.989_	•

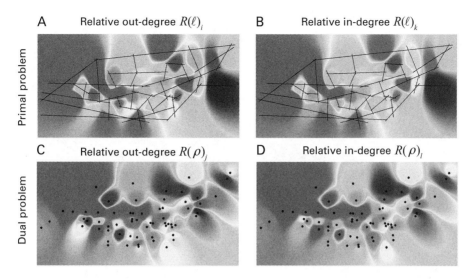

Figure 7.12
Relative proximities for the primal and dual pure syntax analysis of Gassin.

dual and primal problems. As in the previous tests, correlations with the point distributions for the relative adjacency out-degree measures in both examples have low correlations because the out-degrees of the points have hardly any structure, reflecting the near-planarity of each example. Correlations between the relative adjacency in-degrees and out-degrees are quite high. A more graphic demonstration of this is illustrated in the surface representations of these two measures, shown in figures 7.12(a to d) for lines and points in Gassin and in figures 7.13(a to d) in Central Melbourne.

Figures 7.12 and 7.13 reveal that the relative adjacency measures in these two examples are near-symmetric, as might be expected in systems where surfaces are completely covered by streets. In the applications where there is real structure in terms of what is adjacent to what (administrative units such as census tracts, for example), the lack of symmetry is much more significant, but in these examples, this is not the case. In fact, what these proximity indices show is that the proximities of lines in the primal and nodes in the dual problem in Gassin are quite close, apart from the importance of the northern axis, which is stronger in the primal than the dual. In Melbourne there is a little less correspondence between the primal and dual; the line-based primal problem reveals proximity measures that are much more spread over the central area than in the case of the nodes in the dual, where the north-south axes produce striations in the accessibility surface distorting the spread. Unlike the Euclidean distance measures, the position of the rail stations and their

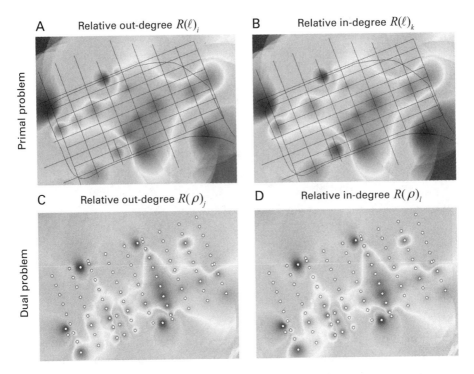

Figure 7.13
Relative proximities for the primal and dual analysis of Central Melbourne.

relative inaccessibility to the street system in terms of their remoteness to lines of sight explains the relative lack of proximity of these points within the central area.

Taking Urban Network Science Forward

What these last two chapters have shown in quite graphic terms is that the accessibility measures associated with axial lines and their intersections are quite different from those that result from measuring physical distance between such points and lines. The two types of measure imply two different problems, but the problems are only separate in conceptual terms, because they are defined with respect to the same network of physical relations: the street network, which contains the axial map. In fact, we have restricted our generation of networks to systems of relations that relate two sets of objects, both of which pertain to spatial networks. The primal and dual problems that are generated from bipartite relations always lead to graphs that can be embedded in different ways in Euclidean space, but the real power of this approach—beginning with bipartite relations—is that only one set of objects might be explicitly spatial, the other social. Or, indeed, strings of bipartite relations might be generated in terms of how one kind of network translates to another. This is the approach we will take in part III, which puts the spatial domain into the background and proposes ways in which networks can be used to manipulate ideas—designs and plans—that ultimately structure two- and three-dimensional space. In short, all we have shown so far is that different conceptions of distance embedded in spatial networks can be used to indicate ways different kinds of spatial network might be coupled, but these different conceptions imply different perspectives. Relating them is a major challenge, which we have illustrated here, but this is only the tip of the iceberg with respect to the ways in which we can form and generate different kinds of urban networks. Indeed, the study of coupled networks represents one of the research frontiers of network science at the present time, and the kind of intractability implied here in joining networks together that are defined for different purposes, often involving different transport media, represents a major challenge (Buldyrev et al., 2010).

In the case where we have a pure space syntax problem—where axial lines are defined solely with respect to lines of sight and where they are then used to interpret physical distance—we have two largely separable problems but with an ability to compare line of sight accessibility with accessibility based on physical distance. How these accessibilities interact with each other to produce the kind of morphologies that emerge in cities is part of the problematic our tools are meant to address and disentangle as we seek to understand the urban forms that we see around us (Hillier, 1996). The questions thus posed are: Which accessibilities should we define to show that this is actually the way cities develop? Is there any combination of them that will achieve this? And what are these multiple accessibilities? We will return to some of these questions in chapter 9, where we link our science to traditional, aggregate spatial-interaction and activity modeling.

The mixed syntax problem complicates the picture even further. In this case it is not possible to see physical and topological accessibilities as being separable. The street network can be based on lines of sight, but there is always a part of the network that does not have lines of sight and where physical accessibility is paramount. We need to compare different step-length accessibilities with physical distance accessibilities when the actual routes defining such accessibilities are themselves mixed. The answer to this probably lies in identifying different types of street network for different purposes and reorienting the analysis this way. The task remains, however, of integrating the measures that are derived for each problem once they have been generated. This then constitutes the next challenge—to actually work out whether accessibility measures for the primal and dual problems can somehow be partitioned to be associated with Euclidean or with line of sight, step-length distances. However, the problem remains how to generate different measures and use them when the lines and points involved are associated with one or the other or a mixture.

Our digression into proximity measures, which clearly correlate highly with traditional space syntax distances based on step length, reinforces the need to look at adjacency rather than geometric measures, and this suggests that the material of these last two chapters, far from providing the last word on distance in complex networks, is just the beginning. Thus, what began with some coherence regarding primals and duals, lines and points, has emerged into a debate about topological, Euclidean, and proximity distances and a mix of these in systems where topology and physical distance are clearly all appropriate, but probably reflect different purposes. It is these distinctions of purpose in characterizing urban morphology that future comparisons of distance and accessibility must address. We will leave these issues as open questions and now switch tack back to location, while at the same time examining urban form and structure, in which network characterizations are never far from the surface. In the next chapter, we will focus on building morphologies around ideas about simplified networks of treelike fractal structures, which will present yet another perspective in our scientific kaleidoscope.

8

Fractal Growth and Form

Our knowledge of the whole complex phenomena of growth is so scanty that it may seem rash to advance even these tentative suggestions.
—D'Arcy Wentworth Thompson, *On Growth and Form* (1917, p. 282)

It has been nearly one hundred years since D'Arcy Wentworth Thompson published his seminal book advancing a science of form, albeit for the biological world, but with strong implications for how the social world might grasp a parallel challenge. His reticence about his understanding at the time has turned out to be rather prescient, for it has taken the last century to push us to the point where we have the sense of a theory of cities that enables us to trace the mechanisms of how cities evolve and grow from the bottom up. In the last half-century, the focus introduced in chapter 1 for systems in equilibrium has slowly given way to dynamics—aggregate dynamics, at first, but more recently to a dynamics of growth and change that is built from the bottom up (Batty, 2005). Now our picture of cities is no longer one of equilibrium but of systems far-from-equilibrium, beyond disequilibrium, in a state where systems can change quite abruptly and with surprising consequences. This involves processes that are complex in that their outcomes and interactions with one another lead to unanticipated effects. There is order in these processes, as indicated in chapter 1, which we will present here as signatures of growth. Yet because cities are still largely observed as if they are in equilibrium, progress has been slow in building ideas about how various urban morphologies evolve and change. Our concern in the last two chapters with respect to networks was entirely focused on their equilibrium properties, and insofar as dynamics has entered the picture so far in this book, it has tended to relate to routine, fast processes such as daily travel rather than longer-term change. Here we will redress the balance and provide another glimpse of cities from the perspective of their longer-term evolution, while ensuring that the signatures of complexity introduced so far are clearly linked to the processes of urban growth.

Although we have barely touched on a dynamics of urban networks or space, many of the concepts that underpin complexity theory have already been introduced.

In earlier chapters, we were at pains to emphasize the notions of modularity, self-similarity, recurrence, and hierarchy that we consider central to the way urban form and structure are organized. Indeed, one of the hallmarks of systems that evolve in relatively stable ways is that they grow in terms of their modules, with each of their components adjusting to those they are most closely related to as they evolve in terms of size and complexity. Alexander (2003) makes the point that complex systems evolve using what he calls structure-preserving transformations. These transformations, he argues, affect any system through local adaptations that gradually converge on configurations fit for the purpose. He writes: "How can a complex system find its way to the good configuration? In a theoretical sense, we may say that the system walks through the configuration space, taking this turn and that, and always arriving at a well-adapted configuration. The huge question of course is how this walk is controlled: what are the rules of the walk, that make it lead to good adaptation?" His answer to this question is as open-ended as D'Arcy's was a century ago. He continues: "Although a few, very preliminary answers have been given to this question, no good ones have yet been given. This is perhaps *the* scientific question of our present era" (p. 19).

The tradition first suggested by Patrick Geddes (1915/1949) in his book *Cities in Evolution* was that small change can lead to large effects—indeed, that what he called "conservative surgery" was the *modus operandi* of how cities should develop. In fact, throughout most of the twentieth century, city planning was composed of large change, although Geddes's mantle was taken up in earnest in the early 1960s by Christopher Alexander (1964) and Jane Jacobs (1961), among others. Their view, which is entirely consistent with Thompson's (1917), has slowly sensitized us to the need to step carefully when intervening in complex systems. Its message is that we plan "at our peril" and that small interventions in a timely and opportune manner attuned to the local context are more likely to succeed than the massive top-down plans that were a feature of city planning throughout much of the twentieth century. This is Darwin's message, too, for biological systems: that life proceeds through a natural selection that slowly but surely preserves the fittest in the population and destroys the rest. This view appears increasingly attractive in explaining the growth dynamics of a variety of nonbiological organizations such as cities. The emergence of order on all scales is the hallmark of complex systems, and it is hardly surprising that with the growth of digital computation, it is now possible to simulate such evolutionary processes, thereby suggesting how "good" designs might emerge among a universe of possible designs.

If we can show that good urban designs can be grown by manipulating this kind of complexity, this promises to provide a much more sensitive, less intrusive way of managing our environments than the blunt instruments that have hitherto characterized planning. City design should thus ascribe to Darwin's message that it is small changes, intelligently identified in the city fabric, rather than massive, monumental plans, that lead to more successful, livable, and certainly more sustainable environments (Hamdi, 2004). To emphasize this new style of planning, we will proceed by analogy, using metaphors about how cities are formed, again outlining the notions of modularity and hierarchy, self-similarity and scale in the physical and functional form of cities, and then presenting ways in which basic functions generate patterns that fill space to different degrees. Cities develop by filling the space available to them in different ways, at different densities, and using different patterns to deliver the energy in terms of the people and materials that enable their constituent parts to function. We will demonstrate a simple diffusion model and then generalize it to grow city forms and structures. We will allude to city plans in history that demonstrate our need to plan with and alongside the mechanisms of organic growth rather than against these processes, which has been the dominant style of planning in the past century.

In this chapter, we will hardly provide answers to Alexander's big question, but we will speculate that local adaptation that makes itself clear in the geometry of form has much to do with the way cities evolve. We will introduce fractal geometry, which is the geometry of hierarchy and self-similarity and entirely consistent with our earlier ideas about size and scale, but here we will develop generative mechanisms that specifically lead to self-similar forms across different scales, in this way evoking a dynamics of growth. In particular, we will implement these ideas using what has become a fashionable tool of development, namely cellular automata, in which cities are conceived as tessellations of cells that change state, dependent on local transformations, thus generating higher-order patterns that resemble processes at lower levels. Thus far in this book we have not introduced specific models of generative processes, apart from some simple demonstrations of how populations evolve in terms of size, but in this chapter, we will introduce mechanisms that lie at the basis of a class of models being used empirically to search for an understanding of growth in cities. In the next chapter, we will present more traditional cross-sectional models and set all of these in context as another way of developing our science. This will be through the ideas of models of systems, rather than through network representations, which we have taken as our starting point here. Nevertheless, in the pragmatism of our approach, we urge readers to think of these tools as exemplars for much wider applications than those we have demonstrated here. To this end, our approach is one of making analogies and fashioning metaphors.

Modularity, Hierarchy, and Self-Similarity

Modular construction is not simply a functional process of ensuring that component parts of a system are stuck together efficiently and sustainably, but a means of actually operating processes that drive the system in an effective way. For example, different functions that relate to how the economy of a city works depend on a critical mass of population, and the more specialized the function, the wider the population required to sustain it. In short, more specialized functions depend on economies of scale such that the size and spacing of various functions produces a regular patterning at different hierarchical levels. The modules are thus replicated in ways that change their extent with their scale.

We can demonstrate this point using some simple geometry that illustrates how we can scale a physical module, producing a fractal that is similar on all scales. Imagine that we need to increase the space required for planting a barrier along a straight path that always has an obstacle at its center. If we divide the first line of the path into three equal segments, we can take two of these segments and splay them away from the path and the barrier so that they touch and form an equilateral triangle in the manner shown in figure 8.1(a), thus bypassing the obstacle. This clearly increases the length of the line L (which has an original length of three units) by one unit, so that the new length of the line becomes $(4/3)L$. We can further increase the length of the line by subdividing each segment into three and displacing the central portion of each of the original segments to form the same equilateral triangle, but at one scale down from the original. If this is done for each of the original four segments, then the length of the second line composed of these four segments increases by 4/3. This in turn is 4/3 the length of the original line L, and

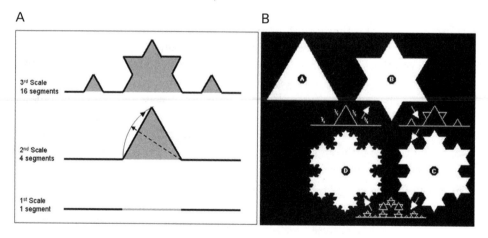

Figure 8.1
Constructing a space-filling curve: The Koch snowflake curve. (a) Successive displacement of the central section of a line at ever-finer scales. (b) Application of the displacement rule to the lines defining a triangle shape called a Koch Island.

the new line is now $(4/3)(4/3)L$. We can continue doing this at ever-finer scales and the length of the line at scale n thus becomes $(4/3)^n L$.

If we now call the initiator line $L(1)$, the length of the line $L(n)$ at any iteration of the recursion is

$$L(n) = \phi^n L(1), \tag{8.1}$$

where ϕ is the composite scaling ratio, in this case 4/3. Clearly, we can vary the scaling ratio. If it approaches 2, then this means the line fills the two-dimensional space. This construction is a recursion of the same rule at different scales, and it generates a pattern which is self-similar in that the motif—the triangular displacement—occurs at every scale and is in a sense the hallmark of the entire construction. The structure grown from the bottom up produces a shape that is a fractal, a regular geometry composed of noncontinuous, sometimes statistically irregular parts repeated on successive scales, which is indicative of the same processes being applied over and over again. The process can be viewed as a hierarchy that is clearly present in the pattern itself, but in terms of the recursive process can be abstracted into the usual treelike diagram shown in figure 8.2.

There are several strange consequences to the process we have just illustrated. If the process of adding more and more detail of the same kind continues indefinitely, the length of the line increases to infinity, but it is intuitively obvious that the area enclosed by the resulting shape, either in the Koch curve in figure 8.1(a) or the Koch island in figure 8.1(b), converges to a fixed value. Second, if the line becomes more and more convoluted in filling the plane, then it would appear that the line, which

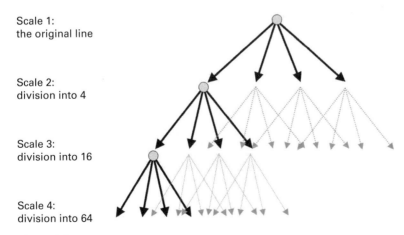

Scale 1:
the original line

Scale 2:
division into 4

Scale 3:
division into 16

Scale 4:
division into 64

Figure 8.2
The hierarchy of composition in constructing a fractal.

has Euclidean dimension of 1, seems to have the dimension of the plane which is 2. This concept of space-filling can be formally demonstrated as being encapsulated in the idea of a fractal dimension. The best way to illustrate this is to count the number of new lines N that are generated for a transition of scale using the formula

$$N = K\varepsilon^{-D}, \tag{8.2}$$

where K is a constant of proportionality that ensures the measure relates to its context, ε is the scaling factor that scales up the current number of lines to the next iteration of N, and D is now a fractal or fractional dimension that ensures this. In fact, for a deterministic fractal such as that in figure 8.1, the scaling is identical from iteration to iteration.

It is easy to see that the fractal dimension can be computed for such a construction at any scale as

$$D = -\frac{\log N - \log K}{\log \varepsilon}. \tag{8.3}$$

If we assume that K is normalized to unity, we can work out the fractal dimension for a variety of shapes where we know how many pieces N are generated from applying the scaling factor ε. In the Koch curve, $N = 4$ and $\varepsilon = 1/3$, and thus $D = \log 4/\log 3 = 1.2619$. You can easily see that if the line does not scale faster than the scaling factor, the dimension remains at 1, which is the Euclidean dimension for a line, but if it scales at twice the rate of the factor, that is $N = 4$ and $\varepsilon = 1/2$, the dimension equals 2, which is a true space-filling curve. If we have a series of numbers N_k of empirically measured line segments that we measure at different scales ε_k—that is, at a series of scaling factors—then we can estimate the fractal dimension from equation (8.2) by least squares as

$$\log N_k = \log K - D \log \varepsilon_k. \tag{8.4}$$

While the Koch curve in figure 8.1(a) has a fractal dimension of about 1.26, a more convoluted line like a fjord coastline has something like 1.7. Rather smooth curves, such as the coastline of southern Australia, have a fractal dimension of about 1.1. In fact, the inventor of the concept of fractals, Benoit Mandelbrot, wrote a famous paper in *Science* in 1967 entitled "How Long Is the Coast of Britain? Statistical Self-Similarity and Fractional Dimension." To cut a long story short, objects that are irregular in the way we have shown and that manifest self-similarity are fractals whose dimension lies between the dimension that they are defined by and the dimension of the space they are trying to fill. In cities, filling the two-dimensional plane with particular forms of development from the parcel to the street line and at different densities suggests that their fractal dimension lies between 1 and 2. Thus this dimension becomes the signature of urban morphology,

A   B

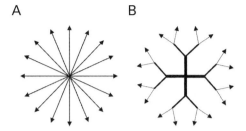

Figure 8.3
Literal hierarchies: Transport from a central source. (a) Each link is separate. (b) Arranging links into a more efficient structure.

which is the outcome of processes that generate fractal shapes. There is a growing tradition in thinking of cities as fractals that originates from the work of the author (Batty, 1985) and that is continually being elaborated in different physical contexts and at different scales (Batty and Longley, 1994; Frankhauser, 1994; Salingaros, 2005; Salat, 2011).

There is, however, a much more literal morphology that is fractal, and this is the shape of an object or set of linked objects that form a tree or dendrite. If you want to transport energy from some central source to many distant locations, it is more efficient to develop infrastructure that captures as much capacity for transfer as near to the source as possible. This is rather easy to demonstrate graphically, for if there are 16 points arranged around a circle, rather than build a link between the source and each of these 16 points, it is more efficient to group the links in such a way that the distance to these different locations is minimized. In figure 8.3(a), assuming each single link is of distance 1 unit, then the length of the routes needed in total to service these locations ("fill the space") is 16, in comparison with the grouping of these routes into 2, then 4, then 8 shown in figure 8.3(b). The total distance of this arrangement in 8.3(b) is something between one-half and three-quarters of the original form in figure 8.3(a) depending on the precise configuration, although the capacity of the links that take more traffic nearer the source are bigger, and this would incur extra costs of construction. Nevertheless, this demonstrates the important point: when resources are to be conserved (which is in virtually every situation one might imagine), space must be filled efficiently. The tree structures in figure 8.3 are fractals. Figure 8.3(b) illustrates this self-similarity directly, while at the same time being a literal hierarchy spread out in space demonstrating quite explicitly the pattern of its construction. This relates to many of the networks discussed in chapter 3, particularly where we examined the allometry of transport networks, as well as the fractal network introduced in the last chapter where we examined its syntax.

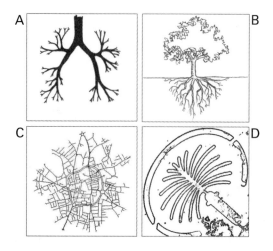

Figure 8.4
Space-filling hierarchies. (a) Top left: a schematic of the human lung. (b) Top right: a schematic of a tree growing into different media, with air above ground and soil below. (c) Bottom left: the road network of Wolverhampton, a mid-sized English town. (d) Bottom right: space-filling in difficult media: Nakheel's Palm Island in Dubai.

There are many examples of such hierarchical structure in the forms we see in both nature and in made-made systems. Energy in the form of blood, oxygen, and electrical signals are delivered to the body through dendritic networks of arteries and veins, lungs, and nerves, as the schematic of the central lung system in figure 8.4(a) illustrates. Plants reach up to receive oxygen from the air and down to draw out other nutrients from the soil, as in figure 8.4(b). Nearer to our concern here and reflecting the discussion of route systems above, figure 8.4(c) shows the network of streets in the midsize English town of Wolverhampton (population circa 300,000 in 2001). It is clear that the traditional street system has grown organically, but the ring around the town center has been planned, imposed from the top down, thus illustrating the notion that what we observe in cities is a mixture of different scales of decision-making. In figure 8.4(d), we show one of the Palm islands off the coast of Dubai developed by the construction company Nakheel. This is a wonderful example of how it is necessary to conserve resources when building into hostile media—in this case by reclaiming land from the sea, where transportation and access become the main constraints on the way the resort is formed.

These examples complement the networks introduced in chapter 3, but now we are impressing the growth dynamics of their evolution, demonstrating that they are built from the bottom up, incrementally, and when they are planned from the top down, the plan is usually a small part of wider organic development. When cities grow, at any point in time, we have little idea of what the future holds with respect to new behaviors, values, technologies, and social norms, and thus it is not surprising that cities grow in an ad hoc manner reflecting the efficiencies and equities that dominate the consensus at the time when development takes place. To illustrate how we can model this process, we can abstract it into two main forces that reflect the desire for space on the part of any individual, developer, or consumer, which is traded off against the desire to live as close as possible to the "city" composed of other individuals so that economies of scale might be realized. This is a simple model, also noted in chapter 1, which captures all the ideas we have introduced so far, and we will now develop it as a hypothetical simulation.

Simulating Space-Filling Growth

Our model is based on two key drivers. First, cities exist as machines for enabling us to divide our labor so that we might realize economies of scale, or agglomeration economies as they are also called. This we argued in earlier chapters, invoking Alfred Marshall (1890), who made the point over one hundred years ago: "Great are the advantages which people following the same skilled trade get from near neighborhood to one another. The mysteries of the trade become no mystery but are, as it were, in the air" (quoted in Glaeser, 1996). Our first principle is that for the city to exist at all, individuals must be connected to one another in terms of their proximity to others, and this means that new entrants to the city must somehow connect physically to those already there. In contrast, individuals seek as much personal space as possible for themselves, and this translates into the notion that they wish to live as far away from others as possible but still reside in the city space. This may translate into living at low densities but, as in Manhattan, large apartments in the sky may be another way of realizing this quest. In addition, there are increasingly innovative ways of meeting this goal in different global locations. In our context here, we will embody this second principle as one in which people wish to live on the edge of the existing city rather than in the center, notwithstanding the great variety in these kinds of preferences. We note also that we need not insist on physical contiguity for the city to be said to be connected, for it is enough for a network of connection to be established, but to best play out the demonstration, we assume physical contiguity or adjacency.

Our model can be constructed as follows. Imagine a trader decides to locate his or her base at the intersection of a trading route and a river where the land is fertile and flat. Many cities have grown from such humble origins, where comparative natural advantages such as these determine the best seed for a settlement. Now imagine another individual seeking a permanent location comes into the vicinity of the lone trader's base. If that trader happens by chance to get to within the neighborhood of the existing trader, that trader may decide to locate there although there may be many traders in the wider hinterland who do not enter into the neighborhood and never find the emergent settlement. However, a certain proportion will find the settlement with a certain probability, and given enough time and enough traders, the settlement will grow. As it grows, the probability of being discovered increases because it occupies more space. From these simple principles, we can demonstrate the form of the growing city. In figure 8.5(a), we show a schematic of the location process. The individuals are arranged around a circle well outside the location of the settlement, which is fixed at the center of the circle with the black solid dot. This is where the original trader locates. Each individual is a solid gray dot, and they begin the movement in search of the location using a random walk. They decide at each step to move up or down or left or right randomly and in this way walk across the locational plane. If they move to a cell adjacent to the fixed solid black dot, they settle; they stop any further walking and turn stippled black, which shows they are now fixed, stable, and no longer in motion. The first one to do so is shown by the stippled black dot adjacent to the initial solid black dot. That is all there is to it. You can see the final form, so you know what will result, but if you had not seen this result, then many would guess that the result would not have been a treelike structure but a compact growing mass.

We show the progression of this in figure 8.5(b). To talk this through, what happens is that as soon as a trader settles next door to the existing black dot, the chances of another trader finding that new trader settlement as opposed to any other one increase just a fraction. As time goes by, the linear edge pattern that is characteristic of the growing tip of the cluster begins to emphasize itself, and increasingly a trader finds it impossible to penetrate into the fissures in the growing cluster. Traders then are more likely to find the growing cluster at its edge, and in this way, the cluster begins to span the space. If this were to produce a growing compact mass, it would have dimension nearer to 2—the Euclidean dimension—but in fact it has a fractal dimension between 1 and 2, about 1.7, as empirical work in many fields has determined (Batty and Longley, 1994). This, like any dendritic structure, is a fractal, and it is easy to see the self-similarity that is contained in its form. Break off any branch and you can see the entire structure in the branch, just as you can usually see the entire structure of a tree or a plant in its leaf. As we increase the resolution of the grid or lattice on which this walk takes place, then we get finer and finer treelike structures in which the fractal structure is readily apparent, as shown in figure 8.5(c).

This form is generated by a process called diffusion-limited aggregation (DLA), which has been found and used extensively in physics to grow crystal-like structures and to examine ways in which one media penetrates another, such as oil diffusing into water (Stanley and Ostrowsky, 1985). You can see similar patterns if you pour concentrated liquid soap into ordinary bath water, and this is also reminiscent of

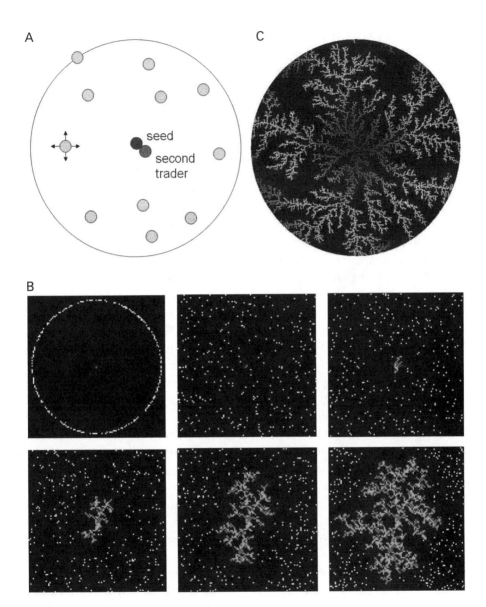

Figure 8.5
Generating clustered city growth using diffusion-limited aggregation.

the way the Dubai palm island resort has been "forced" into the sea. It is a general principle that a substance with a higher density creates such patterns when infused into a substance of lower density. A model, of course, is only as good as its assumptions, but it is possible to tune this DLA model to produce many different shapes, some of which bear an uncanny resemblance to those we find in real cities. For example, we can "tune" the DLA to produce sparser structures if we relax the criterion that the individual who settles must exactly touch the already settled structure. We could, for example, set a distance threshold for this, or we could insist that more than one individual must already have settled. In this way, we can change the density, growing structures that are very heavily controlled in their dependence on what has gone before, or generating compact linear structures in which the degree of control over where traders are allowed to settle is very weak. Some of these extensions are explored by the author elsewhere (Batty, 2005).

These kinds of generative mechanisms give rise to patterns of space filling that tend to be linear rather than compact and in which space is filled by network-like composites. To an extent, this is consistent with the way energy is delivered to the various parts of an organism whose metabolism obeys various allometric laws of scaling, such as those introduced in chapters 1 and 3 (West, Brown and Enquist, 1997). There we illustrated how travel is minimized for monocentric cities, in which treelike networks centered on a central zone of the city deliver all the energy to the center and the assumption is that everyone travels to this center. This is entirely consistent with the DLA model, which generates activity around a central seed site. If we relax this model to incorporate several seeds or even to place new seeds in areas that are not yet served, we build up a polycentric structure that still has network properties. In the same way, the allometric network model can be thus relaxed, and as illustrated in chapter 3, following Samaniego and Moses (2008), the scaling begins to change in a consistent way. In essence, what our discussion reveals is that relatively robust and simple models of urban location in which relations are key to where activities are located are much more likely to generate, not dense clusters of activities, which tend to be the exception rather than the rule, but network structures. In short, contemporary location models that generate the sorts of fractal structures that characterize real cities essentially reproduce the network structures that form the skeletal frame around which all cities are constructed.

Real Cities and Patterns of Complexity

There are many examples at different scales of the way cities are structured along dendritic lines mirroring the lines of energy that serve their distant parts. We glimpsed an idea of this in figure 8.4(c), which abstracted the street network of

Wolverhampton, but cities are not pure dendrites. Different networks are superimposed on one another for different kinds of transport, ranging from different modes requiring different networks to social and electronic networks that underpin the way people trade and communicate. In figure 1.4 in chapter 1 and figure 3.10 in chapter 3, we showed several maps of transport networks in London that essentially organized the hierarchy of routes and modes according to traffic volumes and accessibilities, which in turn are highly correlated with the energy used in transporting people and materials. In figure 8.6, we show the street transport network map of inner London. The streets are colored according to the energy they transport using the proxy of road traffic volumes, which give some index of both capacity and congestion or saturation. This is also highly correlated with patterns of accessibility, which mirror the proximity of places to each other.

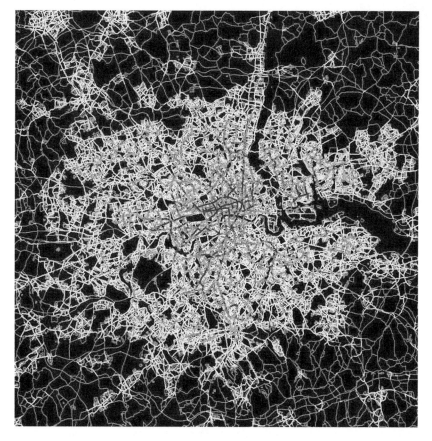

Figure 8.6
The organically evolving network of surface streets in Greater London classified by traffic volume.

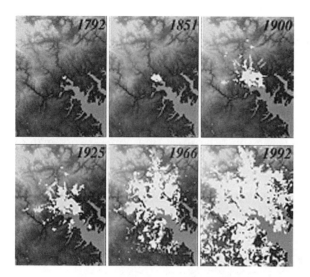

Figure 8.7
Two hundred years of urban growth in Baltimore (from Acevedo, Foresman, and Buchanan, 1997).

As was clear in the last two chapters, street networks are excellent examples of how cities grow from the bottom up, for they represent the skeletal structure on which all else in the city hangs. As we can see from the way cities grow, transport and land use are intimately related. Indeed, in the 1930s, as urban sprawl first became significant in Britain, "ribbon" development became the pattern of transport linked to land use that was the subject of fierce control in the quest to contain urban growth. It is possible to see this connectivity in many patterns of urban growth. The picture of the way the eastern US city of Baltimore in Maryland has grown over the last two hundred years shown in figure 8.7 is a clear illustration of the way development proceeds along radial routes from the traditional center, particularly at the turn of the last century when streetcars, bus, and railways systems dominated (Acevedo, Foresman, and Buchanan, 1997). Although this pattern is breaking down as cities become more polycentric and specialized in their parts and as new kinds of central business districts such as "edge cities" become established, it is still significant.

These patterns recur at different scales, although the notion of them being faithfully reproduced at every spatial scale needs to be tempered with the obvious fact that individuals are diverse in their tastes and values and thus heterogeneous in their actions. Moreover, the sort of similarity that occurs in cities is statistical self-similarity rather than the rather strict self-similarity we saw, for example, in the construction of the Koch snowflake curve in figure 8.1. In fact, although the pattern of transport routes in cities is generally radial, focusing on significant hubs, and organized according to a hierarchy of importance that mirrors different transport technologies, capacities, and speed of transmission, street systems illustrate the space-filling principle quite clearly. At the local level, there is more conscious planning and design of street systems, particularly in developments that are self-contained for purposes of the actual construction as well as financing and sale. For example, residential areas are often formed as small, single-entry streets of houses arranged around cul-de-sacs for purposes of containment and traffic management as well as security. We will return to these ideas in the next section when we speculate on how such structures can be formed in a more conscious sense through explicit design and planning, but it is important to note that these patterns do recur across different scales, as can be seen in their statistical distributions as well as their physical self-similarity.

The obvious question is, how far can we get with the DLA model of the last section in generating simulations of real structures that we see in the transport development of London in figure 8.6 and urban development of Baltimore in figure 8.7? This kind of model is, of course, a demonstration of how two principles or forces interact to produce a structure that resembles certain features of the modern city. It is not intended as anything other than a graphic way of highlighting the notion that bottom-up, uncoordinated change leads to highly ordered structures—fractals—which emerge from this comparatively simple process. One can begin to illustrate how one might make this more realistic, but it is a far cry from the kinds of operational models that are used routinely for strategic planning by government and other agencies. The model is made more realistic simply by planting it into a space or terrain that has real features. In figure 8.8, we show four different simulations of development in the town of Cardiff, Wales, which is contained by the coastline and rivers defining that area. We set two seeds, one at the historic center and one at the dockside, and let the DLA model operate in the manner we have shown in figure 8.5. From this, we realize quite quickly that the river cutting the town in two makes a difference to the rate of growth in parts of the town, while the fact that Cardiff has two centers shows how difficult it is to generate a pattern that gives the right historical balance to each. This is not surprising, because none of the factors that affected this competition between the two centers is contained in the model. A more detailed discussion of the simulation is presented by Batty and Longley (1994). The model, however, can only simulate patterns that are a consequence of its assumptions. Yet these kinds of simulations also provide a means for demonstrating and testing various future hypotheses about urban form. To explore how such structures might be generated empirically, it is now worth presenting how models of these processes can be designed and implemented in such a way that they can be used to help generate effective designs.

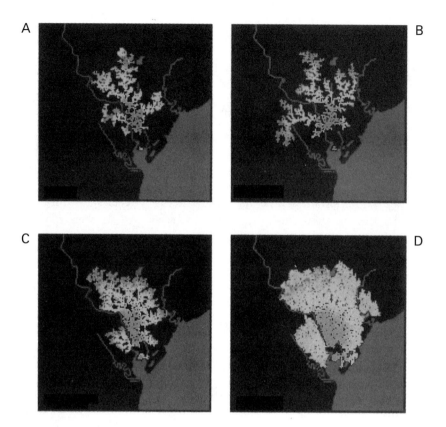

Figure 8.8
Simulating growth using diffusion-limited aggregation, or DLA, in the spatial landscape centered on the city of Cardiff. The control over the development is tuned to decrease systematically through the simulations, from (a) top left to (b) top right to (c) bottom left to (d) bottom right.

Generating and Simulating Cities Using Cellular Automata Models

The Mechanism of Cellular Automata

We need a greater degree of control over our simulation process than that provided by the DLA model or its variants. In fact, we generated the previous clusters by using a generative algebra that lies at the basis of many pattern-making procedures called automata. An automata is usually defined rather generally as a finite state machine driven by inputs that switch the states of the machine—the outputs—to different values. The outputs from the machine may then be used as inputs to drive the process of state transition through time, and this generative process can be tuned to replicate the sorts of patterns we have introduced here. For example, the input to the DLA model is an individual who moves in a cell space, and if certain conditions in the space occur, the individual changes the state of the cell from undeveloped to developed. This, of course, is done in parallel for many individuals. The idea that the space might be characterized as a set of cells simply gives some geometric structure to the problem, and although we have taken for granted the fact that cities are represented in this way in these simulations, for automata in general, and spatial automata in particular, they can be any shape and in any dimension.

The automata we use here to generate physical development are called cellular automata (CA). We assume a regular lattice of (square) cells in which development takes place by changing the state of each cell from undeveloped to developed as long as certain rules apply. The elements of CA are, thus, a set of *cells* which can take on one of several states, in this case developed or undeveloped, extendable to different kinds of development; a *neighborhood* of 8 cells in the N-S-E-W-NE-SE-SW-NW positions around each cell in question; and a set of *transition rules* that define how any cell should change its state dependent upon the configuration, state, and possibly attributes of cells that exist within the neighborhood of the cell in question. Now, start with the initial condition of one cell in the center of the lattice being switched on—developed—and apply the rule that if there exists one or more cells in the neighborhood of any cell, this will generate a diffusion around the initial cell that mirrors the process of successive spreading of the phenomena, just as a physical substance with some motion might diffuse. The diffusion is square because the underlying lattice is square, but we can easily develop versions where the diffusion is near circular if we so configure the lattice. We show this diffusion and the implicit rules generating the automata in figure 8.9.

We will develop these models formally, for they are by far the simplest of any urban model in that they entirely merge their populations with their environment. In essence, the components of the environment are identical to the objects comprising

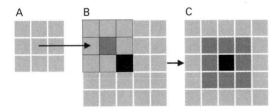

Figure 8.9
Cellular automata: How cells are developed. (a) Left: an 8-cell neighborhood around a central cell in question is applied to (b) Middle: each cell in a lattice. If one or more cells in the lattice is of a particular state, in this case developed (black), the state cell in question in the neighborhood (dark gray stipple) changes to developed. If this rule based on one or more cells in the neighborhood is applied to every cell in the lattice, the result is (c) Right: the set of cells around the central cell (black) becomes developed (black hatch).

the population, in the sense that the locational spaces defining the environment at any point in time are equivalent to each element in the population. Now each cell i in a CA model can take on more than one state A_{it} at time t, which means that the population object can vary in its attributes. Again, the simplest form is that a cell i at time t can take on one of two states—it can be switched on or off, which in urban terms might be compared to the cell being developed or not developed. This is often represented as

$$A_{it} = \begin{cases} 1 & \textit{if } i \text{ is developed} \\ 0, & \text{otherwise} \end{cases}.$$
(8.5)

In slightly more complicated CA models, there may be more than one population object in each cell. If a cell has one population object only but that object can take on different attributes or changes in state, then this is still a CA model. In short, when a cell can take on more than two states, this is usually used to reflect changes in land cover, such as land use types, but it could also be associated with different changes in the population object, such as its level of income, age, and so on. The formulation is entirely generic.

As we have already shown in figure 8.9, CA models in their strict sense have no action-at-a-distance except in the most restrictive sense. A cell is deemed to influence or be influenced by its nearest neighbors, where "near" is defined as physically adjacent if the application is to some spatial system. This is the only way emergence can be charted in such models, since if the field of influence is wider than nearest neighbors in a regular sense, then it is impossible to trace any emergent effects on the ultimate spatial structure. In this manner, CA is used to implement procedures that lead to fractal structures, in which patterns repeat themselves at different scales and only emerge when the system in question grows and evolves. We can illustrate strict CA in the following way. Assume that the set Z_I is the set of immediate neighbors on a regular square lattice. The usual neighborhood is defined as the Moore neighborhood—all cells at the eight compass points around the cell in question as in figure 8.9, or the von Neumann neighborhood, which are the cells N, S, E and W of the central cell. Then we define a function F_{it} as the concatenation of effects in the Z_I neighborhood. If this function takes a certain value, this generates a change in state of the cell in question, cell i. Imagine that the rule—and there can be many, many different rules—is that if this function is greater than a certain threshold Ψ, which is a count of the developed cells in the neighborhood, then the cell changes state. In the simplest case, it is developed if it is not already developed, or it stays developed if already developed. Using the definition in equation (8.5), then

$$F_{it} = \sum_{j \in Z_I} A_{jt}$$
(8.6)

and

$$if \quad F_{it} > \Psi \quad then \quad A_{it+1} = 1. \tag{8.7}$$

It is very easy to show that this process leads to a regular diffusion starting from a single cell. If we assume that the threshold $\Psi = 1$, all the cells in original Moore neighborhood around the seed cell get developed first, then all cells around those that have just been developed, and so on, with the recursion simply leading to the growth of a square cellular region around the starting cell. In fact, in this case, space and time are collapsed into one, which is the key criteria of regular physical diffusion. These ideas are developed in more detail in Batty (2005), to which the reader is referred for many illustrations of such strict CA models and their elaboration into agent-based model forms.

If the CA models are slightly more complicated in terms of neighborhood rules, then various geometric fractals result, while there can be key spatial orientations and biases introduced into the structures that are generated. However, it is common in CA modeling for the neighborhoods, the rules, and the process of generation to be entirely uniform. As soon as the notions of varying neighborhoods over space and varying rules over time are introduced, the models are no longer CA. In fact, many urban applications are not strictly CA models at all but cell-space models, motivated by physical land development problems and raster-based GIS (geographic information systems) map algebras. These do not generate emergent patterns in any recognizable form, and they usually relax the constraints placed on both size of neighborhood and uniformity of cell transition rules. In figure 8.10, we show three typical CA models generated using the Moore neighborhood. The first is the simple diffusion from a source, where any development in any adjacent cell spurs development of the cell in question; the second is simple diffusion from a source using a fractal generating rule, where the pattern of cells developed determines the rule; and the third is based on a more complicated pattern of cells in the neighborhood that steers growth, which in this instance is stochastic in a given direction. These are the kinds of structures that form the basis of such automata, and all applications to real systems contain mechanisms of recursion built along the same lines as those used to generate the patterns in figure 8.10.

Applications of Cell-Based Models

There are several ways in which the strict CA model has been relaxed in developing spatial applications. First, it is easy to control the growth of developed cells by imposing some sort of growth rates with respect to different cells. If growth is one unit cell, then various external constraints can be used to control the growth, but as in all cases where the homogeneity rules are relaxed, then the CA can no longer generate emergent patterns in quite the same simple way as those in figure 8.10.

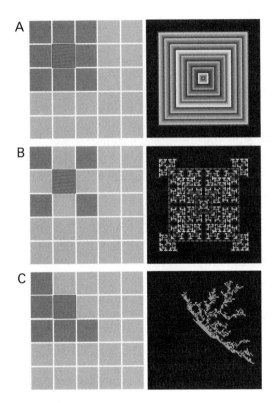

Figure 8.10
Classic CA models. (a) Nearest neighbor physical diffusion on a grid. (b) Koch-like fractal diffusion. (c) Oriented diffusion-limited aggregation.

Moreover, to introduce variety and heterogeneity into the simplest models directly, sometimes the cellular count or concatenation of cells performed in the neighborhood is converted to a probability function, which is then used to condition the development using a random number generator. For example, the structure in equations (8.6) and (8.7) now becomes

$$P_{it} = \sum_{j \in Z_I} A_{jt} / 8, \quad 0 \le P_{it} \le 1 \tag{8.8}$$

and

$$if \quad rand(\Psi) < P_{it} \quad then \quad A_{it+1} = 1, \tag{8.9}$$

where P_{it} is a probability of development and $rand(\Psi)$ is a random number between 0 and 100, say, which if less than the probability (percentage) implies the land should be developed. There are many adaptations that can be made in this manner, but the most significant is related to relaxing the strict neighborhood rule and replacing this with some sort of action-at-a-distance. For example, replacing F_{it} in equation (8.6) with the gravitational expression for accessibility leads to

$$F_{it} = \sum_j A_{jt} / d_{ij}^2, \tag{8.10}$$

and this provides a model that can predict development in proportion to accessibility:

$$A_{it+1} \propto F_{it}. \tag{8.11}$$

This almost converts the CA model to an accessibility potential model, which lies at the core of spatial interaction theory and was first developed for these purposes at the very inception of land use transportation modeling from the mid-1950s (Hansen, 1959). We examined such models in chapter 2, and we will return to them in the next chapter. The open question, of course, is how such a model might relate to the extensive tradition of land use transportation interaction models. These are far superior in their explanatory and predictive power than the kinds of CA model discussed here, but they are limited in their treatment of dynamics. Fusing the two traditions is difficult, and probably counter to the pluralistic conceptions of how we should think about cities that we espouse here.

One of the major developments of these cellular models is to specify different cell states in terms of different land uses, which we will disaggregate and notate as k, A_{it}^k being the appropriate land use k in cell i at time t. In several models, these land uses relate to one another as linkages that determine, to an extent, the locational potential for a site to be developed. Then we might write the change in state of the cell in question as a function of several land uses in adjacent cells, where we use a functional notation to simply indicate that the change in question has to be specified in more detail once the model application is implemented. Then the new state of cell i at time t would be

$$A_{it+1}^k = f(A_{jt}^\ell, d_{ij}) \quad \forall \ell, \tag{8.12}$$

where $j \in Z_l^k$ is a neighborhood defined entirely generically and the field over which distance is defined is again specific to the zone in question. In fact, this relaxes the strict CA quite dramatically and is characteristic of many applications (for reviews, see Batty, 2005, and Liu, 2008). It is worth noting that the rules defining land use transitions generally vary the definition of the neighborhood from the strict no action-at-a-distance principle to the gravitational one. This links different land use states and their densities and types to each land use in question, and also relates these links to different action-at-a-distance effects. These rules also pertain to constraints that are hard and fast on whether a cell can be developed and up to what level, thus defining how land uses do or do not relate to one another. Rules extend to the development of transport links in cells that ensure land uses are connected, and structure the regeneration of cells according to various life cycle effects. All of these rule sets are featured in CA models, and they are central, for example, to the *SLEUTH, DUEM, METRONAMICA,* and related model packages (Batty and Xie, 2005; Batty, 2012).

A more generic CA-like structure, which is a lot closer to the differential model that dominates the dynamics of physical phenomena at much finer scales, is based on a reaction-diffusion structure that might be written in the following way:

$$A_{it+1} = \alpha A_{it} + \beta \sum_{j \in Z_I} A_{jt} + (1 - \alpha - \beta)X_{it}, \tag{8.13}$$

where α and β are normalizing parameters between 0 and 1, and X_{it} is an exogenous variable that reflects changes from the wider environment that might be treated as error or noise in the system, but more commonly is treated as an exogenous shock or as an input that is not predictable by the model. To operationalize this structure, it may be necessary to impose various other constraints to ensure that variables remain within bounds, but the essence of the structure is one where the first term on the right-hand side is the reaction, the second the diffusion, and the third the external input or noise. If we assume that $X_{it} = 0$, the evolution or growth is purely a function of the trade-off between how the system reacts and how activity within it diffuses. In fact, this is rather an artificial structure, since change in absolute terms always needs to be controlled. In this sense, external inputs are always likely to be the case. Many CA models do not explicitly adopt this more general structure, and a lot of applications have tended to simply scale the outputs of the developed cells to meet exogenous forecasts rather than introducing such exogeneity in more consistent and subtle ways, as in the reaction-diffusion model in equation (8.13).

The real critique of CA models relates to their highly physicalist approach to urban structure and dynamics. Essentially these are supply-side models, simulating the supply of land based on physical constraints. The notion of demand is foreign to these models, as is the notion of interaction as reflected in transport. By abandoning the principles of uniformity, restricted neighborhoods, and homogeneity of states, which it is often necessary to do once one applies these ideas, then the models often become poor equivalents to land use–transportation interaction and other models that we will introduce in the next chapter. However, in their favor is the fact that they are explicitly dynamic, although dynamic processes other than physical land development do not feature very much in their formulations. Their dynamics is also rather straightforward and if surprising and novel forecasts do emerge, this is more by accident than design, in the sense that these models tend to simulate a relatively smooth dynamics.

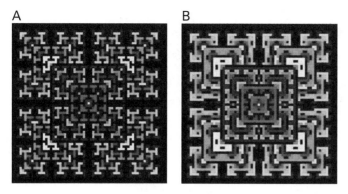

Figure 8.11
Regular diffusion using CA: Patterns reminiscent of idealized Renaissance city plans. (a) Left: only one cell in the neighborhood. (b) Right: one or two cells developed.

If we modify the rules by noting that the number of cells in the neighborhood of an undeveloped cell must be only one, then we generate the kinds of diffusion in figure 8.11(a), and if there are one or two, then the simulation generates figure 8.11(b). There are literally millions of possibilities, and the trick is, of course, to define the correct or appropriate set of rules. Wolfram (2002), in his book *A New Kind of Science,* argues that such automata represent the fundamental units on which our universe is constructed. Although we have more modest ambitions here, this kind of automata can be tuned to replicate many different generative phenomena that characterize many different forms of city.

To generate ideal cities using such automata, it is necessary to begin with a set of realistic rules for transition. Ideal cities are often designed to meet some overriding objective function, such as to minimize density, as in Frank Lloyd Wright's BroadAcre city; to maximize density, as in Le Corbusier's City Radieuse; to generate formal vistas and garden squares, as in Regency London; to generate medium-density new towns with segregated land uses, as in the first generation of British New Towns, and so on. A rather good example that can be generated using cellular automata principles is the plan for the Georgian colony of Savannah in the New World. Developed in 1733 by General James Oglethorpe (Wikipedia, 2012), the plan is shown in figure 8.12; the CA rules might be imagined in analogy to the way we generate development in figures 8.10 and 8.11.

Usually plans for ideal cities are not grown using a generative logic because plans are conceived of all-in-one-piece, so to speak, and the notion of an uncertain future is never in the frame. However, CA allows us to generate plans that evolve through time, and we can continually change the rules so that the idealization is a shifting vision. In a sense, the plans grown in figures 8.10 and 8.11 have stable rules, which

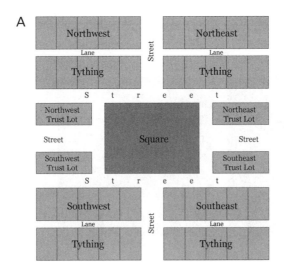

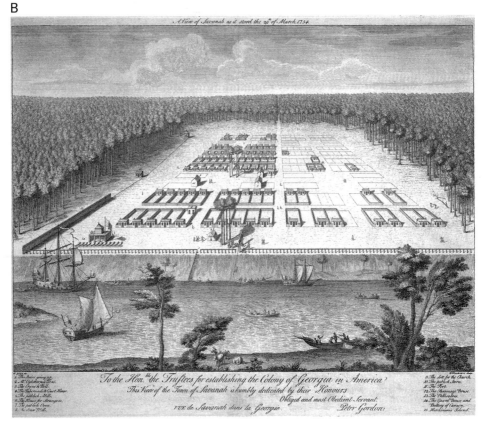

Figure 8.12
The colonial plan for Savannah, Georgia. (a) The original neighborhood plan, from http://en.wikipedia
.org/wiki/Squares_of_Savannah,_Georgia. (b) The plan in 1734, from http://www.torontopubliclibrary.ca/
content/ve/bain/images/111gordon-savannah-cons-large.jpg.

may or may not be considered as ideal objectives to be attained. To conclude our demonstration of this kind of logic and the intrinsic complexity of cities (in that their ideal form is never certain), we will return to the DLA model and tweak the rules a little so that a systemwide objective might be met. Imagine that the agents in our model move randomly in the manner described earlier, that is to all points of the compass; this can be simulated using CA by assuming that where the state of the cell is an agent, then the cell changes state according to the movement. If an agent is at cell i, j, and it moves to cell $i+1, j$ in the next time period, then the cell state switches accordingly, from the cell where the agent is located to the cell where it is newly located. Our first rule, then, is simply cell state switching from the place where the agent was located to its new location. But we also have a rule that says that if the agent is located at cell i, j and there is another agent fixed at a cell in the neighborhood of i, j, then the agent remains fixed and the cell on which it sits changes to the stable state. Note in this version of the CA, the cells contain mobile or fixed (stable) agents or have no agents within them at all. The cells have three possible states, which are appropriately coded, but this is still a CA with two sets of rules. Many similar variants can be constructed that embody more elaborate rules of growth and development, but the quest, of course, is to select those rules that are observable in the way real cities develop, while keeping such models as simple as possible.

Simulations, Predictions, Designs

There is much still to say about how cities are formed and evolve, how we might best understand and then simulate them, and most importantly, how we should design plans that enable them to function in more efficient and equitable ways. Much of what we have done in this book so far is to define representations that apply largely to structures at an instant of time, although throughout we have emphasized the dynamics of how these structures change. We have not organized our science around different sectors of the city in terms of theories or models that can be used to simulate structure and dynamics, for our focus has been much more on relations and networks than on the processes that give rise to the cities we see around us. In this sense, our science is not comprehensive, for our purpose here is to shift the emphasis from locations to relations, and in doing so to emphasize interactions rather than actions per se. Nevertheless, it is important that readers grasp that the field can be organized quite differently, and at this point it is worth saying a little about how one might model cities, either in their cross-section as static structures or as dynamically flowing processes.

Half a century ago, urban researchers began to build models of the cities in their cross-section. Urban structure composed of different sectors—residential, industrial, retail, commerce, and so on—each had clearly different locational requirements and interactions with one another, and the first attempts at building models were based on generating consistent functional patterns for each of these sectors in terms of the locational activities at the cross-section in time. Spatial interaction and its wider context of social physics, as we outlined in chapter 2 and will continue with in the next chapter, constituted the focus (Batty, 1976). These were aggregative models that assumed cities were in equilibrium, but as we have argued, this paradigm was found lacking in that cities are clearly far from equilibrium, while dynamics and innovation are key to their survival and growth. Hence the paradigm changed toward more disaggregate, micro, individual-based models reflected in individual and group processes of interaction and location. CA models represent this trend in its simplest form, but as we have argued, these models are simplistic at best and represent pedagogic ways of thinking about dynamics and evolution rather than structures that can be made strongly operational. Moreover, as the quest to disaggregate models in terms of attributes, populations, and temporal dynamics continues, the kinds of models that have resulted are much less easy to apply and validate. The current generation of models is much richer than the static equivalents we will deal with in the next chapter, but models in which space is represented by cells, populations by individuals or agents, and actions and interactions by dynamically varying spatial behavior are less easy to implement and quite problematic to validate (Batty, 2012). Indeed, cities and their models are so complex that the notion of validation in general is under severe scrutiny, as indeed is the wider question of how we use science in human affairs.

We have also broached here the idea that cities evolve into an unknowable future that is always uncertain. Therefore, any goals we might have for the future city are contingent on the present, hence continually subject to revision and compromise. In the past, cities have been designed in a timeless future where sets of objectives have been defined to be achievable as if the city were cast in timeless web, and it is of little surprise that few cities have ever achieved the aspirations set out in their plans. Complexity theory broaches the problem of the unknowable future and the way cities evolve from the bottom up, incrementally, as the products of decisions that might be optimal at any one time but are always subject to changing circumstances. This would appear to be a far more fruitful and realistic way of generating cities that meet certain goals, since the city emerges from the product of decisions that might be optimal in the small but whose global effects are unknowable in the large until they emerge. As we argued at the start of this chapter, effective intervention and small, incremental design might lead to large and effective changes that go with the flow and do not fight against the grain. Such planning through incremental evolution has not been the history of most city plans hitherto, but our science is evolving to meet this challenge, particularly, we hope, through using the ideas we are developing here.

9

Urban Simulation

Abstraction today is no longer that of the map, the double, the mirror or the concept. Simulation is no longer that of a territory, a referential being or a substance. It is the generation by models of a real without origin or reality: a hyperreal.
—Jean Baudrillard, *Simulations* (1983, p. 2)

Simulation is a generic activity that has been introduced implicitly throughout this book. The notion of representing cities as formal mathematical functions, logical sets of relations, and digital descriptions is core to our science of cities, and in previous chapters, although we have hinted at various models that tie many functions of how cities evolve and are linked together, we have not yet introduced large-scale simulation, which enables this science to be tested in empirical and practical contexts. Nevertheless, from the time when digital computers were invented over half a century ago, the idea of simulating large-scale, extensive systems such as cities became prominent. Computer models of their form and structure as patterns of location and of interactions or flows providing the glue that binds various economic, social, and land use activities together were first proposed in the 1950s. The kinds of theory invoked then, on which the design of such models was predicated, was and still largely is based on the social physics of movement and potential that we presented, albeit briefly, in chapter 2. It embraced central place and location theory (Isard 1956; Fujita, Krugman, and Venables, 1999), transportation flows in analogy with gravitational models, and the urban economic rent theory underpinning the mechanisms used to develop land, housing, and various commercial activities that form the spatial structure of large cities (Alonso, 1964). We have not provided an exhaustive catalog of these theories in this book and nor shall we (although we hinted at this in the introduction to part II) but we will introduce enough of this material to give some credence to the kinds of urban simulation that constitute the focus of this chapter.

The Logic of Simulation

As we alluded to in the last chapter, simulation models of urban systems have evolved from aggregate representations of spatial interaction and locational activities that reflect the spatial structure of cities at a cross-section in time to much more detailed disaggregate dynamic models of individual behaviors that reflect much richer decision-making processes relating to individual spatial behavior and land development. This reflects the shift from land use–transportation interaction models to agent-based and cellular automata models similar to those briefly presented in the last chapter. Urban simulations have also moved from more thematic, abstract representations of activity and interactions between zones or census tracts to much more detailed point locations where buildings, land parcels, and streets form the focus, thus reflecting the shift from geographic to geometric representations. The first attempts at urban simulation models were developed half a century ago against a background of belief that good and accurate predictions could be made for systems as complex as cities regarding the impact of urban growth and new transportation infrastructure on their form and functioning. This experience was marred in that early models were judged to be either "too simple" or "not simple enough" to grapple with the daunting complexity of cities (Brewer, 1973; Lee, 1973); and so began a long period of reflection, extension, and reworking of model structures in the quest to make such models more applicable and relevant to policy making. Although progress has been made, many problems remain (Timmermans, 2006).

Two key themes have dominated model development since then. Urban models by their nature tend to treat the city system "comprehensively," and there has been a long line in developing models that treat ever-more detailed representations and functions, largely through disaggregating activities and adding new sectors. This has involved the representation of markets, which balance the simulated demand and supply of urban activities at ever-finer scales of disaggregation to the level of firms and households. Such models now fashion explicit links to quasi-independent transportation models or embed these models directly within their structure. They remain largely cross-sectional and equilibrium-seeking, but tend to simulate urban change between two or more points in time, often using combinations of model types ranging from spatial interaction to microsimulation. These models still appeal to the tradition of being "large-scale" in that they are complicated to set up, take time to run, and are the product of teams of analysts rather than the work of individuals. The reasons for their continued development revolve around the notion that the complexity or variety of any particular application requires equal complexity or variety in the model system, while policy makers usually demand the kind of detail that such models are able to supply. Good reviews of the state of the art are provided by Timmermans (2006), Hunt, Miller, and Kriger (2005), and Iacono, Levinson, and

El-Geneidy (2008). Typical of current developments is the series of models developed by Echenique (2004). It is important to note that these models differ quite massively from those introduced very briefly in the last chapter based on articulated physical development processes using ideas about automata, bottom-up decision processes, and nonequilibrium conditions in cities. These are more pedagogic than those introduced in this chapter, which are more testable and operational in focus.

Another contrast to this tradition of making simpler models more complex has been a less organized quest to develop models that are simpler than their predecessors. This has proceeded by decoupling submodels and developing these in more detail, or by fashioning different elements of comprehensive models into individual models that are used as elements in a tool box of techniques. Many planning support systems are constructed in this fashion (see Brail, 2008), although the quest to develop simpler models is dwarfed by the wider trend of extending models to embrace new developments in information technologies and better and richer sources of data. Perhaps a clearer way of impressing this difference between simplicity and complexity, between the small and the large scale, is to adopt Bankes's (1993) distinction between "consolidative" and "exploratory" models (or modeling styles). The consolidative style tends to focus on the construction of models that might ultimately provide accurate or focused predictions in contrast to exploratory models that will never do so, but are used to define salient characteristics and to "inform" the debate over particular problems. The large-scale tradition in urban modeling very definitely relies upon the former, while the notion of using simpler models over and over again, often modifying their structure in countless ways, accords to the latter, more exploratory viewpoint.

It is not our purpose here to provide a review of land use–transportation interaction models, sometimes called urban models, but to simply provide the reader with a glimpse of this tradition. Accordingly, the model we will present is clearly in the newer exploratory tradition, but it originates from earlier large-scale modeling efforts. It is based on simulating interactions from work to home through a residential location model with four transport modes, where the modes in question are based on road, heavy rail, tube and light rail, and bus networks. In this sense, it is another approach to dealing with coupled networks, though through the medium of their flows. It sits squarely in the tradition of aggregate spatial interaction modeling as a singly (origin), semi-destination-constrained model, in the jargon of chapter 2. Flows of workers to residential zones across the four competing modal networks determine the population that locates in each of the destination zones, and in this sense, it is both an interaction and a location model. This version was first developed as one stage in an integrated assessment of climate change for the London region based on a series of coupled models, beginning with a national-regional input-output model (Hall et al., 2009). These employment forecasts were then scaled to small areas of the urban region, feeding to the residential location model, the subject of this chapter. These residential populations were then reduced to an even finer spatial scale using a model reflecting physical constraints on land development as incorporated in GIS, reminiscent of more physically based urban development models in the cellular automata tradition introduced in the last chapter (Batty, 2009b). This enables these predictions of population to be tested with respect to flood risk derived from hydrological models associated with predicted sea level rises in the Thames and its estuary, consistent with forecasts from the UK Climate Impacts Programme for the next fifty and one hundred years (Dawson et al., 2009).

To illustrate these simulations, we will develop the model in a way that makes its use in assessing the impact of abrupt changes in energy and travel costs immediate, before one's very eyes, so to speak, and this will directly relate to how switches between different modal networks take place. This immediacy is a major requirement for communicating the modeling outcomes to a range of stakeholders who are nonexpert in the particular model design. To this end, our model system is visually-driven so that the greatest amount of information about the model and its predictions can be communicated as effectively as possible to diverse audiences. We first define a series of strict criteria that the model must meet before we outline its structure. We then examine the implicit dynamics of the model, which is a cross-sectional equilibrium structure, before presenting its derivation using entropy and utility-maximizing, drawing on tools introduced in chapter 2. This sets up the method for consistently calibrating, validating, and evaluating the model, where we focus on how the model handles energy use in the urban system. We then reach the position where we can examine the impact of abrupt and rapid change in energy costs, which are encapsulated in fast changes in interaction patterns through mode shift across the networks and much slower changes in location patterns reflecting redistribution of the population. We finally evaluate these changes using changes in the energy-entropy balance consistent with the model's structure, and this sets the context for generalizing this method for the rapid assessment of future scenarios.

Requirements for Urban Simulation

Rapid Execution and Visually Driven Predictions in the Dialogue with Stakeholders

The predictions this model addresses are posed as the outcomes to "what if" questions. These assume either an immediate reaction or impact on the system of interest, or a reaction that takes place over an unspecified time; that is, very short or very long time horizons for which the outcomes in either case assume the city system will adjust to some equilibrium state that is the focus of interest. Thus accurate predictions are not the goal of this kind of model, for the predictions may take many years to realize in terms of generating a long-term equilibrium. Perfect accuracy, in fact, can never occur due to unforeseen changes and adaptations that will take place on the trajectory toward this state. In this sense, the model predictions are designed to inform the debate and engender learning among stakeholders. As Epstein (2008) so cogently argues, this style of model can "discipline the dialogue about options and make unavoidable judgments more considered" (paragraph 1.7). In terms of the style of model developed, this is quite consistent with evaluating predictions for 50 or 100 years that embody radical impacts (such as those due to climate change), but as we shall see, the model is also capable of examining much more immediate change and its consequences.

The model must be capable of being used over and over again so that a rapid dialogue can be maintained between model builders and users. This puts an upper limit on the time required to run the model, which must be in an environment that generates predictions in a matter of seconds or at most minutes, suggesting a desktop or Web-based media in which outcomes can be communicated through visual analytics such as maps, bar charts, tree diagrams, flow networks, and so on, all of which might be represented in different dimensions and through various animations. Again, it is not our purpose in this book to particularly stress visualization, for we assume that the complexity of cities is such that state-of-the-art ways of communicating digital data and predictions should be routine, but these requirements are essential to a diverse community of stakeholders and we need to be clear about this. In using the model for integrated assessment, which consists of chaining different models together across different spatial and temporal scales, different kinds of expertise are required, and visual media makes it easier to communicate model structures and outcomes to other scientists involved in models that require different disciplinary and professional expertise. Our focus then extends to stakeholders, professionals involved in policy making of various sorts, from those trained in cognate professional and scientific disciplines but not involved in developing models per se, all the way through to policy analysts and advisors. Finally, we consider the more visual products used to present model outcomes should also be intelligible to those involved in the policy process who are nonexpert but at least informed, affected by, and instrumental in resolving the problems in question.

Models whose outcomes can be generated quickly and disseminated rapidly must also be capable of being reconfigured to embrace different features of the problem-solving context that become important during analysis. This suggests that these kinds of models should be modular in some sense. Although the model structure developed here is relatively simple, lacking any extensive modularity per se (although its structure can be easily replicated for other subsystems of the city system), the manner in which its outcomes can be communicated involves extensive modularity with respect to the tool kit of visual analytics available. Modularity is also essential in integrating different model types into sequences of predictions that are coupled over different spatial scales with different kinds of science being used at each stage, and this again reinforces the need for a common medium of communication between different models and model-builders. Visualization is by far the most effective medium in which to communicate different kinds of model outcomes, thus posing additional requirements about the need for rapid and quickly repeatable model runs that can be generated *in situ* in the presence of relevant scientists, decision makers, and stakeholders.

In terms of interpretations, outcomes and hence model structures should be intelligible enough to associate causes with effects. In complex models, predictions may be the result of emergent processes that, some argue, may not be explicable in terms of our knowledge of how the model works. But here we impose the requirement that whatever the outcomes, these must be traceable to changes in input values, notwithstanding counterintuitive impacts. In short, such counterintuition must be ultimately explicable in terms of the model's functioning, for the essence of informed debate is to recognize when and why such effects can occur. In terms of the role of the model here in integrated assessment, the organizational environment in which these kinds of models are developed is one that is highly fractured, with different scientific expertise located in different places. Thus the need to stitch models together, and even to assemble elements of the same models that are built in different places, requires common media of communication. This we believe is more likely to be the rule rather than the exception in applying our science of cities, since resource constraints and different expectations dictate the need for simplicity in design and communication. All this implies fast, simple, visual, and accessible models.

Dynamics and Comparative Statics: Equilibrium in Terms of Fast and Slow Change

The argument that cities should be treated as equilibrium structures is based on wide agreement that most cities display a similar generic spatial structure and morphology, notwithstanding distinct differences at more detailed levels. Such structures also appear to persist over decades and longer, giving power to Harris's (1970) point that such clear evidence of an equilibrium should provide the prime focus for simulation. This view does not reject the notion that urban change must also be a focus, but asserts that models that attempt to replicate urban form and structure must simulate this equilibrium prior to any additional features such models may take on. This is given added weight in that idealized plans for cities often redefine this equilibrium in ways quite different from that which is observed, and thus the first stage in understanding must be to simulate what already exists.

Models that do not reflect an explicit dynamics simulate what is observed at a cross-section in time and make the assumption that whenever a prediction is made, the outcomes from the model reflect the fact that the system will have moved to a new equilibrium within the given time period. Lowry (1964) in his model for Pittsburgh referred to this as an "instant metropolis" (p. 39) and suggested that forecasts made with such models must be seen "as 'quasi-predictions' of the emerging spatial structure" (p. iv). How appropriate this is depends on the time period in question as well as the processes of change involved. In contrast, the alternate view is that cities are forever in disequilibrium, and consequently simulation must focus not on replicating a static urban structure but on changes to this structure, thus reflecting dynamic processes that unfold in time and destroy any equilibrium that might be assumed to exist at a more aggregate level. In this respect, dynamic models tend to be more complex than cross-sectional, in that processes of change are integral to the model design.

Equilibrium models can deal with both short- and long-term change if the intricate dynamics of the way this change works itself out in the city system do not need to be explicit. Very long-term change over periods of fifty years or more as, for example, those that relate to climate change, lead to a new equilibrium that is clearly only one of a multitude of possible future states. The long-term outcome that is predicted is purely notional, in that the sheer scale of adaptation that would take place between the current and future prediction dates would be such as to destroy any idea that this outcome would ever take place. In these instances, forecasting with such models simply provides a perspective that "informs the debate" about the long-term future, which is the case with any radical change. Very short-term change, however, shows what might happen immediately if the prediction could be borne out, assuming no other constraints on the outcome. But this, too, is unlikely for there are many constraints that only become explicit when an outcome is emerging. Adaptation usually happens even in the very short term, and it is often unclear how this works itself out.

There is a distinction between slow, medium, and fast processes of change in urban systems first formally noted by Wegener, Gnad, and Vannahme (1986). The slowest changes relate to infrastructure, particularly transportation networks and the built environment; the medium to demographic, economic, and related processes; and the fastest to mobility, ranging from local migration to flows across many different scales and types of networks. This continuum can be further elaborated from changes in physical structures, including land use, which are slow, to changes in population and labor markets through redistribution and migration, which are faster. In terms of the spatial interaction-location models developed here, Wilson (2008) identifies fast change in interactions, in this case, the journey to work, which is a diurnal cycle, in contrast to population change in terms of the supply of housing, which is more likely to take place over years. Spatial interaction-location models are usually formulated in cross-sectional terms and when used in a predictive context, it is assumed that the flows generated begin to change immediately, while the ultimate locational redistribution takes longer to work itself out. In fact, this process of working out is implicit and the ultimate equilibrium that occurs is a product of both fast and slow processes, which have no explicit time scale. The assumption is that the predicted outcome would take place if all other conditions were kept the same, thus representing an ultimate steady state that would only occur under the idealized conditions of no other change. As Wegener, Gnad, and Vannahme (1986) so persuasively note, buried within this mix of slow and fast processes are strong contradictions as to the best ways in which these intricate processes might be modeled.

If the model is constructed for simulating changes in the demand for interaction and location, then the new equilibrium that results is one that assumes demand is met with entirely elastic supply. We know that this will never be the case in real systems, and this is thus another way in which predictions made with such equilibrium models represent an idealized future state. In most instances, changes in demand will be moderated by supply, and the ultimate equilibrium will be composed of a complex process of demand and supply adapting to one another and to other exogenous constraints. It is in this sense, then, that predictions with this model are to be used in wider processes of planning support to inform the debate and to pose immediate answers to "what if" types of questions. To this end, we require fast and accessible models of the kind that we will now describe.

The Generic Location Model

Specification and Derivation Using Entropy Maximization

We first cast the model in the most parsimonious form possible, explaining flows between workplaces (origins) and residential areas (destinations) as a function of strictly physical quantities. In short, the variables we wish to model—flows (or trips)—are measured in terms of persons, but we explain them entirely with respect to physical quantities that are determined by the size and scale of the system itself. This is phrased in terms of the technological limits on how people are able to interact, which relate ultimately to the geometry of the system, albeit expressed in units of cost of travel, and in terms of the land area associated with these flows. The model is in the tradition of spatial interaction introduced in chapter 2 but will be expressed in an explicit energetic framework, consistent with its application here.

Flows defined as T_{ij}^k are movements from origin zones i, 1, 2, ... , I to destination zones j, 1, 2, ... , J with respect to the mode of travel k, 1, 2, ... , K. The numbers of zones is in the hundreds, here 633 for both origins and destinations, in contrast to a handful of modes, four in all comprising road, heavy rail, tube and light rail, and bus. We will derive the model in terms of the density of trips T_{ij}^k / A_j destined for a particular zone j with residential land area A_j, but it will be expressed in terms of trip volumes. The model will be subject to two physical constraints, the first based on the total cost of travel by each mode C^k, defined as

$$\sum_i \sum_j T_{ij}^k c_{ij}^k = C^k, \tag{9.1}$$

where c_{ij}^k is the energy expended, measured in terms of travel costs using the modal technology k associated with the network k used to move between i to j. The second constraint is on the origin activity measured as the number of jobs E_i, which provides the overall dimensioning of person activities in the system:

$$\sum_j \sum_k T_{ij}^k = E_i. \tag{9.2}$$

The total number of trips in the system T is fixed implicitly by equation (9.2), which can be written as

$$\sum_i \sum_j \sum_k T_{ij}^k = \sum_i E_i = T. \tag{9.3}$$

It is worth noting the particular structure of this model. The modal costs in equation (9.1) are constrained so that each mode is distinct in terms of the energy it uses, whereas this is not the case when the trips are summed across modes with respect to their origins. This implies that the model simulates competition between modes, an essential criterion for handling switches between the modal networks. As the basic model is a singly or origin-constrained spatial interaction model, besides the flow matrix, the main predictor from the model is activity destined for each residential location, which is working population P_j, derived as

$$\sum_i \sum_k T_{ij}^k = P_j, \tag{9.4}$$

but other volumes can be predicted such as employment and population by mode at origins and destinations.

To derive the model, we follow the well-established method of defining and maximizing the entropy S of the distribution associated with $\{T_{ij}^k\}$. In fact, we use a more consistent definition for entropy than that used originally by Wilson (1970), which is the discrete approximation to the continuous form proposed by Batty (1974b, 2010), given as

$$\begin{aligned} S &= -\sum_i \sum_j \sum_k T_{ij}^k \log \frac{T_{ij}^k}{A_j} \\ &= -\sum_i \sum_j \sum_k T_{ij}^k \log T_{ij}^k + \sum_i \sum_j \sum_k T_{ij}^k \log A_j \end{aligned} \tag{9.5}$$

We next perform a maximization of this entropy by forming a Lagrangian L using equations (9.1), (9.2), and (9.3) as constraints on its form. Then

$$L = -\sum_i \sum_j \sum_k T_{ij}^k \log T_{ij}^k + \sum_i \sum_j \sum_k T_{ij}^k \log A_j$$
$$+ \sum_i \lambda_i \sum_j \sum_k \{T_{ij}^k - E_i\} + \lambda \sum_i \sum_j \sum_k \{T_{ij}^k c_{ij}^k - C^k\}' \tag{9.6}$$

which we set equal to zero for the maximization condition. This leads to

$$\frac{\partial L}{\partial T_{ij}^k} = -\log T_{ij}^k - 1 + \log A_j - \lambda_i + \lambda^k c_{ij}^k = 0, \tag{9.7}$$

from which the model can be easily derived. Note that we can incorporate the -1 term in the multipliers λ_i without loss of generality. We will do so, but will not define new variables. The model can be stated first in log form and then in its normal form as

$$\left.\begin{array}{l} \log T_{ij}^k = -\lambda_i + \log A_j - \lambda^k c_{ij}^k \\ T_{ij}^k = \exp(-\lambda_i) A_j \exp(-\lambda^k c_{ij}^k) \end{array}\right\}. \tag{9.8}$$

There are two properties that are worth noting. First, we can produce an interesting form for the normalizing equation (9.2) if we substitute the model in equation (9.8) into this constraint. From this we derive the value of λ_i as

$$\lambda_i = \log\left\{\frac{\sum_j \sum_k A_j \exp(-\lambda^k c_{ij}^k)}{E_i}\right\}, \tag{9.9}$$

which is a log sum accessibility that appears extensively as a measure of benefit in consumer analysis, particularly related to transportation. As we shall see, this is related directly to the free energy in the system. An equivalent expression, however, cannot be easily produced for the modal cost parameters λ^k. Second, if we compare any two modes as modeled from equation (9.8), then these produce a particularly simple form of competition. Taking the ratio of the relevant model equations for, say, $k = 1$ and $k = 2$, then

$$\frac{T_{ij}^{k=1}}{T_{ij}^{k=2}} = \frac{\exp(-\lambda^{k=1} c_{ij}^{k=1})}{\exp(-\lambda^{k=2} c_{ij}^{k=2})}, \tag{9.10}$$

and this implies that the modal split in logarithmic form is a direct function of the ratio of the relevant costs of travel $\lambda^{k=1} c_{ij}^{k=1} / \lambda^{k=2} c_{ij}^{k=2}$. These issues are important here, for one of the main developments of this model is for purposes of comparing changes in energy costs by mode c_{ij}^k.

There is one last point that will change the detail of these equations, but we will not show this formally here. In versions of the model, constraints on destination activities—in short, on population densities—have been imposed. This turns the model from an origin into an origin-semi-destination constrained model in which the following constraint is imposed,

$$\sum_i \sum_k T_{ij}^k \leq P_j^{\max}, \tag{9.11}$$

where $P_j^{\max}$ is the maximum residential population allowed in zone j. Only a subset of zones are so constrained, for in many, this constraint is purely notional in that there is so much space in some zones that it is unlikely the constraint would be breached. However, in inner cities and areas of denser population, equation (9.11) can be critical. If this is breached, a new parameter λ_j must to be introduced to ensure equation (9.11) is met, and then the model needs to be solved and iterated in a somewhat different fashion. Many of the above and subsequent equations would need to be modified to account for this new constraint, but in the example that follows, we report the case of the pure singly constrained model. An extended version of the model is currently being built that incorporates these constraints and much else, to which the reader is referred (Batty et al., 2011).

Calibration, Validation, and Evaluation

There are several ways of determining the parameter values, all of which revolve around the fact that the two key constraints in equations (9.1) and (9.2) must be met. We first write the model in its full form, making explicit the origin constraint as

$$T_{ij}^k = E_i \frac{A_j \exp(-\lambda^k c_{ij}^k)}{\sum_j A_j \sum_k \exp(-\lambda^k c_{ij}^k)}, \tag{9.12}$$

from which it is easy to see we could find the modal parameters λ^k starting with some reasonable estimates such as $\lambda^k = 1.5/C^k$, and then checking how close the predicted value of these costs are to the observed. The iteration is then performed by changing the values of these parameters with respect to the differences between the predicted and observed costs until convergence. This is akin to solution of the model using maximum likelihood or by actually maximizing the entropy equation directly. In some respects, this is simply dimensioning the model to the data, in that no attempt is made to actually optimize the goodness of fit other than ensuring that the total travel costs for each mode generated by the model are the same as those that are observed (for an extended discussion of these issues, see Batty, 1976).

There are a number of indicators that can be generated from the model that pertain to the energy used in spatial interaction. First we will write the entropy from equation (9.5) by substituting equation (9.8) into the standard form, which results in the following simplifications:

$$
\begin{aligned}
S &= -\sum_i \sum_j \sum_k T_{ij}^k (-\lambda_i + \log A_j - \lambda^k c_{ij}^k) + \sum_i \sum_j \sum_k T_{ij}^k \log A_j \\
&= \sum_i \lambda_i E_i + \sum_k \lambda^k C^k \\
&= \log \sum_j \sum_k A_j \exp(-\lambda^k c_{ij}^k) + \sum_k \lambda^k C^k
\end{aligned}
\qquad (9.13)
$$

In equation (9.13), it is clear that the land area term cancels from the maximization, indicating that the way this is introduced into the equation is purely for purposes of dimensioning the distribution. Entropy S in generic terms has a structure that associates it with unusable energy in the system, and in this context it is usually assumed to be equal to the actual energy C less the free energy F. In fact, equation (9.13) is not quite in this form, since the values of the parameters from the maximization are assumed to be negative, and of course in terms of the normalizing constraint on origins, this is likely to be positive. Hence equation (9.13) might be interpreted as $S = -F + C$, from which it is clear that total energy is $C = S + F$, unusable energy plus free energy (Atkins, 1994). This links to more formal analogies between urban structure and statistical thermodynamics (Wilson, 2009; Morphet 2010).

The real value in thinking in terms of different measures of energy becomes significant when changes in the input variables—specifically costs and employment—are made. Note that changes in land supply have no effect here, but if land area is instrumental in bringing energy benefits or costs to bear on the system, then these need to be formulated as constraints in the manner of equations (9.1) and (9.2). If we assume only changes in travel costs—let us say each travel cost for each mode can change by an increment or decrement Δ_{ij}^k as $c_{ij}^k(2) = c_{ij}^k(1) + \Delta_{ij}^k$—then we can compute the change in entropy as a change between free and actual energy; that is, $\Delta S = \Delta C - \Delta F$. Using the above definitions, the actual computation can be written as

$$
\left.
\begin{aligned}
\Delta S &= S(2) - S(1) \\
&= \log \sum_i \sum_j \sum_k A_j \exp(-\lambda^k c_{ij}^k(2)) \\
&\quad - \log \sum_i \sum_j \sum_k A_j \exp(-\lambda^k c_{ij}^k(1)) + \sum_k \lambda^k C^k(2) - \sum_k \lambda^k C^k(1)
\end{aligned}
\right\},
\qquad (9.14)
$$

where equation (9.14) can be further simplified to

$$\Delta S = \log \frac{\sum_i \sum_j \sum_k A_j \exp(-\lambda^k [c_{ij}^k + \Delta_{ij}^k])}{\sum_i \sum_j \sum_k A_j \exp(-\lambda^k c_{ij}^k)} + \sum_k \lambda^k \left\{ \frac{\sum_i \sum_j T_{ij}^k(2)(c_{ij}^k + \Delta_{ij}^k)}{\sum_i \sum_j T_{ij}^k(1)c_{ij}^k} \right\}. \tag{9.15}$$

The free energy term, which is the first on the right-hand side of equation (9.15), is reminiscent of consumer surplus, and there is a degree of intuitive sense in this derivation. It might even be possible to simplify these measures further, but in this form they are probably most useful for our analysis. The crucial issue is to examine actual changes in the three overall measures—ΔS entropy, ΔF free energy, and ΔC actual energy—and all of these measures will be used in the subsequent analysis to show how changes in travel costs have repercussions on accessibility in terms of their benefits as well as the total amount of energy used or distance traveled.

This formulation, which expresses energy in two varieties—"unusable," which is entropy S, and "usable," which is free energy F—is the usual thermodynamic interpretation (Atkins, 1994). In this context, if we increase energy C by adding to transport cost—where it is assumed the energy is "imported" from outside the system, then this will be distributed between usable and unusable and both may increase, since trips are forced onto lower cost modes, but these lower costs in total might be much greater than the costs incurred on the changed mode. It is, however, their relative distribution that is important. If entropy remains the same but energy increases, then this means that free energy increases at the same rate as total energy and the relative distribution of flows does not change. If entropy decreases, then this means the system becomes more concentrated, and thus we can examine the different components of this entropy and free energy as the system changes. When we examine these changes later in the chapter, we need to note that increases in energy will be measured in total trips. If, for example, total costs of energy double, then the average costs will not react in the same way, because the model will redistribute individual flows that are associated with these costs. We thus need to be very careful when interpreting these measures in subsequent discussion.

The Visually Driven Interface

Principles for Visualization

The general principle we ascribe to here is to put as much information used and generated by the model as possible into the display device used to communicate the model's data and predictions as well as its implementation for the user. In this context, our display device is essentially the desktop, possibly the desktop linked to the Internet through a web browser, as well as more conventional media, such as paper products. There are three key stages in the model-building process that most simulation models are constructed around, and these loosely follow the traditional scientific method: first, exploration of the model's input data, which reflects the way the system of interest is articulated in terms of the model, this corresponding to the inductive phase of analysis; second, the calibration or fine-tuning of the model to this data (as well as its interpretation through validation and verification), which is akin to experimentally validating the model; and third, the generation of model outcomes as predictions. In each of these three stages, which drive the process forward in sequential fashion, the same graphics tools are used to display information visually in the form of maps, flows, networks, tree diagrams, and so on, as we illustrate below. This is a similar process to that we have developed before using more primitive graphics (Batty, 1983; Batty, 1992).

The interface is organized as one main window controlling key operations and displaying key outputs and two different toolbars; the main toolbar strings each stage of the modeling process together in an implied sequence of progression from data to prediction, and a second toolbar launches at each stage when graphical outputs of various sorts are required. The main toolbar begins with data input, normalization, and then exploration. The last of these launches the second toolbar for display, which is central to the process of exploratory spatial data analysis. Then the particular model variant can be chosen, and this leads immediately to a window that is launched in which the model is fine-tuned or calibrated to observed statistics. This is followed immediately by the second toolbar, from which the model outcomes at calibration can be explored visually in terms of their goodness of fit. Finally, predictions with the model can be activated from the main toolbar: scenarios can be either imported from file or constructed on the fly, which involves altering locational data concerning employment, floorspace, network characteristics, and travel costs for the various modes on the desktop. Once the scenario is built, predictions are generated, and then these can be once again explored through the second toolbar, which provides similar graphical display capability as at the data input and calibration stages.

The graphics tools accessed through the second toolbar mainly display media in the form of two-dimensional thematic maps; maps that produce "desire lines" recording interaction from origins and destinations in proportion to their flow volumes; histograms or bar maps showing activity volumes; tree maps that display the hierarchy of activities in proportion to their volume in small areas and the next level of hierarchy, which in this case is the London boroughs; and scatter graphs of trips and travel costs. All these data can be displayed as either counts or densities, and there are several derived maps that are built from comparing one activity to another in ratio form. To enrich the analysis by comparing model data and predictions with spatial data that cannot be imported into the model, we have enabled all the data produced by the model in map, flow, or histogram form to be exported on the fly to Google Earth, where it can be compared against network data such as roads and rail lines and a variety of raster-based data such as topographic and climate layers. Data is exported to Google Earth in XML format, which enables KML files to be constructed from the vector data the model generates. This is accomplished in real time when the model is running. The user can thus display data in 3D, flying through it to gain a rich and detailed impression of how the data used and generated by the model compares to other features of the region input to Google Earth.

Figure 9.1 illustrates the basic template—the main window and the two toolbars, where the window in this case is the panel that displays the data input. From this panel the user can explore the data numerically, query the map for the location of zones and higher-level units, and get some sense of the correctness and dimensionality of the data in context. On launching the model, the splash screen first occupies

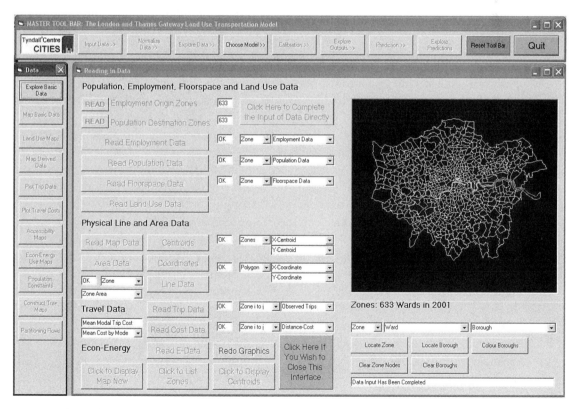

Figure 9.1
Windows comprising the basic interactive model template.

this window, and once the data input has been displayed, a window controlling the normalization of the model data and thence the choice of the model and its calibration are launched within the template, always leaving the location map to the right on screen to keep the user oriented. Once calibration is finished, the main window is refreshed to enter the stage of defining and thence generating predictions, which we illustrate in the next section.

Exploratory Data Analysis, Calibration, and the Generation of Scenarios
The main toolbar drives and directs the user to the sequence of stages defining the model-building process from data through to prediction, but the second toolbar controls the graphics and is launched at each of the three key stages. We will examine the tools that are available at the data analysis stage, but these in fact are similar to those used for the model's calibrated predictions and scenario predictions at the second and third stages. The toolbar contains ten key display types, seven of which are maps of various kinds: histograms showing activity volumes by location, thematic maps showing the same volumes by area, and flow maps showing different flow volumes from any origin to all destinations or vice versa. These displays can also be queried with respect to individual locations and individual flows in interactive form, although for the most part, the data is completely mapped each time a map is drawn. One key distinction is between count and density data, so, for example, population $\{P_j\}$ is plotted as an absolute count, which can then be compared against its density $\{P_j/A_j\}$, where A_j is land area devoted to residential use in zone j. All other variables can be so defined as densities from the land use area data that is available for different land use types.

The first button enables the user to query individual location and flow data for population, employment (densities and counts), and trip data $\{T_{ij}^k\}$ from each origin or destination over any mode of travel k. The second button enables the user to plot these as complete maps, which can then be exported for display within Google Earth. Buttons displaying each land use as a thematic map, and then derived data (for example, activity rates such as P_j/E_j) follow, and then the user can plot trip data such as $\{T_{ij}^k/E_iP_j\}$ against travel costs $\{c_{ij}^k\}$ as scatter graphs. Buttons activating maps as cost surfaces from any origin to all destinations for any mode and vice versa, followed by detailed accessibility surface maps based on potential, consumer surplus, and related indices, again for specific origins or destinations by mode, can then be displayed. The final maps reflect wages, as well as housing price data not utilized in any of the applications here but built into extensions of the model as equivalents to trip flows, and then population constraints are displayed in terms of land availability. Last but not least, we let the user plot any of this location data as tree maps, which effectively represent the volumes of activity in each zone tagged to each higher-level unit—the borough in this case—in terms of proportional rectangles.

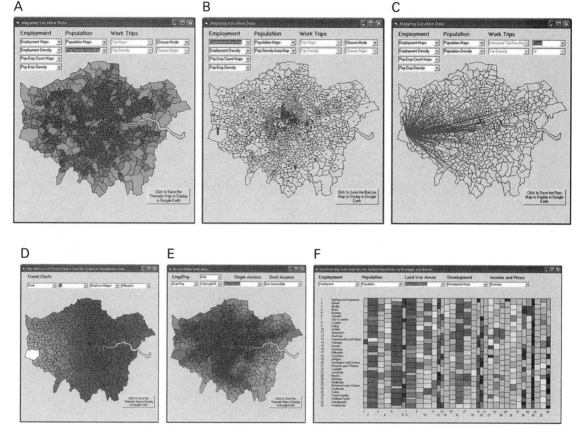

Figure 9.2
Small multiples of graphic output from exploration of the model data. (a) Population density. (b) Employment counts. (c) Road trips from zone 6, Heathrow. (d) Travel costs from Heathrow. (e) Accessibility potential on the tube system. (f) Tree map showing residential areas at borough and ward levels.

In figure 9.2, we present a small collage of these displays showing population density, employment counts, road trips from the western airport in zone 6, travel costs, accessibilities, and the tree map for residential land area. We use these maps to learn about the region in terms of its structure, which, from figures 9.2(a) and 9.2(b), is strongly monocentric with respect to employment densities and counts. Note how the congestion charge zone is picked out in terms of the road accessibility from the airport but is not featured in any way in the travel cost map from the city center in terms of the tube and light rail network. To supplement this visualization, for a number of these layers we can export the input data to Google Earth. In figure 9.3, we show how we can plot employment counts as histograms, population density

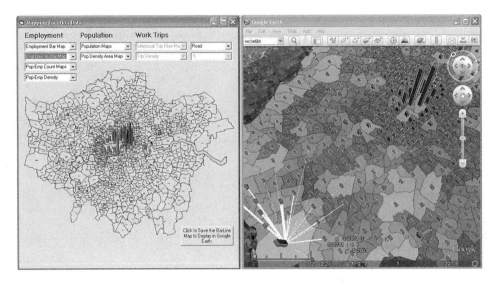

Figure 9.3
Visualizing thematic map layers, flows, and histograms using Google Earth as an external viewer linked to the desktop interface.

in the thematic map layer, and the flow data over the road from zone 6 in 3D, successively updating this visualization as we continue to generate the data from various displays produced from the model. There is much else we can explore using this capability, but figure 9.3 gives some sense of this potential.

We can initiate the same learning cycle with respect to comparing the predictions from the calibration to the data using similar map layers as well as direct comparisons of deviations between observed and predicted activities. Similar displays are available for exploring the impact of various activity and interaction-network changes. However, in driving the process forward, we first need to normalize the travel cost data to relate it to other costs, and then we have the option of choosing the attractor for the model, which is a function of the land area A_j. In the model demonstrated here, we simply use the raw variable as implied in equations (9.8) and (9.12) above. We initiate the calibration using an iterative method to ensure that the trip lengths in equations (9.1) are reproduced, and we use a damped Newton-Raphson (hill-climbing) method to ensure that this convergence takes place as quickly as possible (see Batty, 1976). In terms of the fit of the model we report here, this is quite modest, since some 62% of the variation in the population and 43% in terms of the overall trip matrix are explained. As the purpose here is to solely introduce a typical framework for urban simulation and not comment on the substantive results, we show the simplest possible unconstrained version of the model,

A   B   C

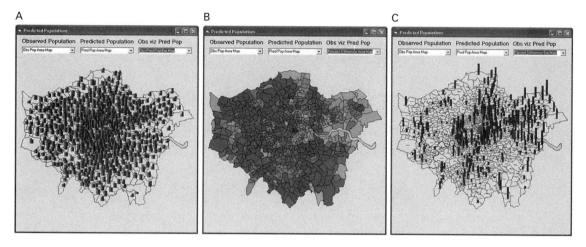

Figure 9.4
Predictions of residential population from the model. (a) Predicted and observed populations. (b) Percentage differences (gray overprediction, black underprediction). (c) Bar graph of percentage differences (gray, over; black, under).

not the one that generates the best goodness of fit, which occurs when the model is extended to deal with wages, prices, and residential land use constraints, which we note later. In figure 9.4, we show the typical fit of the model in terms of population counts and densities.

The last stage in this process involves testing the impact of scenarios. These are framed in terms of changes to the input variables—employment, travel costs, land area constraints, and so on. Users can import new data files that contain these scenarios or can develop them directly in the visual interface screen activated in the main window. We show one such screen in figure 9.5, which illustrates how the user can add to the network by drawing a new transport route—in this case, a line from the western airport to one of the central rail stations, which when input to the model enables the shortest routes to be recalculated for that mode. Many such changes defining a scenario can be input for any set of locations and modes, and in this way, the future that is to be tested is assembled. We are then able to generate the relevant predictions and explore these using the second toolbar, which activates the graphics. We now have three possible sets of comparisons to make: between observed and calibrated data, between observed and scenario predictions, and between calibrated and scenario predictions. In fact, the appropriate comparison is the latter one since any errors introduced by the model need to be factored in, so it is wiser to compare calibrated rather than observed with predicted when generating some measure of change or impact.

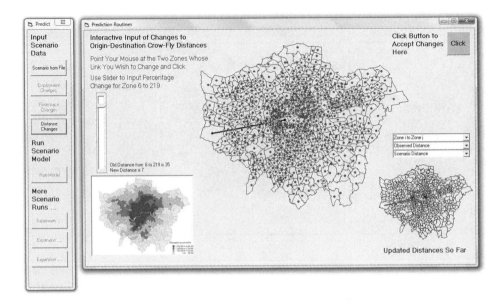

Figure 9.5
Building scenarios on the fly: Inputting a new heavy rail line from the western airport at Heathrow to the West End of the central business district.

Slow and Fast Change: The Impact of Urban Energy Costs

Although this genus of model essentially simulates a world in equilibrium, using the model to predict future change implies a variety of dynamics that are assumed to work themselves out completely by the time the new equilibrium is established. Lowry's (1964) original idea of the "instant metropolis" was predicated on the basis that contained within the existing equilibrium were emergent structures that would be revealed when the model was calibrated, but would only truly show themselves when predictions to a future state were made using the model. Thus the notion of comparing the calibrated against a future state would be one of comparing the implied equilibrium of the present (not the same as the actual present) with a future equilibrium once new changes embodied in the scenario had worked themselves out. The meaning of these predictions in terms of simulating relevant changes thus turn on the processes that are implicit in the internal dynamics of such models.

There are clearly an array of different dynamics that are involved in any changes in location and trip-making decisions. In terms of trip-making, if costs increase, the response is likely to be rapid in that trip-makers will switch to lower-cost network modes. If the capacity of the mode that appears more cost-effective is limited, switching to such a mode might increase congestion, thereby increasing costs to the point where the original mode switch is discounted. This process might take time to work itself out, but it is likely to be a lot faster than other types of locational change. Changing locations with response to such costs clearly takes longer as there can be no immediacy in making a residential move, and it is likely that changes in jobs (which are not part of this model) have a longer time dynamics. The real issue is those effects that are second order, third order, and so on, which are in fact longer-term adaptations that we know little or nothing about because they are often obfuscated by other changes.

We can strictly differentiate here between changes in modal split and changes in location. Changes in both, however, ultimately translate themselves into changing built infrastructures, which tend to be slow in contrast to fast changes that involve people using the same infrastructure but in different ways, generating different volumes of activity (Wegener, Gnad, and Vannahme, 1986). For example, our model was originally designed to examine the impact of very long-term changes in climate, specifically rises in sea level on the locational pattern of the population in Greater London over the next 100 years. There will be substantial adaptation to and indeed mitigation of these effects over this period which would clearly lead to a future state very different from the equilibrium that this model would predict.

In fact, the equilibrium model is important so that the changes that will inevitably occur can be filtered out. Most urban simulation models of this genus are classic "what-if" types of instruments in that they are used to pose and answer questions of the kind, "Assuming everything else remains the same and X changes, what is the effect on the system of interest?" In short, the model can be used to generate the causal chains that are exposed by this usage in terms of definite and explicit impacts, in this case on interaction and location. Although the model can define how much change is due to interaction versus location, the balance in fact cannot be attributable to anything other than the structure of the entire model, which contains many such causal chains. Changes in infrastructure, however, are harder to gauge because individuals switching transport mode, route, and location in response to such changes clearly take place over much longer intervals. Our key example here, in fact, will be changes in cost, not infrastructure, and we will examine one from many such possibilities. What we will do here is examine the impact of a doubling of the cost of road transport relative to all other modes: that is, we will increase the cost of road travel uniformly over the system, doubling the unit cost and keeping the costs of the other three modes—heavy rail, tube and light rail, and bus—constant. The residual mode, which caters to all other users, remains as a residual to the analysis, for it is not formally modeled. The change from state (1) to state (2) in transport cost for the car mode $k = 1$ is written as $c_{ij}^{k=1}(2) = 2c_{ij}^{k=1}(1) = c_{ij}^{k=1}(1) + c_{ij}^{k=1}(1)$. If we substitute this into the modal split comparator equation (9.10), we can see that the relative shift in trips between any mode is a simple function of the previous time step, that is

$$\frac{T_{ij}^{k=1}(2)}{T_{ij}^{k\neq1}(2)} = \frac{T_{ij}^{k=1}(1)}{T_{ij}^{k\neq1}(1)} \exp[-\lambda^{k=1} c_{ij}^{k=1}(1)], \tag{9.16}$$

where it is clear that as the cost gets greater for the mode with the increased cost, the percent shift gets greater. This is logical, given that trips decline exponentially with respect to travel costs.

When we double costs in this manner, the model predicts shifts across all network modes as travelers seek to travel on more cost-effective routes, and since the model is singly constrained, there will also be shifts with respect to their residential locations. There are two key indicators—first, the total average travel costs, which we would expect to rise for road travel (although modal switches are likely to mean these road costs will not rise by the exogenous change of 100%), and second, modal split. These statistics are presented in table 9.1, where it is clear that the overall average trip costs rise by 17%, of which by far the largest component of this is the increase in road trip costs by 27%. Rail and tube only rise between 2% and 3%, while the average cost of bus travel drops by slightly less than 2%. These changes are almost the inverse of shifts in modal split, where car ridership decreases by 46%, in contrast to bus transport, which increases by some 42%. Heavy rail and tube ridership also increase substantially by 35% and 21%, respectively. Such large shifts might also be expected to be associated with shifts in residential location activity, which we will now examine. In reality, we must note that such large shifts would not actually occur, for they would require major increases in rail infrastructure and a massive extension to the bus fleet. However, they are indicative of the pressures in the system, and in this sense, quite consistent with the idea of using equilibrium models to explore such future possibilities.

Table 9.1
Changes in average trip costs and modal split

Mode	Observed mean trip cost[†]	Predicted mean trip cost	Percent difference	Observed modal share	Predicted modal share	Percent difference
Road	38.668	49.157	27.124	0.389	0.210	−45.899
Heavy rail	77.780	79.591	2.328	0.122	0.165	34.997
Tube	59.662	61.196	2.570	0.331	0.400	20.988
Bus	14.659	14.428	−1.576	0.158	0.224	41.926
All modes	47.600	55.621	16.851			

[†]The observed mean trip costs and modal shares are the same as the calibrated values, for the model is calibrated to meet these constraints.

Changes in trip volumes between the existing and new states (1) and (2) lead directly to changes in activity at residential destinations through equation (9.4). In difference terms, these changes can be portrayed as

$$\sum_i \sum_k T_{ij}^k(2) - \sum_i \sum_k T_{ij}^k(1) = P_j(2) - P_j(1), \tag{9.17}$$

where it is clear that as the total number of trips is conserved by definition as $\Sigma_{ijk} T_{ij}^k(2) = \Sigma_{ijk} T_{ij}^k(1)$, then the sum of the differences between residential activities across all locations is zero, that is

$$\sum_j P_j(2) - \sum_j P_j(1) = 0. \tag{9.18}$$

In short, changes in costs simply lead to a redistribution of existing activities, and the new equilibrium predicted by the model—which is clearly a first-order equilibrium, in that there is nothing in the model to predict further-order effects—is composed of these locational changes and the shifts in mode split shown in table 9.1.

We can compute two rather graphic illustrations of these locational shifts. First, we compute the absolute proportion of all population moving as

$$\Phi = 100 \frac{\sum_j |P_j(2) - P_j(1)|}{\sum_j P_j(1)}, \tag{9.19}$$

and we can also partition the system at any point into two sets of zones called Z_1 and Z_2, where the entire set of zones is $Z = Z_1 \cup Z_2$:

$$\sum_{j \in Z_1} [P_j(2) - P_j(1)] = -\sum_{j \in Z_2} [P_j(2) - P_j(1)]. \tag{9.20}$$

Equation (9.20) means in any partition of the system into any two sets of zones, we can examine the flow from one subset to another. In this way, we can examine if the effects of locational constraints are spatially biased toward any locations in the system, specifically in this case either toward the inner city or the outer suburbs, which would reflect major differences in terms of car usage (Batty, Hall, and Starkie, 1974).

The most surprising prediction from the model is that the percentage of the working population shifting residential locations Φ is only 2.4%, which involves some 110,736 persons. This is extremely low, and it is a measure of the resilience of the system to changing transport costs. However, overall costs might rise substantially, as we will show below, but actual second-order costs due to potential shifts in residential location are likely to be much less than expected. To an extent, this result is simply indicative of the fact that there are many more degrees of

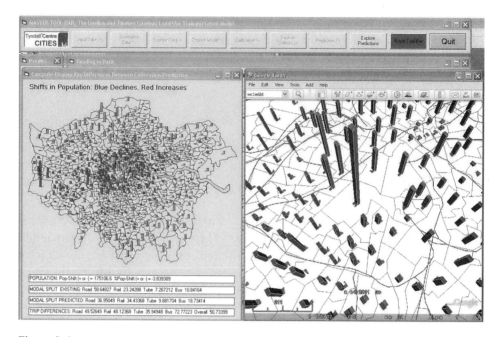

Figure 9.6
The impact of a doubling of road travel costs for private car on location (gray, increase in population; black, decrease in population).

freedom with respect to potential changes in interactions than in locations. Figure 9.6 shows the pattern of these shifts in location, from which it is clear that the city tends to compact slightly with loss of population from many of the suburban areas all around the city with the exception of the relatively vibrant western corridor, which attracts some population. This is probably due to the configuration of employment in the west, in and around the airport, and the relatively prosperous belt of commuters with good access to transport infrastructure in west and southwest London. In figure 9.7, by trial and error we show a noncontiguous partition of the system that leads to a high flow of population across these boundaries. The user can choose any partition by clicking on the zones at the fine or broad scale—wards or boroughs—and in terms of the wards, there are 633! possible combinations to consider. Clearly, some intuition about the workings of the model and the structure of the spatial system is required to use this tool.

The energy equations introduced earlier also provide a useful, if somewhat oblique, perspective on these results. The units in which energy and entropy are measured bear no resemblance to the units in which the data for the model is input, which is minutes of travel time. This is because entropy measures in equations (9.13) to (9.15) are computed in terms of total trips, not probabilities, as formulated in

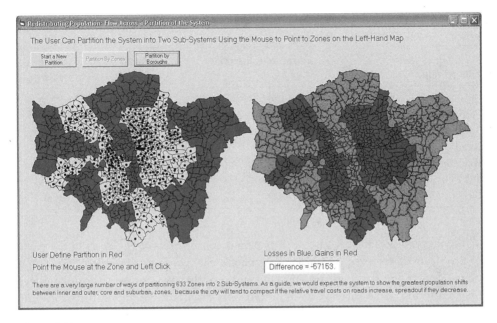

Figure 9.7
A noncontiguous partition of the system leading to population relocation (loss of population in black, gain of population in gray).

traditional versions of entropy-maximizing derivations of these models (Wilson, 1970). Moreover, as equation (9.13) makes clear, travel costs by mode are normalized by the appropriate travel parameter and then summed to produce a composite cost. However, the relative weighting of these measures gives some sense in which the system changes from the first state (1) to the scenario state (2). We show all these values in table 9.2, where the percent changes are indicative.

The critical issue is that the total energy in cost terms massively increases due to the external imposition of the 100% change in road costs, leading to an equivalent increase in free energy while the entropy increases only slightly, which implies a slight decrease in the concentration of all trips across the system. The scale of these increases in total energy and in its free energy component is largely due to the fact that when the road costs increase by 100%, there is a dramatic redistribution of trips onto other modes, which increase total trip costs on these modes massively. Clearly the travel costs by mode sum to the total costs, and this can be seen in the disaggregations. We have not explored the disaggregation of the entropy equation by different modes, because this is distorted by the fact that the free-energy equation cannot be so broken up, since it contains the coupling mechanism needed to ensure the model acts as one. However, this is an emerging and active area of research, and

Table 9.2
Changes in entropy, energy, and costs

Energy value	Calibrated-observed state (1)	Scenario state (2)	Percent change
Entropy S	4550379	4551015	0.014
Free energy F	9243178	50642320	448
Total trip costs $\Sigma_{ijk} T_{ij}^k c_{ij}^k$	4692799	46091305	882
Road costs $\lambda^1 C^1$	3657600	1654163	−55
Rail costs $\lambda^2 C^2$	3142503	6351686	102
Tube costs $\lambda^3 C^3$	1343076	10268510	665
Bus costs $\lambda^4 C^4$	1100000	32367970	2843

all the more important because of our current concern for energy costs in the wake of problems of resource depletion and climate change. The substantive interpretations of energy in these entropy-maximizing models have remained dormant since their inception some forty years ago, and only now is there any effort to ground these concepts in real measurements (Batty, 2010). Various interpretations of the values in table 9.2 are possible, and the reader is referred to the recent papers by Wilson (2009) and Morphet (2010). A science of cities should, of course, point the way to how cities transform energy spatially and how the morphology of cities, in terms of their urban form and networks, enables energy to be distributed with respect to its costs. In this sense, the ideas of this chapter are suggestive of a way forward.

To complete the picture, we will examine the changes in accessibility occasioned by this 100% rise in cost of travel by road. In figure 9.8, we show changes in accessibility based on computing the standard log sum term, which is the first component of the entropy in equation (9.13) and apply this to the accessibility of origins, meaning that this is accessibility to the location of employment. We can restate the change equation for each destination as follows,

$$\Delta V_i^k = \log \frac{\sum_j A_j \exp[-\lambda^k c_{ij}^k(2)]}{\sum_j A_j \exp[-\lambda^k c_{ij}^k(1)]}, \tag{9.21}$$

but it is clear that no changes are communicated from mode to mode through this form of accessibility: for three of the modes, the travel costs are the same for both the before-(1) and after-(2) states. However, we can examine changes in the road accessibility, and in figure 9.8, we show the before, after, and ratio of these two sets of accessibilities as computed from equation (9.21) with $k = 1$. These two

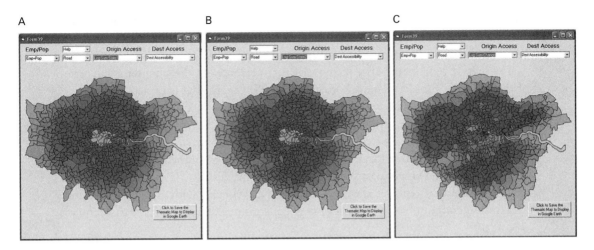

Figure 9.8
(a) Before and (b) after accessibility by road, and (c) their ratio.

accessibility surfaces are mapped in rank form, with the highest accessibilities as the darkest color and the lowest the lightest. The surface tends to contract a little between the two states—that is, the surface tends to draw itself close to the center—but a better illustration is the ratio of the two surfaces, as in figure 9.8(c). This shows that there is a loss of relative accessibility in the southwest, and the congestion charge area becomes much more accessible, largely because if costs are increased uniformly across the board, then the higher-cost areas (Central London) become more advantageous even though they still have the highest unit costs.

Extending Urban Simulation

We are currently extending the model in terms of the number of sectors modeled and the number of zones defining the size of the urban region. Our criterion for visually accessible, rapid operation of the model is still a major objective in its development, but to achieve these changes in scale, we are developing a slimmed-down desktop version of the model and a much faster Web-based version built with state-of-the-art software. The extension of the residential model to link with retail and local services location models mirrors developments of more integrated models elsewhere (Batty, 2009b). These include disaggregation by activity type as well as mode and are interfaced with various capacity constraints on location and on the transport network. The new model includes the assignment of trips to the various networks and the assessment of related capacity constraints reflecting cost of transport. In fact, we have not dwelt on the nature of the transport networks used to underpin the four modes of travel in this model, but beneath the thematic surface used to structure the various visualizations here lie these networks, particularly in terms of the geometry introduced in chapters 1 and 3.

The biggest potential change to the structure presented in this chapter is the form of the residential location model. Wegener (2008) argues that each sector modeled in the urban system is likely to be subject to very different explanations in terms of economic dynamics. Retailing, he argues, is a process of rapid response to changes in demand and supply that can be seen largely in terms of travel costs and accessibility, while residential location is based much more on the trade-off between housing prices and travel costs, which depend on wages. We have a version of the current model that replaces the constraints on travel costs with a budget equation that links incomes to costs through a new constraint. Equation (9.1) thus becomes

$$\sum_i \sum_j T_{ij}^k (w_i - c_{ij}^k - r_j)^2 = \sigma^k, \tag{9.22}$$

where w_i is the average wage earned at location i and r_j is the average housing price (or rent suitably discounted to the appropriate time period) at location j. Equation (9.22) is, in fact, a variance that ensures the majority of trips will cluster around the mean value based on the difference between wages earned and the costs of travel and housing. This assumes that the probability of where people will locate increases as they move closer toward exhausting their budget. The model generated using this constraint and the usual constraint on origins in equation (9.2) thus becomes

$$T_{ij}^k = \exp(-\lambda_i) A_j \exp[-\lambda^k (w_i - c_{ij}^k - r_j)^2]. \tag{9.23}$$

There are many variants of this model, since we can formulate the budget equation in different ways, making it mode-specific if data is available, and of course disaggregating the equation to deal with different employment groups and housing types. This is the form we are taking forward in a three-sector, enlarged model in which local service employment and retailing are treated using specific and different models from the residential one (Batty et al., 2011). Moreover, in such models, the physical volumes of trips made are also paralleled by money flows with respect to rents, travel costs, and retail expenditures.

When we examine changes to travel costs in the current model, the shift in population locations is of an order of magnitude less than the shifts in modal split. Our casual knowledge of urban systems and the way people react to such changing costs suggests that the order of the locational shift—some 2.5% of the working population—is too low. This shift is almost the first-order change that might take place, although the model makes no such distinctions. The new model in equation (9.23), which we have already experimented with a little, gives much larger shifts as travel costs are directly compared to housing prices. If travel costs increase by

100% on road journeys, which was the scenario tested here, then using the model based on equation (9.23), the shift in population is of the order of 12%. This is directly due to the fact that as travel costs increase for road users, they have less to spend on housing and consequently seek cheaper residences, which is a direct locational effect. In the extended model, this kind of structure is central to such locational behavior. Thus, the impact of changing costs is much more realistic. In fact, in future applications of this and the extended model, we propose to explore many different but related scenarios so that we can examine sensitivity to changing travel costs as well as actual impacts. As the model can be run rapidly, hundreds of such scenarios can be generated, which raises the prospect of some means of disciplining such actions so that the user can explore future solution spaces easily and effectively (Batty et al., 2011).

Our last foray into future developments concerns the underpinning of this and similar models using the entropy-energy framework. This needs substantially more effort in making the proper connections and interpretations between city systems and the way energy flows through their spatial fabric. We need a clearer explanation of free entropy and a detailed analysis of the way energy flows in systems are coupled through the modal choice model. One of the problems in measuring such energies is that the entropy-maximizing framework is, for historical reasons, difficult to dimension, and the roles of the travel cost parameters and partition (normalization) functions need to be worked out consistently for such coupled systems. These are all issues that are under active development in the context of the extended model, in which the impacts of changes in energy will continue to be a central motivation for their continued development. In this, we anticipate that the kind of science we have hinted at here will begin to merge with physical morphologies associated with the way energy flows through the city. In this sense, we will begin to enrich the analysis of urban form with respect to notions about how we can reach out to create and design more sustainable cities.

III

The Science of Design

In part III, we will shift rather dramatically from understanding cities to their design. We use the same tools, however, to extend our science. We conceive of design as embedded in networks of relationships between those who have a stake in a problem, and the process of generating new designs as one of communicating and resolving conflicts between different views of the future. Many approaches to a science of cities do not consider their design, with the assumption that cities and their planning are quite separable from one another, science simply informing design in a somewhat passive manner. There is even an assumption that design is rather a modest activity in the evolution of cities and, to a first approximation, it can be ignored. However, one of the most profound changes in our understanding of cities over the last half-century has been the realization that such understanding is impossible without taking account of the very ways in which we intervene in cities through their government, planning, and design. Just as we disturb the universe when we attempt to observe waves and particles at the quantum level, we disturb the city through our planning actions, sometimes up to the point where the very act of measuring urban phenomena changes it. Moreover, our planned actions can and frequently do become part of the problems that future generations are required to deal with. In short, the city and its design are all of one piece and cannot be disentangled and treated without reference to one another. Design is often part of the problematic we seek to understand.

When city planning first became institutionalized and professionalized in Western countries just over 100 years ago, the prevailing view both intellectually and among an informed public was largely anti-urban. The mission of planning in its highest sense was to make cities beautiful, which in turn translated into practical expedients such as reducing pollution, densities, and congestion of all kinds, and of injecting back into cities something of the "green" heritage that had been largely lost by the Industrial Revolution, rapid urbanization, and the rise of cities. New towns, garden cities, segregated zoning of land uses and activities, and the location of new economic activity, decentralized from the core cities, became the order of the day. These instruments were the prerogative of professionals—architects, planners, and social analysts—whose mandate assumed that cities in general were rather passive environments where plans could be defined quite independently of their functioning and their populations. Because of their manifestly technical focus, it became a sine qua non that such plans would simply improve the urban conditions they were designed to alleviate.

This perspective essentially separated our knowledge and understanding of cities from the designs that were to change their function and structure. The kind of rudimentary science that was largely physicalist a century ago was regarded as being central to a professional concern that was quite separate from the city itself. The notion that planning might actually make matters worse was simply not part of this intellectual agenda, until, however, experience with such interventions began to accumulate. By the 1970s, Rittel and Webber (1973) in their review of planning theory were suggesting that many urban problems were what they called "wicked," since intended improvements often intensified the problems they were designed to solve. In short, they argued that cities were so complex that it was near impossible to trace all the repercussions and impacts of proposed solutions, which often ended up making the original conditions more problematic than if no plans had been devised and implemented in the first place.

In our first chapter, we outlined the traditional model of city systems based on the analogy between cities and machines in which, alongside the system, some form of controller steered the city cybernetically toward a future desired state. This model was never any more than a metaphor, for no one supposed that cities could be controlled in the manner of a servomechanism. Nevertheless, although the model was soon relaxed, the notion that there was a science of cities that could be used to inform a technical problem-solving process was one that continued to lie at the basis of how city planning was conceived. This rational decision model was loosely structured around the idea that plans could be produced using some informal process of optimization in which analysis drove the process of generating alternative plans, these alternatives being evaluated against pre-set goals. This decision model is best summed up in Simon's (1960) sequence of "Intelligence-Design-Choice," which encapsulates this rationality.

This model no longer lies at the heart of city planning. Insofar as our science is being used to inform planning, it is now part of a much wider dialog in which many different perspectives—many different sciences, if you like—are brought to bear on urban problem solving. Furthermore, our models of cities have now become much more bottom up, as parts I and II of this book have emphasized. In this sense, all of us are now regarded as agents in an evolutionary process that casts the role of planning alongside all other roles. In short, the focus has shifted much more to the actors or agents who form part of the process of planning. In this perspective, models of how agents generate designs are much more central to the way we might think of how plans and policies are created. Indeed, design might be defined generically as the process of producing plans and then making decisions. What we need are models of the design process and the way decisions spin off from these. As Boulding (1975) has said: "The world moves into the future as a result of decisions, not as a result of plans. Plans are significant only insofar as they affect decisions" (p. 12). What we require are good models of design that lead to decisions. In part III, we will start to explore how we might fashion such models as forms of dialogue between relevant stakeholders.   *good*

In chapter 10, we will begin with models of design that lead very quickly to models of decision making. Throughout part III, we will be concerned with design and decision making as processes of resolving conflict or reaching consensus between relevant stakeholders. The idea of averaging initial opinions—of opinion pooling— will be central to our concerns, while the notion of weighting the importance of opinions will arise over and over again in the models we introduce. Weighting opinions will be simulated in terms of the relationships employed by actors and agents who interact with one another, relationships that are a function of the networks they use to communicate and disseminate their opinions. Location and spatial interaction are still in the frame within these models, for actors identify with locations, and often their decision making is motivated with respect to what goes on in particular places and across different networks.

Our basic model of averaging is mirrored in a first-order Markov process, which leads to a steady state—a consensus—for networks of communication that are strongly connected; that is, where every actor can communicate directly or indirectly with every other actor. We develop this model in chapter 11, while at the same time emphasizing that our focus is shifting in part III from product to process, and that design is as much about the process as the product, the plan, or the eventual outcome.

In chapter 12, we move our focus on design to processes of collective action by illustrating a theory of exchange between actors. At this point, we invoke the idea that models of both the way cities function and the way actors design plans involve networks that can be predicted from more elemental relations. The bipartite graph will again be useful in demonstrating a network of relations between, say, actors and the locations that they aspire to control or have interest in, or between actors and problems, actors and policies, or any two sets of distinct objects relevant to the design process. We can predict the unipartite networks that relate to interactions within each set of objects from these bipartite graphs. We exploit this widely in a theory of collective action that generates a series of relations or weights defining the importance of each set of objects in their steady state, when consensus is reached. The models remain Markovian in structure. The implication is that such bipartite relations might be extended to tripartite and beyond, blowing up the dimensionality of the problem massively. But the requisite calculus does not yet exist, and we must content ourselves to remain prisoners of a two-dimensional world where this richness can be partly captured by strings of bipartite sets.

In chapters 13 and 14, we develop examples of exchange and communications that lead to changes in the interest that actors have in land and the control they have over that same land. We develop various semi-real applications, first to a problem of conflict over land development in Central London and second with respect to committee decision making involving budgetary control in a typical municipal context where resources are required to be allocated to various services. These examples point the way to how our models might be further developed in real-world contexts, coming back full circle to models of cities. There are many ways forward in this endeavor, but this is unfinished business, as we will elaborate in our conclusions.

10

Hierarchical Design

The artist does not illustrate science; . . . [but] he frequently responds to the same interests that a scientist does, and expresses by a visual synthesis what the scientist converts into analytical formulae or experimental demonstrations.
—Lewis Mumford, "The Arts," in *Whither Mankind: A Panorama of Modern Civilization* (1928, p. 296)

The tools we have so far introduced have largely been focused on developing a science of cities analytically, but as we will see, these same tools can be fashioned to generate an understanding of cities synthetically. In this sense, analysis involves deconstructing the city into its component parts and interactions, in contrast to synthesis, which is putting the city back together again. Analysis tends to be top down while synthesis is bottom up, but there is a twist to this argument, in that deconstruction tends to focus on extant forms, construction on new forms, deconstruction on analysis and science, construction on synthesis and design. The tools we have introduced so far have been focused on explaining what exists—on actualities—while those we will introduce in this last part of the book are focused on designing what might exist—on possibilities. Although the tools have great commonality in the way we treat interactions and relations, the substantive orientation of the methods we use to reconstruct and design new city forms involves a shift in perspective, from a concern for the past and the present to a concern for the future. As Mumford (1928) so cogently argues in the opening quote above, design as art is contrasted against science in a comparison of visual synthesis with analysis and experiment.

Our implicit theory, which has driven the way we have introduced tools to elaborate our science, is based on defining variables we assume to be independent of outcomes. In this sense, we assume that these variables are not functionally dependent on those outcomes, in that an outcome y is driven or generated by a variable x but not vice versa. Both the dependent and independent variables are observables, and it is the task of science to figure out how good the relationship is between x

and *y* in theoretical as well as empirical terms. Good theory must be empirically falsifiable, robust to such attempts at destruction, and of course plausible with respect to our reflections and intuitions about how the world works. Defining appropriate outcomes and independent variables is a skill that is not science, per se, which implies that there can be as much design in science as science in design. Yet the way science proceeds involves defining hypotheses as relationships that can be tested through experiment or computer simulations in quasi-deductive manner, or are inferred or induced from sets of observations that set the hypothesis-testing cycle in motion. Perhaps the most important test of good science is whether or not a theory enables robust predictions borne out through observations that are truly independent of the way the hypotheses have been generated in the first instance. To an extent, this goal, which was the cornerstone of classical science in the nineteenth and twentieth centuries, is now under intense scrutiny as the complexity of the systems that we deal with increases.

Just as science is driven by the search for causal relationships between one set of variables and another, design can be seen as following the same structure but with a significant difference of perspective and intent. Often design can be construed as the operation of a series of independent variables that when synthesized or integrated form an outcome we can call a "plan" or "design." The independent variables are factors that determine in some way the solution to a problem, and the act of their synthesis is one of aggregating or merging these factors in such a way that the solution or outcome reflects some optimal resolution of the factors in question. The solution or design can also be seen as a plan *y* determined by a series of factors that define desirable but often conflicting states *x*. These factors are usually defined by the designer, and in spatial problems are sometimes considered as "layers" that are combined to generate a solution. In fact, the process of science and design can be likened to the composition of explanations or solutions that are clever combinations of these layers, and the analogy is particularly fruitful in spatial systems where the layers might be considered as spatial maps. However, unlike scientific variables, there may be no consensus among different designers about their importance, and this makes their particular combination unique to any design. These factors or variables may not be independent of one another, for they are selected by the designer as being relevant in some way to his or her intuition. Combined with the process of synthesis that produces a solution, this process in and of itself is also unique to the designer.

As we will see, the processes we define are highly formalized but, nevertheless, given the subjective nature of how the factors influencing a design are defined, every outcome will be unique. Whether a design is good or not depends on how it is evaluated, and often this is a process of testing the outcome or solution against a series of goals or objectives that measure the fit of the solution to the problem as defined in terms of those goals. Unlike the scientific method, evaluating designs against preset goals does not involve taking the design elsewhere and testing it against an independent set of goals. This would be logically equivalent to the process of testing a theory in science, where the goals in that context are defined in terms of the observations, but the process of design does not involve any such independence. In short, the design is tested against the very same goals that are used in the process of synthesis that generates the design in the first place. There is an inevitable and intrinsic circularity, and there is probably no true test of a design, since consensus is very often not the purpose for which the design was produced. In fact, in this context, consensus will be important, but this is never a requirement for good design.

In one sense, all science and design involve optimization. The process of fitting or fine-tuning the independent or explanatory variables in science to those that are explained invariably involves searching for ways of fine-tuning the explanation by introducing various parameters that define the relative importance and significance of the independent variables. At various points in part II, we used linear analysis to assess the weights of such variables using the minimization of some function such as least squares to produce a best-fitting model that minimized the errors between observed and predicted outcomes—in short, finding sets of weights w_k that minimize functions such as $\Phi = \Sigma_i(y_i - \Sigma_k w_k x_{ik})^2$. The same kind of structure has sometimes been used to define optimal solutions to design problems, usually based on some objective function such as $\Phi = \Sigma_k w_k x_{ik}$, where the weights w_k are now costs. We then choose variables in the solution y_i, usually measured in terms of benefits, profits, etc., to optimize Φ, which is usually subject to a series of constraints on the limits the solution variables (and often the independent factors) can take on. In fact, in the early days of land use transport modeling, models such as those we explored in the last chapter were set within such optimization frameworks, although it was quickly realized that such formal attempts at design tended to be somewhat artificial, producing rather blunt-edged solutions that ignored crucial qualitative variables clearly important to design.

Optimization for any but the most tractable and simple systems involves some form of feedback that moves the solution toward what is optimal. To an extent, this looping is key to the scientific method in that hypotheses are successively improved as more is learned about their nature and as observations are refined in the quest to produce better and better predictions. Much the same is true of design in that even in formal optimization, the solution space within which the optimal solution might lie is successively narrowed in the quest to hone in on the best outcomes. In the various models we will examine in this and subsequent chapters, such processes of interaction are key to the way in which conflicts are resolved and consensus— which we often take as optimality—reached. Many of our models, in fact, will be linear in structure and in some instances, as for example in statistical analysis, it is possible to produce solutions directly when the convergence properties of such interactions simplify to tractable forms.

Our first foray into formal design is through generating solutions as a linear synthesis of factors that affect solutions to the problem of finding best locations for various land uses. Simple methods of linear weighting of the predefined set of factors used to structure the problems are introduced. We will then examine the structure of relationships between these factors, using network representations showing the relative importance of these factors one to another using ideas based on connectivity introduced in part I. This provides us with methods for organizing the factors influencing the solution into various hierarchies of subproblems that provide structured synthesis and in which differential weights are implicit. We then generalize this to a form of sequential averaging, which we then formalize as an algebra for design. This is akin to a first-order Markov process whose convergence to a steady state represents another weighted solution to the problem. We will explore this method in detail in the next chapter, where we then push the design problem one stage backward, posing it as one of choosing the best design machine—that is "designing the design." We then illustrate its application to a traditional problem of locating a highway in a landscape, a well-known application first posed by Alexander and Manheim (1962a) and also articulated many years ago by McHarg (1969) in his seminal book *Design with Nature*. These methods are generic in the design of land use plans with contemporary applications involving the new science of geodesign, formalized by Steinitz (2012). Overlay design of the kind demonstrated in this chapter has been developed in many forms of map-based geographic information systems technologies that are increasingly used in land use planning (Carr and Zwick, 2007).

Articulating Design Problems

Problems of design tend to manifest their structure in qualitative rather than quantitative terms, and thus classical research designs based on statistical or numerical analysis are often inappropriate. Of particular importance in solving any policy problem is the definition of alternative solutions where the different solutions pertain to objectives or constraints that the designer considers of differing degrees of importance to the problem at hand. There are many ways of articulating such problems, but here we will consider any solution to a problem to be related to any other. The task of the designer is thus to synthesize these "partial" or subsolutions into an acceptable solution in which the importance of each factor in determining the final solution is clear. In this spirit, the relationships between the elements of the problem in terms of different solutions will be formalized using qualitative relations based on a series of structural models that build on various graph-theoretic and network representations we also exploited in part II of this book. These models are essentially aids for thinking about the process of plan-generation. As such, they involve simple structures that can be used as pegs on which to hang ideas about solutions to planning problems. The models to be introduced were first formalized many years ago as design methods; some are still in common usage in practical design and planning problems, particularly in land use and landscape analysis, while others are derivatives and reformulations of well-known methods.

Our typical problem in thinking about future cities is how their land use and activities might be organized to meet specific objectives that determine a context in which changes in location and urban morphology might generate more efficient and equitable conditions for urban living. Typically, problems of where to locate mass housing and transportation fall into this domain, but so do many problems of urban regeneration. The kind of problem we will explore in this chapter is a relatively straightforward one of finding the "best" locations for a particular land use that meets certain predefined objectives agreed upon by those involved in preparing the plan. The easiest way to visualize this is in terms of some spatial extent, mappable as areas that might be defined as being the best locations and those that are not. Thus a solution to the problem might be cast as a spatial array $\bar{A}_i$ where there are m locations i and where the value might be either binary, with 1 is a best location and 0 not a best location, but it may also vary over a range, say from 0 to 1, which gives the suitability of that location for the particular land use in question.

If we suppose any solution to the problem, not necessarily the best, is defined as a spatial extent A_{ik} where there are $k = 1, 2, \dots , K$ partial or subsolutions, then the design problem might be framed as one in which the best solution is some combination of these partial solutions, that is

$$\bar{A}_i = f(A_{i1}, A_{i2}, \dots, A_{iK}).\tag{10.1}$$

Each partial solution is based on an issue of importance to the best solution, which in turn might reflect some objective or constraint. For example, A_{ik} might be a solution to the problem based solely on defining best locations with respect to maximizing accessibility or nearness to existing retail facilities, something that few would dispute as being relevant to a location problem involving land uses or activities that were needed to be serviced by those retail facilities. Another issue might be based on the constraint that all land to be developed must meet certain minimal restrictions on potential for flooding, and in such a case, A_{ik} would define those areas that met this criterion and those that did not. In this sense, these partial solutions would reflect "all or nothing" development or a continuum of development potential. In short, then, we will call these issues *factors influencing development,* which in turn reflect individual objectives or constraints on what is desirable or otherwise.

Of course, the key to defining a best solution is specifying the ways in which the combination of partial solutions or objectives implying such solutions occur in the generic relationship in equation (10.1). One obvious way would be to partition the set of factors into objectives that would imply some form of benefit-cost or desirability and into constraints that form some limits on the solution, and to then solve the problem as one of optimization. This is often extremely difficult to achieve because, even though a constraint might individually appear hard and fast, when considered with other constraints, there is some variability in design. For example, most residential development is cost-effective when located on flat to gently sloping terrain, but occasionally inspired designs break this constraint and locate activity on tortuous terrain. This, to an extent, reflects factors that are impossible to quantify and are part of the inspiration and intuition of the designer. Moreover, setting up problems like this as formal procedures for optimization assumes a degree of quantification that is hard to achieve. In fact, a much more obvious way would be to provide some procedure for taking account of these objectives and constraints in some order that could be achieved manually, with the designer considering, reflecting, and pausing at each stage as the solution is evolved. This is the focus we will adopt here. In short, we will provide structural methods for synthesizing each of the factors in turn, progressing toward a solution in a manner that the designer can reverse, changing the process of synthesis at any stage.

To this end, we will introduce a "toy" problem in which we will locate land suitable for residential development in a small English town. This example is based on Macclesfield in North Cheshire, which has a current population of around 50,000 in the town itself and another 20,000 in the wider region that we define as the space in which new development might take place. We have identified $K = 12$ key objectives that we formulate as factors critical to the problem; these are listed in table 10.1. The task of plan-generation is to make use of these factors in generating solutions to this problem. This set of factors is almost certainly incomplete, and thus the problem is somewhat hypothetical. Nevertheless, it serves as a useful example on which to demonstrate the first main class of design methods introduced in this chapter. All these factors can be presented in spatial terms, and each have varying implications for different locations in the spatial system under consideration. We show the set of 12 in hard constraint terms in figure 10.1, where 1 (or black) implies that the location in question cannot be developed and 0 (or white) that the location can be developed. In figure 10.2, we also plot these as degrees of spatial desirability. In essence, the methods we will now introduce are ones where the order in which

Table 10.1
Key factors affecting residential development

A_{i1}	Accessibility to existing urban services
A_{i2}	Costs of spatial congestion
A_{i3}	Accessibility to recreational amenities
A_{i4}	Areas of acceptable microclimate
A_{i5}	Areas of water catchment and poor drainage
A_{i6}	Institutional constraints imposed by government
A_{i7}	Accessibility to external urban markets
A_{i8}	Subsidence and extensive industrial pollution
A_{i9}	Areas of suitable topography
A_{i10}	Rural amenity areas
A_{i11}	Historic urban areas
A_{i12}	Conservation of high-quality agricultural quality

these factors might be compared with one another reflects the relative importance of each within the final "best" solution. It is important to note the degree of fuzziness in this definition, for it implies that the order of comparison we generate is simply one the designer is encouraged to follow, with the assumption that at any stage the designer's intuition can be in full rein.

Traditionally, planners have presented such factors spatially in terms of their desirability or suitability (or otherwise) for solving the problem in hand, and the process of plan-generation and design has been one in which a synthesis of the factors is accomplished by resolving conflict between them by methods involving some kind of averaging. Probably the best-developed method we will illustrate below was first detailed by Christopher Alexander (1964), who argued that each factor relevant to development, such as those listed in table 10.1, usually implies a less than optimal solution to the problem. Then the designer's job is to examine each of these subsolutions and try to reconcile what they imply for the best location with every other subsolution. With 12 factors, there are 144 (12^2) possible paired comparisons, but for each factor, producing some synthetic solution relative to all other factors would simply imply that there would still be 12 different solutions. Alexander (1964) argued that some form of priority order for their consideration is needed, and it is on this basis that grouping the partial solutions or factors according to the degree to which they are correlated or related would provide a good method for generating a synthesis with the least conflict between all 12 factors. This is the procedure we will develop in the next three sections before we finally develop an iterative method that synthesizes the subsolutions in such a way that a true compromise or "best" average is the result.

A

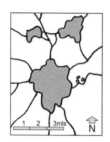

Each factor influencing land development in the Macclesfield urban area shown to the left a) is mapped below. Areas which are suitable for development are shown in gray while those unsuitable for development are white. The criterion of suitability is fixed at a particular threshold in this figure but in Figure 10.2 is shown as contour surfaces. The list of factors is given in Table 10.1.

B

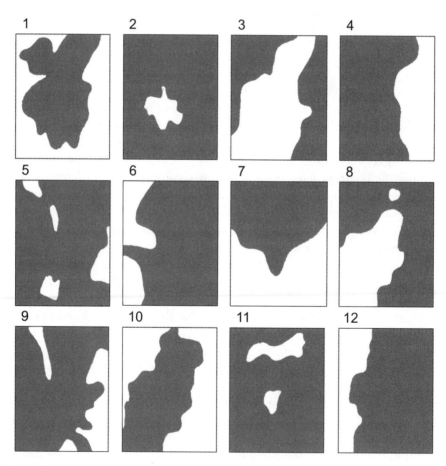

Figure 10.1
Factors represented as partial solutions to the location problem.

Figure 10.2
Factors represented as desirability surfaces for best locations.

Design as Linear Synthesis

These sets of subsolutions or objectives and constraints, defined from whatever perspective the designer has on the problem, must be synthesized in some way to produce a "best" solution. This solution must be in the same form as the subsolutions, that is mappable as a set of land suitability values that fall within some fixed range. This requires, as we have noted, the factors A_{ik} to be commensurate, or comparable over some range that without loss of generality we can normalize to $0 \leq A_{ik} \leq 1, \forall i, k$. The default synthesis is a simple average that we form as

$$\bar{A}_i = \frac{\sum_{k=1}^{K} A_{ik}}{K},$$

(10.2)

where we can assume that the factors can be represented as constraints or objectives. If the factor is a constraint we can define it using the superscript *c*, whereas if it is an objective we use the superscript *o*. This is consistent with these factors taking on values

$$A_{ik}^c = \begin{cases} 1, \ constrained \\ 0, \ unconstrained \end{cases} \quad or \quad 0 \leq A_{ik}^o \leq 1. \tag{10.3}$$

Equation (10.2) is applicable to any mix of factors represented as constraints or objectives because they are commensurate over the same range of values.

The default synthesis assumes that each factor is of the same importance, or weight. This can be easily generalized to differential values such that the final solution is a weighted average of the form

$$\bar{A}_i = \sum_k w_k A_{ik}, \tag{10.4}$$

where the weights of course must be normalized so that these reflect the relative importance of each factor. The usual form is to set these as if they were probabilities of the subsolution or factor determining the final solution, where

$$\sum_k w_k = 1. \tag{10.5}$$

As in previous chapters, we will drop the full summation over the range when it is obvious, as in these instances.

This method of averaging to produce a solution has been widely used in various kinds of practical land use planning where locations are being sorted. Sometimes the factors are referred to as "potentials for development" if they are represented in the objective form and can thus be interpolated as contour surfaces, as in figure 10.2. Defined in this way as far back as the 1970s in a succession of subregional and metropolitan planning studies (see Wannop, 1971), these methods are now firmly established as part of geographic information systems (GIS) technology. Explicit functions are available that enable users to produce and visualize composite solutions, and practical methods such as LUCIS (land use conflict identification strategy) are now widely available that build on these ideas (Carr and Zwick, 2007).

In fact, these methods reflect some of the core tools in geodesign (Steinitz, 2012), but they date back to very early studies in landscape planning, which were dominated by physical concerns. Clearly, their limits lie in the fact that for the set of factors to be commensurate they must all imply some form of physical suitability for development, and this determines their manifestly physical bias to problems of location. In short, whatever issues need to be taken account of, they must be translatable into spatial terms, and although this is often possible, some ingenuity is required in the representation of certain factors. These methods build on a long lineage of what are called "overlay" analyses. Mappable factors are synthesized by overlaying them, one on top of each other, assuming equal weight and physically determining the composite surface either as a visual combination of the various factors or as a count of the areas affected by constraints. The usual process is to represent the factors as constraints and filter out those areas of land not affected by any constraints. A useful history is provided by Steinitz, Parker, and Jordan (1976), and a generalization linking overlay techniques to more formal methods is presented by Hopkins (1977). A particularly incisive illustration of these overlay methods lies at the basis of McHarg's (1969) book *Design with Nature,* and a version of his method used in identifying best locations for various kinds of development that operates on factors represented as constraints $\{A_{ik}^c\}$ is described as "sieve mapping." This is a procedure defined by Keeble (1952) in the following way: "It is assumed that all the land under examination is 'passed' through a series of sieves, each of which represents some characteristic rendering land unfit for the particular purpose being considered. Any land possessing characteristics represented by one or more of the sieves is 'caught,' while that which passes through all of them is *prima facie* suitable for the purpose concerned" (p. 234).

A formal way of representing overlay mapping where constraints are used as filters or sieves is by forming the intersections of the constraint factors A_{ik}^c. That is, the final solution is determined as

$$\bar{A}_i = \bigcap_{k=1}^{K} A_{ik}^c.$$ \hfill (10.6)

If this is a null set, which it might be if every cell of land k is affected by some constraint, then it might be necessary to simply count the constraints as

$$\bar{A}_i = \sum_k A_{ik},$$ \hfill (10.7)

which, of course, is equivalent to the non-weighted average in equation (10.2). In figure 10.3, we show three related solutions by synthesizing the various factors in figure 10.1—constraints—and figure 10.2—objectives. These three solutions are based on the use of constraints in equation (10.6), the count in equation (10.7) that reflects the overlays, and the equal weighting of objectives in figure 10.2 using equation (10.2). We have not specified differential weights, which at this stage would be arbitrary, because the rest of this chapter deals with systems and procedures that explore how we can determine these weights based on differential correlations between the factors.

Before we begin to examine the structure of problems such as these, we should note that in situations where land use or activities might be competing for

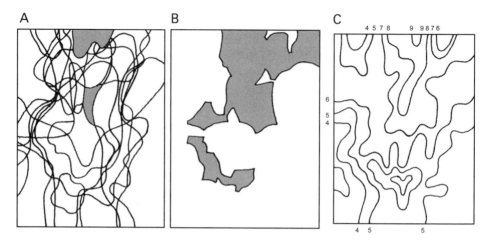

Figure 10.3
A comparison of different linear syntheses. (a) Left: traditional overlay (sieve-style) analysis showing land affected by less than three constraints in gray. (b) Middle: average factor map based on a count of constraints. (c) Right: equal weighting of desirability surfaces.

land—that is, where there are a series of design problems of the kind we have implied, then various techniques for resolving conflicts at higher levels have been devised. Let us now suppose that these methods are used to find best locations for a series of $\ell = 1, 2, \dots, L$ land uses where we might notate the previous weights w_k and factors A_{ik} with this notation, which is then used to derive best locations for each land use as $\bar{A}_i(\ell)$. One way of injecting competition between these different land uses into this process is to specify the relative importance of each solution λ_ℓ, and then to produce new final solutions for each land use as

$$\hat{A}_i(\ell) = \lambda_\ell \bar{A}_i(\ell) + \sum_{j \neq \ell}^{L} \lambda_j [Q - A_i(j)], \quad \sum_\ell \lambda_\ell = 1. \tag{10.8}$$

Q is a suitably defined normalization factor that ensures the comparability between the competing suitability surfaces defining each land use solution is maintained (Grant and Thompson, 1971). This parameter essentially converts each land use solution $j \neq k$ into one that competes with the solution k in question.

Structure and Connectivity of Design Networks

The essential differences between the averaging methods we have just presented and the methods to be presented in this and later chapters revolve around concepts associated with assessing the relative importance of the many conflicting and converging factors that determine the success or otherwise of some solution to the problem of optimal location. In this, we assume that a key argument is that the relative weighting of these factors in any design problem should reflect the problem's intrinsic or generic structure, and this should be extracted and articulated by the designer during the process of design. In short, the problem's structure should be used to guide the design. A central feature of such structures are the relations between each factor which may be specified functionally, causally or in terms of their correlation. In short, the strength of a relation between any two factors depends upon the extent to which basing a design on any one single factor affects the extent to which the design meets the requirements posed by the other. In short, the relationship might be assumed to be strong if a solution makes it easy to meet the requirements of both factors, or weak if it does not. However, the problem might be posed as one in which two factors have a strong connection when it is hard to meet the requirements of both, implying that the designer should worry about these pairs of factors first, since those that are easier to resolve should be left until last. How such relations are measured therefore needs to be reflected in the weights that are used to combine the factors, which in turn can depend on the order in which they are combined. In this exposition, we will generate networks where factors are related if they are very different from one another in terms of the subsolutions that each of them imply, and the network used below is based on such a criterion.

There are various ways in which to explore strategies for weighting. Among the most straightforward, because they tend to be based on structuring qualitative judgments of the importance of one factor with respect to any other, are methods based on paired comparisons (Churchman, Ackoff, and Arnoff, 1957). The method known as the analytic hierarchy process developed by Saaty (1980) has been quite widely applied in practice to derive a set of weights. This proceeds on the basis of making comparisons of the strength of the relationships between any one factor and all others, building up a matrix of such comparisons and then extracting an overall "average" set of weights that reflect the importance of each factor in the final solution. With respect to our problem of extracting the weights associated with the factors $\{A_{ik}\}$, we could proceed as follows. We first assess the relative importance of every factor to every other, assessing this on a scale that is likely to be no more than 7 points and forming the matrix W_{jk}, where this gives the relative importance of factor j to factor k. If j is, say, 5 times as important as k, then $W_{jk} = 5$ and $W_{kj} = 1/5$. Generating the matrix in this fashion may well lead to inconsistencies due to the qualitative nature of the comparisons, but a measure of this inconsistency can be computed and used to establish confidence or otherwise, thus prompting the designer to rework the comparisons in the case of extreme inconsistencies.

Buried in this pairwise comparison matrix is a set of weights, w_j, equivalent to those that we defined earlier giving the relative importance of each factor. If the pairwise comparisons were completely consistent, then these would be the ratio of these weights, that is

$$\frac{w_j}{w_k} = W_{jk},$$ (10.9)

from which it is easy to see that $w_j = W_{jk}w_k$, where for any weight k, this equation should give the weight w_j. It thus seems reasonable in a matrix W_{jk}, which might be inconsistent in terms of its specification, that the average of these values $W_{jk}w_k$ should be equal to w_i, that is

$$w_j = \frac{1}{K}\sum_k W_{jk}w_k.$$ (10.10)

This immediately suggests a scheme for iterating on the matrix $\mathbf{W}$ using $\mathbf{w}(t + 1) = (1/K)\ \mathbf{W}\ \mathbf{w}(t)$. This usually converges to a weight vector that is proportional to the principal eigenvalue of the $\mathbf{W}$ matrix. In fact, good approximations to the weights can be generated using simpler normalized sums of row or column totals of $\mathbf{W}$, and these, as we shall see, are all related to systems that are based on some graph or network of relations that link the factors together in matrix form. We will not take Saaty's method any further here, but there are quite obvious extensions to systems that can be partitioned hierarchically with normalized weights determined at each level and then combined, on the assumption that the weights at each level are independent of those at other levels. The most recent applications explore the networks that underpin such hierarchies (Saaty, 2005).

Many design methods, in particular the method pioneered by Alexander (1964), make use of such relationships in exploring the structure of the design problem. Alexander's thesis, like Saaty's in decision making, is that the pairwise relations between different factors reflect the relative importance of such factors in design. Moreover, Alexander has argued that the central problem of contemporary design involves the search for the true structure of the design problem. He suggests that an objective analysis of the set of pairwise relations will suggest new ways of classifying design factors, and this will ultimately direct the designer onto a path leading to the most relevant solution to such a problem. For example, by defining relationships between factors that strongly conflict and by grouping such factors into subsets or subproblems, it is possible to structure the design process around a method for resolving the strongly conflicting subproblems first, progressing in this fashion toward a final solution. In this sense, subproblems imply subsolutions or partial solutions and thus the process of comparing one subsolution against another and the order involved in this reflects the relative importance of the subsolutions in moving toward a final solution.

The simplest set of pairwise relations between the factors A_{ij}, where we assume the relation for the factors j and k varying across the spatial locations i, is given by the binary graph whose adjacency matrix is $\{a_{jk}\}$ and whose elements are defined as

$$a_{jk} \text{ and } a_{kj} = \begin{cases} 1, \text{ if } A_{ij} \leftrightarrow A_{ik}, j \neq k \\ 0, \text{ otherwise} \end{cases}. \tag{10.11}$$

The graph is assumed to be symmetric. It has no self-loops, since these are not relevant to combining a factor with itself, while we also assume that the graph is strongly connected. That is, for nontrivial design problems, the graph must have the property of strong connectivity; in other words, a path or chain of distinct edges must exist between any two vertices in the graph, thus implying that every factor is related to every other factor, directly or indirectly. In this sense, the problem cannot be partitioned into two or more separate subproblems, at least for purposes of applying the methods presented here. To illustrate these concepts, the 12 factors or requirements comprising the land development problem for the small town illustrated by the factors in figures 10.1 and 10.2 have been related using the simple binary code defined in equation (10.11). The criterion for a positive relation to exist is based upon Alexander's (1964) notion that "two requirements interact (and are therefore linked), if what you do about one of them in a design necessarily makes it more difficult or easier to do anything about the other" (p. 106). In figure 10.4, the graph of its associated matrix is presented diagrammatically. Our task is now to explore various weighting strategies that can be derived from such graphs, and

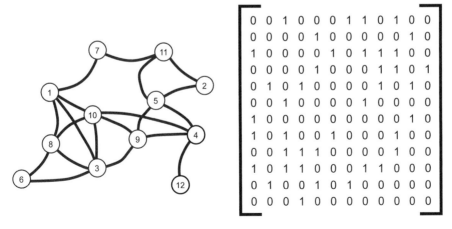

Figure 10.4
The graph of the problem and its associated adjacency matrix.

to then use these in combining the factors in the manner of a linear synthesis but whose structures will imply a much more elaborated order.

The simplest weighting structure which can be derived from the graph or its adjacency matrix is formed by summing the rows or columns of $[a_{ij}]$. Harary, Norman, and Cartwright (1965) first referred to the column sums as in-degrees and the row sums as out-degrees, while Flament (1963) refers to these as degrees of reception and degrees of emission, respectively. This follows the usual definitions in network science introduced in chapter 3 and elsewhere that are widely accepted in the field (Newman, 2010). Some of the results introduced in earlier chapters will be restated here so that readers are up to date on their usage. Clearly, in a symmetric graph, in-degrees and out-degrees for any set of factors $\{A_{ik}\}$ are the same. Formally,

$$a_j = \sum_\ell a_{j\ell} = a_k = \sum_\ell a_{\ell k} , \quad j = k, \tag{10.12}$$

and an obvious set of weights is based on the relative connectivities implied by these in-degrees and out-degrees, that is

$$w_j = \frac{a_j}{\sum_j a_j} = \frac{\sum_\ell a_{j\ell}}{\sum_j \sum_\ell a_{j\ell}}. \tag{10.13}$$

This scheme of weights, however, only accounts for direct relationships within the problem, and it is likely that strategies accounting for both direct and indirect connections are more relevant.

One such scheme is based upon the so-called distance matrix $[d_{jk}]$, which can be defined from simple operations on $[a_{jk}]$ as illustrated in chapter 3 in equation (3.41). Since the graph is assumed to be strongly connected, it is possible to reach any vertex from any other, either directly or indirectly. If each direct link is given the value of 1—the unit distance—then indirect distances can be found as multiples of these unit distances. To compute such distances, we first need to compute successive powers of the adjacency matrix, which can be calculated from the following recurrence relation

$$a_{jk}^m = \sum_\ell a_{j\ell}^{m-1} a_{\ell k}. \tag{10.14}$$

As we discussed in earlier chapters, in a strongly connected graph, the matrix $[a_{jk}^m]$ will always be positive by the power $m - 1$, where m is the number of factors or vertices in the graph. It is possible using equation (10.14) to determine the power m at which any link between j and k becomes positive. When such a link becomes positive, then the power m gives the distance between j and k. Formally,

$$d_{jk} = m, \quad if \quad a_{jk}^m > 0 \quad and \quad a_{jk}^{m-1} = 0. \tag{10.15}$$

A measure of weighting can be defined by taking the reciprocal of the row or column sums of d_{jk}, defined as d_j and d_k, which are

$$\frac{1}{d_j} = \frac{1}{\sum_{\ell \neq j} d_{j\ell}} = \frac{1}{d_k} = \frac{1}{\sum_{\ell \neq k} d_{\ell k}}, \quad j = k \tag{10.16}$$

where it is clear that since the distance matrix is also symmetric, its in-degrees and out-degrees are the same. All this follows from the causal symmetry of the way the design problem is specified. Note that these weights can be normalized to sum to 1 if necessary. They are similar to those defined in chapter 3 in terms of closeness centrality and the various distance measures that are associated with space syntax in chapters 6 and 7.

The main limitation of this weighting scheme, however, rests on the fact that the indirect distances contribute the same amount to the overall weight, as do the direct distances. Therefore, weighting schemes in which the indirect distances contribute less are preferable. A well-known result from graph theory is that the cells of the powers m, $[a_{jk}^m]$ give the number of paths between any two vertices j and k that are of distance m. As the power m increases, the number of paths increases exponentially in a strongly connected graph, and clearly a weighting index such as

$$\bar{a}_{jk} = \sum_{\ell} a_{j\ell}^m \tag{10.17}$$

gives far too much weight to the higher powers of $[a_{jk}]$. Yet it is possible to apply a system of decreasing weights to the higher powers of $[a_{jk}]$. First, the scale effects of the higher power matrices can be ignored by defining a probability matrix $[p_{jk}^m]$ as

$$p_{jk}^m = \frac{a_{jk}^m}{\sum_j \sum_k a_{jk}^m}. \tag{10.18}$$

Then to each matrix $[p_{jk}^m]$, a scalar α^m where $0 < \alpha < 1$ is applied from which a weight for the relevant link can be defined as

$$w_{jk} = Z \sum_{z=0}^{m} \alpha^z p_{jk}^{z+1}. \tag{10.19}$$

Z is a scaling constant, which can be chosen so that the weights sum to any predetermined total. As $[w_{jk}]$ is symmetric, a unique weight can be defined for each factor as

$$w_j = \sum_\ell w_{j\ell} = w_k = \sum_\ell w_{\ell k} , \quad j = k. \tag{10.20}$$

The constant Z can be chosen so that the weights sum to 1, that is

$$\sum_j \sum_k w_{jk} = 1, \tag{10.21}$$

which implies that

$$Z \sum_z \alpha^z \sum_j \sum_k p_{jk}^{z+1} = 1. \tag{10.22}$$

Equation (10.22) can be rearranged so that

$$Z = \frac{1}{\sum_z \alpha^z}. \tag{10.23}$$

Writing out the series and setting this equal to 1,

$$Z(\alpha^0 + \alpha^1 + \alpha^2 + \dots + \alpha^z) = 1, \tag{10.24}$$

leads to a convergent sum for Z, which we can write as

$$Z = \frac{1-\alpha}{1-\alpha^{z+1}}. \tag{10.25}$$

We can now use equation (10.25) in (10.19) to define the weighted matrix as

$$w_{jk} = \frac{1-\alpha}{1-\alpha^{z+1}} \sum_{z=0}^{m} \alpha^z p_{jk}^{z+1}, \tag{10.26}$$

from which weights for each factor can be defined as the in-degrees or out-degrees of this matrix.

A problem with the weighting scheme just presented concerns the rather lengthy series of computations required to produce the set of weights $[w_{jk}]$, although these are relatively trivial with contemporary computation. Nevertheless, a simpler and somewhat more elegant set of weights $[\bar{w}_{jk}]$ can be computed if we first set up the matrix $[p_{jk}]$ as

$$p_{jk} = \frac{a_{jk}}{\max\{a_{jk}\}}. \tag{10.27}$$

Then the weights $[\bar{w}_{jk}]$ can be found from the following series

$$\bar{w}_{jk} = \alpha^1 p_{jk}^1 + \alpha^2 p_{jk}^2 + \alpha^3 p_{jk}^3 + \dots + \alpha^m p_{jk}^m, \tag{10.28}$$

where the term p_{jk}^m is defined from the recurrence relation

$$p_{jk}^m = \sum_\ell p_{j\ell}^{m-1} p_{\ell k}. \tag{10.29}$$

If it can be shown (which is likely) that $\alpha^m p_{jk}^m \rightarrow 0$ as m $\rightarrow \infty$, then the matrix of weights can be derived directly from the matrix equation

$$\overline{\mathbf{W}} = \alpha \mathbf{P}[\mathbf{I} - \alpha \mathbf{P}]^{-1}, \tag{10.30}$$

where $\overline{\mathbf{W}} = [w_{jk}]$, $\mathbf{I} = [I_{jk}]$ is an identity matrix, and $\mathbf{P} = [p_{jk}]$ as defined from equation (10.27). The term in the large brackets on the right-hand side of equation (10.30) is an inverse matrix. This convergence is a standard result of matrix algebra.

The four weighting schemes based on raw connectivities (in-degrees and out-degrees) in equation (10.12), on inverse distance in-degrees and out-degrees in equation (10.16), on the geometric sequence of connectivities in equation (10.20), and on the convergent series in equation (10.30) have been applied to determine the weighting structure of the 12 factor design problems based on the factors in table 10.1. Furthermore, this sort of exploration and analysis is simple enough to be applied manually for small problems by any designer, and thus constitutes a suitable tool to be used in thinking about design problems. At this stage, it is worth comparing the differences in weighting produced by the four methods. Table 10.2 lists the four different sets of weights for each factor. Although the weights have strong overall similarities in terms of rank-order, there are also significant differences. However, such weightings lie at the heart of more structured schemes for

Table 10.2
A comparison of weighting schemes based on connectivities, distances, and path lengths

Factor	In/out-degrees-of connectivities (equation (10.12))[†]	In/out-degrees-of distance sums (equation (10.16))	Series sum of path lengths (equation (10.20))	Convergent sum of path lengths (equation (10.30))
1	0.100	0.090	0.104	0.099
2	0.050	0.068	0.045	0.053
3	0.125	0.094	0.128	0.120
4	0.100	0.094	0.097	0.095
5	0.100	0.090	0.091	0.092
6	0.050	0.064	0.055	0.058
7	0.050	0.076	0.046	0.053
8	0.100	0.082	0.105	0.099
9	0.100	0.099	0.109	0.103
10	0.125	0.099	0.133	0.123
11	0.075	0.076	0.061	0.068
12	0.025	0.066	0.025	0.034

[†]The steady state of the sequential averaging procedure described in the last section of this chapter generates an identical weight matrix to the in-out-degrees of the basic adjacency matrix.

weighted averaging and conflict resolution. In the next section, some of these methods will be introduced.

Hierarchies and Lattices

Hierarchical structure is intrinsic to the way complex systems evolve, as we have emphasized throughout this book. Indeed, in chapter 5, we explored the role of hierarchy in how cities organize themselves in space, but here we will change tack and illustrate how hierarchy can be central to the processes through which we can build systems and design cities. In chapter 1, we recounted Simon's (1962, 1969) iconic example of the two Swiss watchmakers Hora and Tempus, who made identical watches, but Hora developed subassemblies in hierarchical fashion while Tempus constructed his watches all of one piece. The moral of the story, of course, is that the hierarchical method triumphed in a noisy environment because it was robust to such noise, with the implication that evolution, which builds more complex structures from simpler ones, must proceed using some kind of modular design.

This implies hierarchy. In examining a series of factors that influence a design, it is thus important to examine the structure of relations between them and to consider whether or not they can be decomposed and then grouped into subassemblies. These can then be used to enable the problem to be split into stages, in which factors close to one another in terms of the problem's hierarchical structure can be dealt with more easily first. In short, hierarchy can be relevant to the order in which factors are considered in moving to a solution, quite the opposite to the methods in the previous section where, although factors were prescribed different degrees of importance through their weights, their synthesis did not imply any sequence of tasks.

Simon (1962, 1969) has also referred to hierarchy in systems that evolve as the "shape of design." His description of an architecture of complexity built around such modular construction was exploited by Alexander (1964), who argued that design problems should be analyzed formally by identifying their hierarchies. The process of decomposition of the problem into subproblems is analysis, while solutions are synthesized by exploiting this hierarchy in putting the problem back together again, but now in terms of its solution. In short, Alexander (1964) proposed a hierarchical method in which the problem was decomposed into subproblems, and these subproblems were then synthesized back in the given hierarchical order but expressed in a series of partial solutions or subsolutions, ultimately generating a final "best" solution or design.

Alexander's idea for reclassifying design factors by constructing a hierarchy can be accomplished by decomposing the graph or its matrix into a hierarchy of subsets in such a way that at the base of the hierarchy lie the most closely related subsets of factors. If these subsets are defined so that highly conflicting factors form the most elemental level of the hierarchy, which is what we have done so far, the hierarchy can be used as a structure showing the order in which the subproblems can be resolved (from bottom to top). This order of synthesis enables the designer to solve the most conflicting and hardest subproblems at the most elemental level, thus focusing the designer's mind onto the real purpose of design as a method for reaching a best compromise. However, all this depends on how the problem is structured by the designer and indeed what the purpose of design is. If the hardest subproblems are reconciled first, this suggests that the graph should contain links that connect the least-correlated factors or the factors with the largest negative correlations. This is almost the opposite of the way we specified relationships in earlier chapters, when we generated networks focusing on those factors that were most highly connected spatially. The designer might also adopt this assumption, in that those factors with the highest correlations are the easiest to solve and to an extent duplicate one another. Thus these are reconciled first, and as the designer proceeds to reconcile subproblems moving up the hierarchy, the problems get harder rather than easier, with the easiest problems dispensed with first. This is counter to what Alexander (1964) suggested, but what it does indicate is that design problems need to be thought about carefully in terms of what is being resolved against what. As we will see, there are methods that go beyond extracting such structures for reconciling competing and conflicting factors, and that thus avoid this problem.

Assuming that we work with the graph illustrated in figure 10.4, we will present two methods, both of which merge the most closely related (and conflicting) factors first and generate hierarchies where the most strongly correlated factors are synthesized into the solution last. There are literally hundreds of techniques available to decompose a set of pairwise relations into a hierarchy or to build up a hierarchy from such relations. However, most of these methods are dependent upon the pairwise relationships being specified by continuous variables rather than in the binary terms described earlier. For example, in chapter 3, we used single-linkage cluster analysis to generate a hierarchy of differences, and there are many such methods now available as standard packages. Here we will use two basic methods that depend on the weightings generated in the previous section. The first method depends upon using the weighting matrices defined earlier to construct maximal spanning trees, and the second method is based upon using such weights to form a measure of information loss in building up a hierarchy.

A maximal spanning tree can be defined as a path within a graph on which there are no loops or circuits, and that has the greatest total distance of any tree within the graph. Such a tree will always have $n - 1$ edges, where n is the total number of vertices. A well-known algorithm originally devised by Kruskal (1956) to find the minimal spanning tree can be easily adapted to finding the maximal spanning tree. The algorithm is described as follows: identify the greatest distance or weight

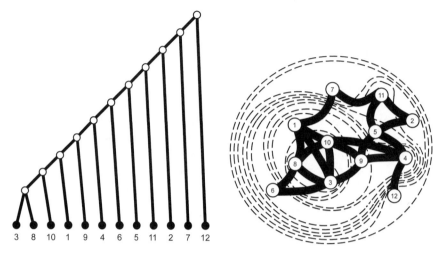

Figure 10.5
A maximal spanning tree and related hierarchy.

between any two vertices j and k and note this edge. Then find the next greatest distance and note this edge, unless the edge forms a circuit within the edges already chosen; if so, disregard it. Continue in this fashion until $n - 1$ edges have been chosen. At this point, the maximal spanning tree has been determined. Using this algorithm, the pair of factors with the greatest weights are linked first and thus form the first level of hierarchy. In figure 10.5, this process of building up the hierarchy is demonstrated diagrammatically using the matrix of weights $[w_{jk}]$ calculated from equation (10.19); the greatest advantage of this method is its simplicity, for it can be applied to quite large problems manually and is thus consistent with the designer continually reflecting on the importance of each set of links. Perhaps its biggest limitation is the fact that it is based on the local structure of the design problem, not upon its more general "gestalt" properties. In fact, as figure 10.5 illustrates, it is hard to compare the order implied by the decomposition with the weights that are implied in the methods based on the different linear syntheses above.

A second method that has been used in deriving a hierarchy is based upon the concept of information first defined by Shannon and Weaver (1949), and extended to problems involving aggregation by Theil (1967). Alexander and Manheim (1962b) used a similar hierarchical decomposition technique based on information-theoretic considerations in their highway design problem, which we will explore in the next chapter. As noted in chapter 9, the information content or entropy H of any structure can be measured by the formula

$$H = -\sum_j p_j \log p_j, \quad \sum_i p_i = 1, \tag{10.31}$$

where p_j is the probability of an event j occurring, which in terms of each factor in the design problem is its relative importance in the range of factors. We assume that this probability is a normalized weight based on any of those compared in table 10.2. If the set of events or factors is then aggregated into a two-level hierarchical subdivision of the set, then equation (10.31) can be decomposed and written as the sum of a between-set entropy and a within-set entropy:

$$H = -\sum_{\ell=1}^{L} P_\ell \log P_\ell - \sum_{\ell=1}^{L} P_\ell \sum_{j \in V_\ell} \frac{p_j}{P_\ell} \log \frac{p_j}{P_\ell}, \tag{10.32}$$

where

$$\sum_{j \in V_\ell} p_j = P_\ell, \text{ and } \sum_{\ell=1}^{L} P_\ell = 1. \tag{10.33}$$

The decomposition is into L mutually exclusive subsets V_ℓ, which are aggregations of the basic factors to the next level of hierarchy. In fact, this decomposition-aggregation can take place successively in generating a hierarchy of sets until the last two aggregated subsets are merged into one. The first term on the right-hand-side of equation (10.33) is the between-set entropy, and the second term the within-set entropy. If the probabilities indicate the importance of resolving the sub-solutions implied by factors that are densely connected, then a process that aggregates factors into subsets by maximizing the within-set entropy would be suitable. One particular heuristic for maximizing this entropy is based on the algorithm proposed by Ward (1963).

Ward's method begins at the base of the hierarchy and aggregates factors or subsets one by one in such a way that at any level in the hierarchy, the factor or subset that is aggregated produces a greater within-set entropy than any other aggregation of factor or subset. Using the weights calculated from equations (10.19) and (10.20) as probabilities, the method has been applied, first with no contiguity constraints on the aggregation and secondly with contiguity constraints based on the positive entries in the adjacency matrix $[a_{jk}]$. Contiguity constraints ensure that factors adjacent in the original network are also "close" in the hierarchy in some sense. In figure 10.6, where the two hierarchies from these applications are contrasted, it is clear that the method involving contiguity constraints produces the most acceptable solution. This solution shows similarities to the hierarchy based on the maximal spanning tree, but it is clear that the method takes greater account of the global properties of the problem, and is thus preferable.

A problem with any mutually exclusive scheme for classification involves the degree to which information is lost on aggregation. Furthermore, there is the ever-present difficulty of choosing the set into which a factor has to be aggregated.

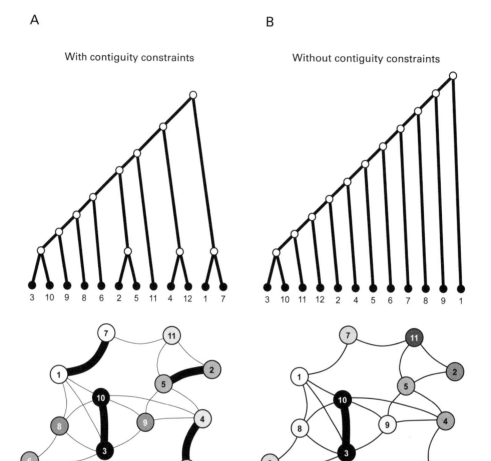

Figure 10.6
Hierarchical structure based on the application of Ward's algorithm with and without contiguity constraints.

Frequently, a factor might be equally important to two or more subsets at the same level of hierarchy. To account for such a possibility, it is necessary to build up a hierarchy based on overlapping sets, thus producing a structure that is lattice-like in appearance and has some of the properties of semi-lattices. In these structures, there is less information lost than in a strict hierarchy, and in one sense, such structures have a measure of redundancy that helps to produce a more realistic averaging strategy.

An algorithm for lattice-building has been assembled using as a basis the set of weights $[w_{jk}]$ defined from the matrix $[a_{jk}]$ in equation (10.19). The principle on which this algorithm rests involves the reduction of the number of subsets or factors one by one at successive levels of the hierarchy. However, at each level of the hierarchy, any factor might belong to two or more subsets subject to the constraint that the total number of factors or subsets be equal to the prescribed number for that level of hierarchy. For example, consider a simple system in which there are 5 factors at the most elemental level of hierarchy. At the next level of hierarchy, 4 subsets or factors are required, and these four subsets could be composed of (1, 2), (1, 3), 4, and 5, or (3, 4), (3, 5), (1, 2), (1, 3), and so on, but not (1, 2), (1, 3), (3, 4) and (1, 4), for here factor 5 is missing. The choice of subsets is based on the matrix of weights $[w_{jk}]$, and subsets are selected by starting with the largest weight $\max_{jk} w_{jk}$ and working down. At each level of hierarchy, the matrix of weights is recalculated by taking the simple average of the previous weights where subsets or factors have been aggregated.

This algorithm generates the hierarchical lattice shown in figure 10.7. The graphical syntheses suggested by the treelike and lattice-like structures of this and the previous section have been worked through and are compared in figure 10.8. There is much less difference between these final solutions to the plan-design problem than in the case of the weighted linear syntheses whose solutions were shown earlier in figure 10.3. The importance of these structures, however, is not solely in their ability to generate planning solutions, but in suggesting to the designer different ways to explore the structure of these kinds of problems. The implications of using these tools in determining best locations is as much in the process of making comparisons and thinking hard about potential subsolutions as in any routine way of generating a final solution. In short, the hierarchical structures that are generated provide an order for design.

There is one inconsistency in these methods that needs to be raised. If we assume that the hierarchy is used to average each of the factors against one another in the given order, then those factors that are closest to one another, and are averaged first, yield overall weights that are lower than those that are averaged with the subsolutions at higher levels in the hierarchy. If we take the hierarchy in figure 10.5 from the maximal spanning tree, then, assuming we follow the order, we can write the averaging process as follows, also assuming that when two factors or subsolutions

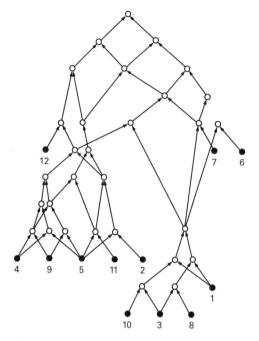

Figure 10.7
The resultant lattice-like structure.

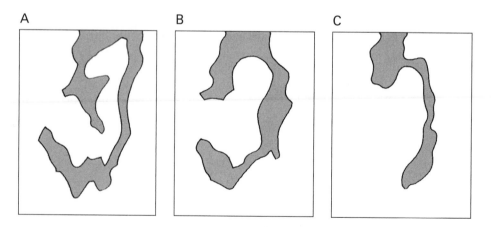

Figure 10.8
A comparison of the syntheses derived from the hierarchical methods. (a) Left: from the maximal spanning tree. (b) Middle: from the cluster analysis with contiguity constraints. (c) Right: from the lattice-like
structure.

are averaged, this is a simple average. From figure 10.5, we can write this process of successive averaging as

$$\overline{A}_{ik} = (((((((((((A_{i3} + A_{i8}) / 2 + A_{i10}) / 2 + A_{i1}) / 2 + A_{i9}) / 2 + A_{i4}) / 2 \\ + A_{i6}) / 2 + A_{i5}) / 2 + A_{i11}) / 2 + A_{i2}) / 2 + A_{i7}) / 2 + A_{i12}) \tag{10.34}$$

It is quite clear that the factor that is least connected, A_{i12}, has the highest weight, and since it enters the hierarchy last, it has a weight of 1/2 when it is averaged with the subsolution based on averaging the other 11 factors in the given order. However, what probably needs to be considered is that the position in the hierarchy should be given a weight, with the order in which the factors are averaged being given an a priori higher weight. However, there has been little research into these various schemes and this remains an area for active exploration.

Design as Sequential Averaging

Apart from the attempt to reduce information loss by building lattice-like structures, there is another method of synthesis in which the entire set of information concerning the problem is utilized. Imagine that the graph shown in figure 10.4 is a communications network in which the vertices represent transmitters or receivers and the edges represent channels. This graph can be redrawn as a directed graph where loops are allowed and the channels are specified in both directions. At each vertex in the graph is a designer who holds a particular view concerning the solution to the design problem; each designer is able to transmit his or her view to other designers and in turn receive other designers' views, but only along the given channels. Based on this power structure is a set of explicit rules governing the pooling of information and subsequent changes in the views of the designers based on such a pooling.

By setting up the problem in this way as one of resolving conflicting factors or views held by different designers, the object of the group dynamic is to reach a consensus between all members by following the rules laid down. Assuming that each designer is able to make a perfect compromise between his view and other views, then the process of conflict resolution begins as follows. In the first time cycle t, each designer transmits a view along the given channels to other designers, and each designer then changes his or her view by taking a simple average of their own view and other views received. Then, in the next time cycle $t + 1$, these new views are transmitted and averaged in the same fashion. This process continues until each designer reaches some overall consensus with the rest of the group. Such convergence to a consensus that is different from all the initial views will only occur in power structures based on strongly connected networks, because only in such structures can the views of any one member permeate to all other members. The

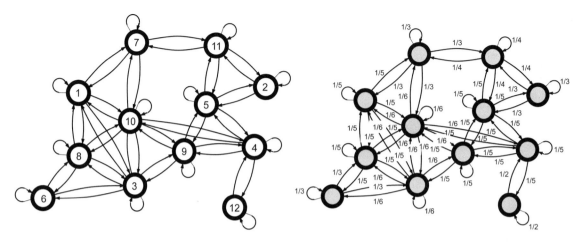

Figure 10.9
The power structure as a directed graph.

continuing transmission of the compromised views gradually reduces differences between the members, but with the ultimate consensus reflecting the network of transmission, which in turn reflects the power structure of the group. This model as a basis for the study of social power was first suggested by French (1956), subsequently refined by Harary (1959), and has become the basis for a variety of approaches to opinion pooling and learning (Golub and Jackson, 2010; Jackson, 2011). We introduced it in chapter 3, and we will rework it again in order to compare it to the various linear and hierarchical syntheses introduced above.

The important feature to note about such a process is that it represents a simple type of averaging procedure. Furthermore, design problems conceived in this way introduce notions concerning realistic procedures involving power, group problem solving, and sequential decision making. In figure 10.9, the nondirected graph of figure 10.4 is redrawn as a directed graph, where the fractions alongside each edge of this graph represent the direct weight each designer gives to the other designers' views. It is not so radical a step to use this formulation as a basis for synthesis in a design problem where each factor represents a partial solution to the problem. However, before this is applied, it is worth examining this process more formally to develop an algebraic interpretation from the theory of discrete-time, discrete-state Markov processes.

Consider the process of averaging or conflict resolution in any time period $t + 1$. This process involves transmitting the factors $A_{ik}(t)$, where the factor is now notated with respect to the time at which some averaging has taken place, with the initial factors being $A_{ik} = A_{ik}(1)$. At time cycle $t + 1$, new averages are formed by averaging the factors associated with the connections from all designers k to each individual designer j by forming the average,

$$A_{ij}(t+1) = \frac{\sum\limits_{k} a_{jk} A_{ik}(t)}{\sum\limits_{k} a_{jk}}.$$ (10.35)

In fact, we can represent the averaging in terms of the weight each designer gives to a factor, which when summed to 1 for each designer gives conditional probabilities defined as

$$p_{jk} = \frac{a_{jk}}{\sum\limits_{k} a_{jk}}, \quad \sum_{k} p_{jk} = 1.$$ (10.36)

Using equation (10.36), we can write the recurrence relation, which embodies the sequential averaging in equation (10.35) as

$$A_{ij}(t+1) = \sum_{k} p_{jk} A_{ik}(t).$$ (10.37)

Here p_{jk} is a transition probability, the matrix $[p_{jk}]$ in which each row sums to 1 is called a stochastic matrix, and the process of first-order updating or averaging is equivalent to a first-order Markov chain (Kemeny, Mirkil, Snell, and Thompson, 1959).

Of course, equation (10.35) and its Markovian form in (10.36) are recurrence relations, or processes in which the average at any stage $t + m$ can be seen as a function of the way the transition matrix embodies all the averaging applied to the initial factors. Then, iterating on equation (10.37) by substituting $A_{ij}(t + 1)$ for $A_{ik}(t)$, we get

$$\begin{aligned} A_{i\ell}(t+2) &= \sum_{j} p_{\ell j} A_{ij}(t+1) \\ &= \sum_{j} \sum_{k} p_{\ell j} p_{jk} A_{ik}(t) = \sum_{k} p_{\ell k}^{(2)} A_{ik}(t)' \end{aligned}$$ (10.38)

where $p_{\ell k}^{(2)}$ are higher transition probabilities and a general recurrence relation can be derived that expresses the average at any time $t + m$ as a function of the factor values at time t,

$$A_{i\ell}(t+m) = \sum_{k} p_{\ell k}^{(m)} A_{ik}(t).$$ (10.39)

This process will converge if the network is strongly connected. This is intuitively obvious in that the initial conditions in terms of the original factors become less and less strong in the final solution, which increasingly embodies the strength of the communication channels in the transmission of information. In fact, it can be shown, and we have sketched this already in chapters 3 (and 6), that $p_{\ell k}^{(m)} \to w_k$ *as* $m \to \infty$, and thus setting the start time as $t = 1$, the limit of equation (10.39) can be written as

$$\lim_{m \to 0} \sum_k p_{\ell k}^{(m)} A_{ik}(1) = \sum_k w_k A_{ik}(1) = c_i, \tag{10.40}$$

where c_i is the average value to which the cell or land area i converges.

The formal theory that underlies this type of process is Markovian, and the Markov chain used to describe the process is called ergodic, for it is possible to reach one particular state from any other. This chain also has the added property of regularity, since there is some power of the matrix $[p_{jk}]$ in which all cells are positive. In fact, there is a complete classification of chains associated with equivalent Markov digraphs, but we will postpone a discussion of these and the full process until the next chapter, when we formally introduce the model once again and demonstrate how it can be used for a problem of highway location.

To conclude, we will use the process represented in equations (10.39) and (10.40) to illustrate how the averaging takes place. We use the iterative structure as specified in equations (10.35) or (10.37) where each factor $A_{ik}(t)$ at iteration t is averaged with respect to those factors to which is linked, with new factors A_{ik} $(t + 1)$ generated from this process. This is continued until convergence. In figure 10.10, we show this process diagrammatically as a set of feedback loops, somewhat reminiscent of the feedback and feedforward loops in neural nets (Gurney, 1997), while in figure 10.11, we show the initial set of 12 factors and how they gradually converge to the same values c_k in map form. We can measure the differences between each of these factors at each iteration t to get a measure of convergence such as

$$\Phi(t) = \sum_i \sum_k |A_{ik}(t) - c_k|. \tag{10.41}$$

In fact, we have computed a measure on the transition probabilities, which is defined as

$$\Psi(t) = \sum_j \sum_k |p_{jk}^t - p_{jk}^\infty|, \tag{10.42}$$

where the limiting matrix can be computed not by direct matrix multiplication but by using other properties of the Markov chain that we will explore in the next chapter. This convergence is fast, and we will illustrate it in a slightly simpler way there. It is also worth noting that the weights associated with generating the final

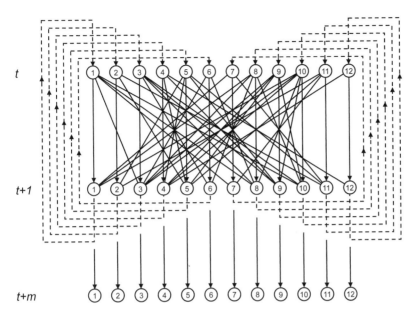

Figure 10.10
The Markov process as sequential averaging.

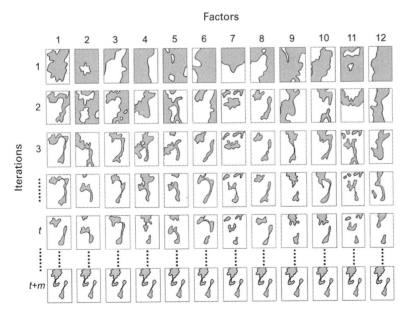

Figure 10.11
The spatial averaging in constraint map form.

average are intimately related to this convergence, and that it is possible to compute this explicitly. These, like most of the weights associated with strongly connected networks in this chapter, reflect the structure of the network, and for completeness we have listed the weights in table 10.2 so that direct comparison can be made with the linear synthesis. In fact, as readers will note, these weights are identical to the in-degrees and out-degrees of the basic connectivity matrix, a property that makes this sequential averaging procedure extremely attractive in that its simplicity is deeply embedded in this method of representing the problem's structure and the averaging process defined upon it. Readers will have to wait until the next chapter to see why this is so, but it is a basic result of working with strongly connected symmetric graphs.

Design Science and City Science

The quote from Mumford (1928) that introduced this chapter implies that scientists and designers are not very far from one another in their quest for understanding, but that designers do not feel they have to test their theories in the same way scientists are mandated to do. Much of what we have presented in this chapter cannot be tested in any conventional sense against data, for this kind of design is about a future that is hidden from view until it is realized. Moreover, as in many other places in this book, the tools, methods, and procedures have been introduced to some extent as "ideal types." These are to be adapted to context and used to inform debate rather than provide firm answers to problems that inevitably will always remain ill-defined, since such is the nature of cities and their planning. These are tools that are useful for thinking about problems of planning and location, not for generating answers or solutions in the conventional sense of the words.

The essence of the methods described here involves the fundamental concept of weighting. Designers and planners have never come to grips with such problems in plan design, and indeed, an awareness of these problems has only developed quite recently. Like all design methods that aid the process of plan design, these techniques involve subjective judgment and are exploratory in nature, but they challenge the designer to think deeply about the implications of problems. Research in this area is difficult and slow, yet the need for research into design processes is urgent if the conceptual and technical advances being made elsewhere in the field of urban research are to become useful and relevant to plan making.

The design methods outlined here are easy to apply, and if computer usage is required, then the necessary time is extremely small. Perhaps one of the greatest advantages of these methods is that designers can apply them "manually." The art + science   arch ≠ u.pl. — social science?

no data per se,   (theory + policy based on data)
decision to build

Markov method outlined above appears to hold some promise in the study of the design process, for not only does it seem capable of relating to existing design methods, but it also brings realistic behavioral implications to the design sciences. We will follow up these ideas in the next three chapters as the idea of sequential averaging is linked to conflict resolution, the generation of consensus, and the whole process of structured argumentation, which have become central to the way physical interventions in cities are now being debated.

Typology
vs.
C. R.
Deductive
vs.
Inductive

designers never have all the information they need or want; They're always playing the odds and their best (intuition = judgement) hunches + instincts. Experience helps. Proven / time-tested Typology helps. patterns + models help.

all the tougher in a faster changing world with greater uncertainty; instincts may become more important, not less important. (Big Data? may help understand past but not some of future!

— Other than scale dwgs + models, they don't get to test it (like a auto prototype or Ind. D.)
— one of the few custom items left in life.

— ITERATION, endless iteration! 'imaginary' testing with eye + mind
— Site info / site specificity vs. typology
— Role of beauty, aesthetics in evolution, in sustainability

Theory

Art
Culture.   Sci, Tech.

practice . (mud)

a bldg. can be   A city cannot
a work of art.   be a work of
                            art — JJ

11

Markovian Design Machines

A "machine" is essentially a system whose behaviour is sufficiently law-abiding or repetitive for us to be able to make some prediction about what it will do.
—W. Ross Ashby, *An Introduction to Cybernetics* (1956, p. 265)

Our approach to design outlined in the previous chapter depends on identifying the structure of the problem with respect to the network of factors that influence its solution. Such a network contains within it the strength of relationships between the parts of the problem, and in general, their resolution with respect to any solution cannot be accomplished in one step. As we illustrated in the previous chapter, the process of reconciling competing or conflicting factors, which was likened to a process where designers have conflicting solutions in mind, can be envisaged as one of averaging in which the structure of the network determines what is averaged. In structures where every factor is related directly or indirectly to every other factor, this process leads to a series of successive compromises in which the initial factors are reconciled with one another and a solution is progressively developed.

We suggested that this process was equivalent to a rather well-developed theory of the general system known as the Markov process. The behavior of such a system is assumed to depend only on its previous states, and in effect the system lacks a memory. Such a pattern of dependence is clearly only a crude analog of the design process, but the importance of developing this approach rests on the implications of this analogy for design. As we have argued throughout this book, our tools for exploring a science of cities and a science for their design are illustrative. They provide analogies that point the way. Howard (1971) reveals this point of view quite cogently when he says "no physical system can ever be classified absolutely as either Markovian or non-Markovian—the important question is whether the Markov model is useful. If the Markovian assumption can be justified, then the investigator can enjoy analytical and computational convenience not often found in complex models" (p. 4). There is another role for such analogies in design that has been elaborated by Churchman (1971). In an exploration of the design process, there must always be some alternative design system against which the real design system can be compared. Progress in design can only occur if such comparisons are continually being made and the processes refined; the theory to be further elaborated here attempts to emulate this role.

As in the last chapter, we use the term "machine" rather than "model" for several reasons. As design can be a particularly private process, it is felt that the term machine rather than model implies that the designer has a greater control over the process and that the machine is simply an aid to imagination—a tool to amplify our intelligence, but also sometimes our creativity. Machines can be used sensibly or badly, they can be misused or abused, and thus the pitch of this definition gives a slightly less exalted tone to the argument. Furthermore, it is no coincidence that the Markovian machine developed here is analogous to a certain class of mathematical machines in the theory of finite automata, although the presentation here will be along conventional lines. The sorts of cellular automata we introduced to model the development of cities in chapter 8 are somewhat broader than the machines introduced here, but the first-order dynamics is still the same. This is yet another example of how networks and dynamic processes used to simulate the form and structure of city systems have parallels in terms of the methods used to design those very same cities. This, indeed, is a feature of the tools in this book: that networks and dynamics are important in examining cities, from any perspective.

This chapter is in three parts. In the first part, we will introduce an algebra for design based on the Markov process. We will more or less repeat the basic equations of sequential averaging presented in the last chapter, but then continue by generalizing these to the more usual form of Markov chain. Of particular importance are the design solutions generated by the machine, which imply a set of weights corresponding to the factors affecting the problem. A classification of design machines serves to delimit the class of problems to which these methods can be applied. We hinted at this classification in chapter 3, but here we will make it explicit. The second part of the chapter is concerned with the choice of machine, from a large number of possible machines, according to certain criteria of selection. The problem of choosing such a machine, which we refer to as the "design of design machines," is formulated as a problem in dynamic programming and solved by an iterative method. Finally, the machine is applied to a problem of design based on a highway location problem first posed and solved by Alexander and Manheim (1962a). This provides a comparison with the hierarchical design method originally used to solve the problem and introduced in the last chapter, and it also acts as a control against which to compare several different solutions generated by the design machine.

An Algebra for Design

Sequential Averaging Restated

All nontrivial design problems involve synthesizing a solution from a set of competing and conflicting requirements expressed as factors; design is thus a process of compromise and conflict resolution. Moreover, design is regarded by some as a process involving invention—something more than mere selection—although Churchman (1971) defines the process in somewhat looser terms as "thinking behavior which conceptually selects among a set of alternatives in order to figure out which alternative leads to the desired goal or set of goals" (p. 5). This is highly resonant with the idea that any solution is a compromise between alternative solutions or subsolutions, which was the way the problem was framed in the last chapter. In this sense, design is a choice between conflicting alternatives.

We will begin by restating the averaging equations introduced in the last chapter, which show how a set of factors that imply different solutions to the problem can be reconciled and compromised with one another. In fact, in chapter 10, we define a factor in terms of a spatial extent, and we will continue this here, but for the moment it is easier to think of a factor as a single numerical value, which will let us illustrate in a little more detail the process of convergence to a weighted average. In chapter 10, we defined a factor k as a spatial extent—a map, A_{jk}—where here we will drop the spatial extent—given by the cells i—and simply define a factor as A_k. Starting with the set of factors $A_k(t) = A_k$, which are initial conditions at time cycle t to $t + 1$, and using a strongly connected binary matrix of links $[a_{jk}]$, new averages are formed as

$$A_j(t+1) = \frac{\sum_k a_{jk} A_k(t)}{\sum_k a_{jk}}. \qquad\qquad (11.1)\ [10.35]$$

We can normalize the network as a transition matrix where $p_{jk} = a_{jk}/\Sigma_k a_{jk}$. Using these definitions, we write the averaging equation (11.1) as

$$A_j(t+1) = \sum_k p_{jk} A_k(t). \qquad\qquad (11.2)\ [10.37]$$

Then, iterating on equation (11.2) by substituting $A_j(t + 1)$ for $A_k(t)$, we get

$$A_{i\ell}(t+2) = \sum_j \sum_k p_{\ell j} p_{jk} A_{ik}(t) = \sum_k p_{\ell k}^{(2)} A_{ik}(t). \qquad\qquad (11.3)\ [10.38]$$

A general recurrence relation can be derived that expresses the average at any time $t + m$ as

$$A_\ell(t+m) = \sum_k p_{\ell k}^{(m)} A_k(t). \qquad\qquad (11.4)\ [10.39]$$

We stated without any proof in chapter 10 that this process will converge if the network is strongly connected. It is obvious from the first set of factors that any averaging across any subset will reduce their range, and if one factor can influence any other factor directly or indirectly, the original factors become less and less strong in the final solution. If the process continues, then in the limit $p_{\ell k}^{(m)} \rightarrow w_k$ *as* $m \rightarrow \infty$, and thus setting the start time as $t = 1$, this limit can be written as

$$\lim_{m \rightarrow 0} \sum_k p_{\ell k}^{(m)} A_k(1) = \sum_k \pi_{jk} A_k(1) = \sum_k w_k A_k(1) = c, \qquad (11.5) \ [10.40]$$

where c is the equilibrium or steady-state value of the final factor solution, and $[\pi_{jk}]$ is the steady-state matrix.

It is worth illustrating in visual terms how this method actually resolves differences producing a weighted average. In figure 11.1(a), we present a hypothetical network between five designers, actors, or factors—whatever interpretation the reader wishes to put on the problem—and we show a strongly connected digraph of the links. In figure 11.1(b), we show the symmetric binary matrix, and in 11.1(c) the transition matrix based on its normalization. Note that this matrix reflects the weights used to form an average of the factors, to which the various nodes are linked. In figure 11.1(d), we show an iterative structure that shows the feedforward and feedback links that generate the sequential averaging. In 11.1(e), we show the convergence of the initial values specified for each factor, which fall in the range 3 to 10. The process starts with the values of each factor as they are noted alongside the nodes in figure 11.1(a). These form the vector

$$A_k(t = 1) = [9 \quad 7 \quad 10 \quad 3 \quad 5] \qquad (11.6)$$

and successive iterations lead to

$$\left. \begin{array}{l} A_k(2) = [6.75 \quad 5.00 \quad 8.00 \quad 6.33 \quad 8.00] \\ A_k(3) = [7.27 \quad 5.66 \quad 7.58 \quad 6.02 \quad 7.58] \\ \quad \bullet \\ \quad \bullet \\ \quad \bullet \\ A_k(\infty) = [6.93 \quad 6.93 \quad 6.93 \quad 6.93 \quad 6.93] \end{array} \right\} . \qquad (11.7)$$

As we will see below, this vector of values can be generated directly from some simple properties of the network.

It is worth noting that there is a similarity between this process and that version of a neural net called a Hopfield net (Gurney, 1997), in which the issue is to find weights on the links that reflect some target set of values the factors must meet—in short, with a list of input factors. The neural net involves finding the values of the weights on the network links that translate these factors into a set of known

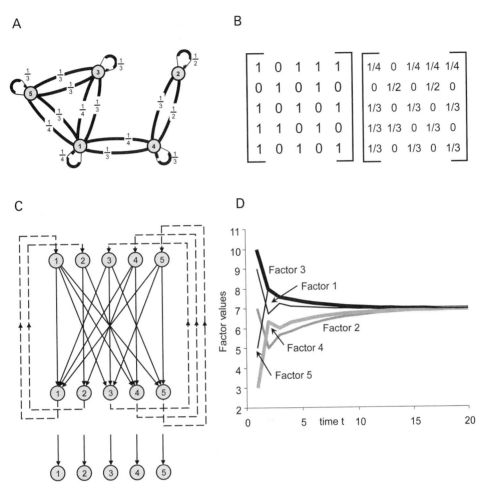

Figure 11.1
An illustrative example of a communications network and the convergence of opinions. (a) Top left: the network. (b) Top middle: the adjacency binary matrix. (c) Top right: the transition probability matrix. (d) Bottom left: the feedback-feedforward method. (e) Bottom right: the convergence to the steady state.

factors—a pattern, for example. In this sense, the neural net problem is quite different from the design problem because the training of the net is in regard to a known pattern of factors, and the input factors are used to determine how these are translated into this pattern. In design, no such training is required, and the only criterion for ending the iteration is convergence to the limit, which in the cases examined here reflects a consensus. Although we will not take this analogy with neural nets any further, readers are referred to this literature, which contains many suggestions for compressing multivariate information into more convergent forms that have some similarities with the design methods developed here.

The Kemeny-Snell Proof of Convergence

In equation (11.5), $[\pi_{jk}]$ is the limiting form of the matrix $[p_{jk}^{(m)}]$, which clearly must exist for the process to converge. If in the limit $A_j(t + m) = A_k(t + m) = c, j \neq k$, then it is also clear that the solution is composed of a weighted average of the initial factors $A_k(t = 1)$, where the weights w_k reflect the contribution of each factor to the final solution. There are many proofs of the conjecture given in equation (11.5), and since such a proof is of central importance to this method and its elaboration in later chapters, it is worthwhile presenting in detail. The form of proof used was originally suggested by Kemeny and Snell (1960), and its popularity is due to their exposition.

In essence, the proof demonstrates that differences between $A_j(t + m)$ and $A_k(t + m)$ decrease as m increases. First let us define x_t and y_t as the maximum and minimum values, respectively, of the factors $A_k(t)$, $k = 1, 2, \ldots, n$ and in analogous manner, x_{t+1} and y_{t+1} as the maximum and minimum values, respectively, of the factors $A_k(t + 1)$. To facilitate the proof, we rearrange $A_k(t)$ and the associated matrix $[p_{jk}]$ so that $A_1(t)$ is the minimum value y_t and $A_n(t)$ is the maximum value x_t. Next we define a set of factors A_j in which $A_1 = y_t$ and $A_k = x_t$, $k \neq 1$. Then, from these definitions, it is obvious that

$$\sum_k p_{jk} A_k(t) \leq \sum_k p_{jk} A_k. \tag{11.8}$$

From equation (11.8), the right-hand side can be expanded and rearranged as

$$\sum_k p_{jk} A_k = p_{j1} y_t + \sum_{k=2}^n p_{jk} x_t = p_{j1} y_t + (1 - p_{j1}) x_t = x_t + p_{j1}(x_t - y_t). \tag{11.9}$$

As $p_{j1} \geq \varepsilon$ where ε is the minimum element of the matrix $[p_{jk}]$, it is clear that

$$x_t - p_{j1}(x_t - y_t) \leq x_t - \varepsilon(x_t - y_t). \tag{11.10}$$

Therefore, using equation (11.8),

$$x_{t+1} \leq x_t - \varepsilon(x_t - y_t). \tag{11.11}$$

An analogous form of reasoning using $-y_t$ for x_t and $-x_t$ for y_t leads to

$$-y_{t+1} \leq -y_t - \varepsilon(-x_t + y_t).$$ (11.12)

Then, addition of equations (11.11) and (11.12) gives

$$x_{t+1} - y_{t+1} \leq x_t - y_t - 2\varepsilon(x_t - y_t) = (1 - 2\varepsilon)(x_t - y_t).$$ (11.13)

Equation (11.13) is a recurrence relation, and thus it can be demonstrated that

$$x_{t+m} - y_{t+m} \leq (1 - 2\varepsilon)^m (x_t - y_t).$$ (11.14)

Since $\varepsilon \leq 1/2$, it is clear that, in the limit, equation (11.14) converges to

$$\lim_{m \to \infty} \{x_{t+m} - y_{t+m}\} = 0.$$ (11.15)

In addition, since the difference between $A_j(t + m)$ and $A_k(t + m)$ tends to zero, the process converges and $A_j(t + m) = c$ for all j. Thus the conjecture in equation (11.5) is proven:

$$\lim_{m \to 0} \sum_k p_{\ell k} A_k(t + m) = \sum_k p_{jk}^{(m)} A_k(t) = c.$$ (11.16)

However, of equal importance from the Markov process viewpoint is the behavior of the matrix $[p_{jk}^{(m)}]$, and this can also be resolved from the proof. Imagine that $A_k(t) = 0$, $k = 1, 2, \dots, n - 1$, and that $A_n(t) = 1$. Then, by using equation (11.16), it is obvious each cell that is the nth column of $[p_{jk}^{(m)}]$ is constant. Repeating the argument for $A_k(t) = 1$, *any* k, then as each column cell is constant, then each row must be equal. This can be formalized as

$$\lim_{m \to 0} p_{jk}^{(m)} = \pi_{jk} = w_k.$$ (11.17)

In equation (11.17), $[p_{jk}^{(m)}]$ is a stochastic matrix, and this implies that

$$\sum_k w_k = 1.$$ (11.18)

Thus the terms w_k are weights in the true sense of the word and reflect the relative importance of each factor in the final solution.

The Standard Markov Process

We have not yet developed this approach to design machines using the conventional Markov process, which was originally developed as a branch of probability theory. In fact the probabilistic interpretation is not as relevant in this context as that of sequential averaging, but it is possible to gain some oblique insights into the process from probabilistic considerations. The process of averaging operates in the opposite way from the probabilistic process in that averaging is achieved by linearly weighting factors across each row of the transition matrix. Computing the probabilities of what state the Markov chain is in is accomplished on the columns of the matrix. In fact, this latter process could be called a forward process and the averaging a backward process, although this is a casual observation and simply relates to the way these processes run. It does not relate to any formal usage or development.

Essentially the standard process—the forward process—computes the probability that the system will be in its different states, which are equivalent to the nodes of the design network or the subsolutions to the problem. These probabilities are relative weights $w_k(t)$, and as the process continues, we will show that these weights become stable—converge to equilibrium weights w_k—which in fact give the proportions of time during which the system will occupy each state. This might then be thought of as the probability that the state of the system might take on a particular subsolution, if left to roam between all possible states which are the complete set of initial subsolutions. The implication is that if we then average the subsolutions according to these weights, we would get a single solution, which is produced directly from the backward process of sequential averaging.

First let us define the probability of solution to the system at time t as the weight $w_k(t)$. This is updated by applying the transition probabilities $[p_{jk}]$, which are in fact conditional probabilities of state j changing to k, using the standard recurrence relation from chapter 3, which is

$$w_k(t+1) = \sum_j w_j(t)p_{jk}, \quad \sum_k w_k(t+1) = \sum_k w_k(t) = 1. \tag{11.19}$$

Recursion on equation (11.19) leads to

$$w_k(t+m) = \sum_j w_j(t)p_{jk}^{(m)}. \tag{11.20}$$

From the conjecture in equation (11.5) and its consequent proof, the matrix $[p_{jk}^{(m)}]$ converges to the stable matrix $[\pi_{jk}]$, where each row is the stable vector $[w_k]$. Thus

$$\lim_{m \to \infty} w_k(t+m) = \sum_j w_j(t)\pi_{jk}, \tag{11.21}$$

and substitution of the stable vector for $[\pi_{jk}]$ in equation (11.21) leads to following steady-state relation

$$\lim_{m \to \infty} w_k(t+m) = \sum_j w_j(t)w_k = w_k. \tag{11.22}$$

Of more importance, however, is the fact that equation (11.22) implies that the convergence to w_k is independent of the starting probabilities $w_j(t = 1)$, and this particular property is often used to argue that the Markov process has "no memory."

The primary significance of this to the theory suggested here is not to introduce the probabilistic argument per se, but to demonstrate a convenient method for calculating the steady-state probabilities or weights w_k. In the limit, equations (11.20) to (11.22) suggest that

$$w_k(t + m + 1) = \sum_j w_j(t + m)p_{jk} = \sum_j \sum_\ell w_\ell \pi_{\ell j} p_{jk}, \tag{11.23}$$

and substituting equation (11.22) into (11.23) leads to the classic steady-state equation

$$w_k = \sum_j w_j p_{jk}. \tag{11.24}$$

This implies that the steady-state equations can be calculated directly from the transition-probability matrix without recourse to iteration. However, the equation system associated with equation (11.24) is linear and homogeneous, and thus can only be solved directly by taking the first $n - 1$ equations from this set and adding the identity $\Sigma_k w_k = 1$. A soluble equation set of the following form can then be derived:

$$
\begin{bmatrix} 0 \\ 0 \\ \bullet \\ 0 \\ 1 \end{bmatrix} =
\begin{bmatrix}
p_{11} - 1 & p_{12} & \bullet & \bullet & \bullet & p_{1n} \\
p_{21} & p_{22} - 1 & \bullet & \bullet & \bullet & p_{2n} \\
\bullet & & & \bullet & & \bullet \\
p_{n-1,1} & p_{n-1,2} & \bullet & \bullet & \bullet & p_{n-1,n} \\
1 & 1 & \bullet & \bullet & \bullet & 1
\end{bmatrix}
\begin{bmatrix} w_1 \\ w_2 \\ \bullet \\ w_{n-1} \\ w_n \end{bmatrix}. \tag{11.25}
$$

The above set of linear equations can now be solved for $[w_k]$ using any of the standard methods.

Several techniques are available for evaluating the behavior of the Markov process as it moves toward the steady state. In particular, methods involving the spectral decomposition of the transition matrix into its characteristic roots or eigenvalues help in determining the rate of convergence and the damping effect of different states (Bailey, 1964). A related method in engineering treats the Markov process geometrically using Z-transform analysis but, as most of these methods reflect more complicated mathematical considerations, they will not be pursued here. Suffice it to say that convergence of the Markov process can be analyzed pragmatically using formulas that relate the higher-transition probabilities to the steady-state probabilities. For example, following Bhat (1972), we define a measure of convergence as the difference between the steady-state transition π_{jk} and the transition probability matrix $p_{jk}^{(m)}$, as in equation (10.42) in the last chapter, while other methods involve a comparison of prior and posterior probabilities. Theil (1972) uses the well-known "information for discrimination" statistic from Kullback (1959) to measure convergence as

$$I(m) = \sum_j \sum_k \pi_{jk} \log \frac{\pi_{jk}}{p_{jk}^{(m)}}. \tag{11.26}$$

Equation (11.26) is in fact like a chi-square statistic. These formulas are easy to apply and have been used in analyzing the convergence exhibited in the applications of the design machines outlined below.

A Classification of Design Machines

To determine the class of problems to which the theory presented above pertains, it is worthwhile to develop a classification of design machines, first in terms of the characteristics of the Markov process, and second using the theory of graphs. A useful concept in discussing types of the finite Markov process or chain relates to whether or not different states are accessible to one another. States that are accessible, that is states that can be reached from one another, form an irreducible chain—one that cannot be decomposed and whose states are said to be persistent (Feller, 1957). In contrast, states that are nonaccessible are called transient, in the sense that once the process leaves these states, it never returns. If only one state is persistent and all the others are transient, then that state is said to be absorbing and the chain is referred to as an absorbing Markov chain.

An intuitively more appealing classification of chains is through the theory of graphs. It is well-known that systems of equations can be represented as directed linear graphs in which the variables are represented as nodes or vertices of the graph, and the coefficients of the equations as arcs or line segments linking the nodes. Indeed, in certain instances it is easier to solve systems of equations using concepts from graph theory. As we implied in the last chapter, it is possible to represent the Markov process given in equations (11.2) or (11.19) by a directed graph such as that illustrated in figure 11.1. Harary and Lipstein (1962) call this structure a Markov digraph, and these authors also demonstrate that the theory of graphs can be used to classify Markov chains. Furthermore, the Markov process illustrated by using graph theory can be likened to a communications system in which the nodes represent message transmitters and receivers and the arcs represent channels of communication. Analogies between the design process described here and social power structures originate from this interpretation (French, 1956; Harary, 1959; Lambiotte et al., 2011), and the theory of graphs also serves to relate this process to previous design methods, as in chapter 10.

The concept of accessibility in Markov chains, alluded to above, can best be seen in terms of the graph in figure 11.1. If as in figure 11.1, there is a possible path, direct or indirect, from one node to another, then this node or state is accessible to the other. This accessibility is in fact a measure of connectivity as we have used it throughout this book and the best-developed classification of graphs, hence Markov chains, uses these concepts. The class of irreducible chains, in which all states are persistent and accessible to one another, can be divided into *completely connected* and *strongly connected* chains. In completely connected chains, all states are directly accessible to one another; that is, the matrices $[a_{jk}]$ and $[p_{jk}]$ are positive in every cell. Strongly connected chains, on the other hand, contain states that are not directly accessible but are indirectly accessible and for which, at some power of the matrix $[p_{jk}^{(m)}]$ for $m < n$, each $[p_{jk}^{(m)}]$ becomes positive.

The class of reducible chains contains three types, which can also be discussed in terms of connectivity. In *unilaterally connected* chains, one chain can be partitioned into one irreducible set with persistent states, and another set of transient states. Weakly connected chains can be partitioned into two or more irreducible sets, and finally disconnected chains are completely separable into two or more chains, which can then be classified in the above manner. In figure 11.2, a diagrammatic presentation of this classification is given, demonstrating visually the meaning of these notions of accessibility and connectivity.

In terms of the Markov process, irreducible chains produce solutions that reflect a true compromise of all the factors, whereas in reducible chains, all factors are not

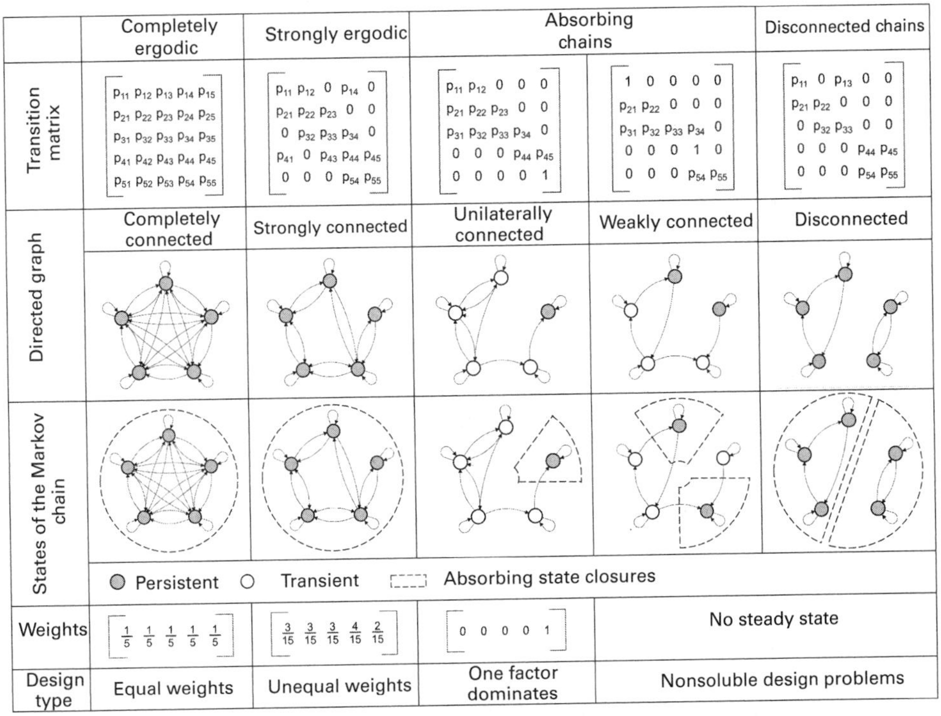

Figure 11.2
A classification of design problems as Markov digraphs.

compromised. In unilaterally connected chains, one factor or set of factors dominates, and the factors associated with the transient state have no effect. In weakly connected chains, however, no single solution will result, for two or more sets of factors show separate solutions that, because of inaccessibility, will never converge. Genuine compromises will only result using irreducible chains and thus, in the rest of this chapter, only strongly and completely connected chains will be considered; these are the only chains that lead to nontrivial solutions. It may be that, in reality, reducible chains match many unresolvable or only partly resolvable design problems, but in essence, a genuine compromise can only be achieved if every node or actor is connected either directly or indirectly to every other.

Symmetric Design Machines: A Dramatic Simplification

There is a special class of irreducible chains that deserves separate and serious consideration before this argument is developed further. These are the chains in which the original set of relationships between the factors $A_j(t)$ and $A_k(t)$ associated with the matrix $[a_{jk}]$ are symmetric. In such matrices, the steady-state weights w_k are identical, proportional to the in-degrees (or out-degrees) a_k (in equation [10.12]); that is, $w_k \propto a_k$. The weights can be calculated directly as

$$w_k = w_j = \frac{\sum\limits_j a_{jk}}{\sum\limits_j \sum\limits_k a_{jk}} = \frac{\sum\limits_k a_{jk}}{\sum\limits_j \sum\limits_k a_{jk}} \quad , \quad j = k. \tag{11.27}$$

If we substitute this weight into the steady-state equation (11.24) and write out the transition matrix in the basic form as $p_{jk} = a_{jk}/\Sigma_k a_{jk}$, we get

$$w_k = \sum_j \frac{\sum\limits_k a_{jk}}{\sum\limits_\ell \sum\limits_k a_{\ell k}} \frac{a_{jk}}{\sum\limits_z a_{jz}} \quad . \tag{11.28}$$

Equation (11.28) is immediately proven because of the symmetry implied by equation (11.27), and it has some important implications for a theory of design machines.

Where the relationships between factors in a design problem are symmetric, the above result means that the solution can be reached in one step from

$$c = \frac{\sum\limits_j A_j(t) \sum\limits_k a_{jk}}{\sum\limits_j \sum\limits_k a_{jk}}. \tag{11.29}$$

Several of the design methods introduced in chapter 10 were based on the assumption that the relationships between factors were binary and symmetric, that is, $a_{jk} = a_{kj}$ and $a_{jk} = 1$ *or* 0. Indeed, the example to be presented later was originally developed using a binary set of relationships. For such binary matrices, the above result implies that the weight of each factor in the final solution is proportional to the number of positive relationships between that factor and others in the set. This fact is intuitively appealing, and it suggests that problems whose structure can be described in simple binary terms have correspondingly simple sets of weights.

The extreme of the binary problem is interesting in this regard too because, as each factor is connected to every other factor, the weights of each factor in the final solution are equal to one another. In such a case, $p_{jk} = 1/n$, and it is clear that the transition-probability matrix is already in the steady state. Substituting for p_{jk} in equation (11.24) leads to

$$w_k = \sum_j w_j \frac{1}{n} = \frac{1}{n}, \tag{11.30}$$

and it is interesting to note that such a structure for the Markovian design machine corresponds exactly to the structure, used in physical design problems, that involves an immediate compromise using the technique of "sieving" introduced in chapter 10. At this point, the formal structure of the design machine has been elaborated in sufficient detail to twist the logic of the argument and extend the problem to consider the choice of a particular machine from a larger class of machines. In the following section, this problem is formally posed and solved in preparation for the application of such machines to real problems.

The Design of Design Machines

Markovian Decision Problems

As the design process is solved sequentially using the Markovian machine, it is likely that rewards or penalties are incurred as averaging takes place. For example, if it is necessary to achieve a compromise of conflicting factors, then some loss or gain can result, which in turn may affect the form of the process. In another sense, the reward-penalty characteristics are even more obvious if the problem is seen as one in which designers, holding different attitudes toward the problem, attempt to resolve their conflicts. The process already described can be modified to account for such behavior, as first suggested by Bellman (1957a) and elaborated in some detail by Howard (1960). The following results pertain only to irreducible chains.

Consider a process in which the designer has to choose between different sets called q of transition probabilities $[p_{jkq}]$. These imply different sets of rewards $[r_{jkq}]$ so that some criterion of value is optimized where the subscript q refers to these alternative sets of relationships and rewards that might be chosen. The problem can be formally set down as one in which the expected value of each factor j at time $t + 1$ called $v_{jq}(t + 1)$ is determined by

$$v_{jq}(t+1) = \sum_k p_{jkq}[r_{jkq} + v_{kq}(t)]. \tag{11.31}$$

As the first term in equation (11.31) is independent of the stage reached by the process, we define g_{jq} as

$$g_{jq} = \sum_k p_{jkq}\, r_{jkq}, \tag{11.32}$$

and equation (11.31) can be written as

$$v_{jq}(t+1) = g_{jq} + \sum_k p_{jkq} v_{kq}(t). \tag{11.33}$$

Note that the value of $v_{jq}(t + 1)$ not only depends upon the constant gain at each stage of the process g_{jq} but the value gained so far, $v_{jq}(t)$.

The task of choosing the design machine from q alternatives is one of optimizing the process at each stage or, in this case, of maximizing $v_{jq}(t + 1)$. This problem is stated as

$$v_{jq}(t+1) = \max_q \{ g_{jq} + \sum_k p_{jkq} v_{kq}(t) \}, \tag{11.34}$$

and in this form it is a classic problem of dynamic programming as considered by Bellman (1957b). Bellman's principle of optimality suggests a method of backward iteration to solve the above problem from time $t = 1$, but this method is mainly used to find the value of the process at any particular time. In the case presented here, of more interest is the average gain of the process as it continues indefinitely, and before this can be determined, it is necessary to investigate its limiting behavior.

Limiting Behavior of the Decision Process

First substitute $v_{jq}(t + 1)$ back into equation (11.33) to find the value of $v_{jq}(t + 2)$. Then, suppressing the superscript q for the moment,

$$v_\ell(t+2) = g_\ell + \sum_j p_{\ell j}\left[g_j + \sum_k p_{jk} v_k(t) \right] = g_\ell + \sum_j p_{\ell j} g_j + \sum_k p_{\ell k} v_k(t). \tag{11.35}$$

If the recurrence relation in equation (11.33) is applied indefinitely, then

$$v_\ell(t+m) = g_\ell + \sum_y p_{\ell y} g_y + \sum_z p_{\ell z}^2 g_z + \dots + \sum_k p_{jk}^{(m)} v_k(t). \tag{11.36}$$

Equation (11.36) can be written more concisely as

$$v_j(t+m) = g_j + \sum_{s=0}^{m-1}\sum_k p_{jk}^{(s)}g_k + \sum_k p_{jk}^{(m)}v_k(t),$$

(11.37)

where $p_{jk}^{(0)} = \delta_{jk}$, the Kronecker delta, so that $[p_{jk}^{(0)}]$ is the identity matrix.

It is possible to define the matrix $[p_{jk}^{(s)}]$ in terms of the steady-state matrix $[\pi_{jk}]$ and an error matrix $[E_{jk}^{(s)}]$. Then, using the following definition,

$$p_{jk}^{(s)} = \pi_{jk} + E_{jk}^{(s)},$$

(11.38)

and noting that

$$E_{jk}^{(0)} = \delta_{jk} - \pi_{jk},$$

(11.39)

equation (11.37) can be simplified in the following way. Substituting for $[p_{jk}^s]$ from equation (11.38), we obtain

$$v_j(t+m) = m\sum_k \pi_{jk}g_k + \sum_{s=0}^{m-1}\sum_k E_{jk}^{(s)}g_k + \sum_k \pi_{jk}v_k(t) + \sum_k E_{jk}^{(m)}v_k(t).$$

(11.40)

Then passing to the limit and noting that $E_{jk}^{(m)} \to 0$ for large m, we have

$$\lim_{s\to\infty} v_j(t+m) = m\sum_k \pi_{jk}g_k + \sum_{s=0}^{m-1}\sum_k E_{jk}^{(s)}g_k + \sum_k \pi_{jk}v_k(t),$$
$$= m\sum_k w_k g_k + \sum_k e_{jk}g_k + \sum_k w_k v_k(t)$$

(11.41)

where the summation of the error term $E_{jk}^{(s)}$ over s is defined as

$$\lim_{s\to\infty}\sum_s E_{jk}^{(s)} = e_{jk}.$$

(11.42)

However, equation (11.41) demonstrates that in the limit the process only depends upon the value of m and is thus linear. This result is seen more clearly if the following definitions are used:

$$v_j = \sum_k e_{jk}g_k + \sum_k w_k v_k(t),$$

(11.43)

and

$$z = \sum_k w_k g_k.$$

(11.44)

The substitution of equations (11.43) and (11.44) into (11.41) leads to

$$v_j(t+m) = v_k + m\,z,$$

(11.45)

which is clearly linear for each state k with intercept v_k and slope z. In fact, equation (11.44) defines the average gain of the process at each stage, and equation (11.45) serves to show that in the limit this gain dominates. Thus the optimization strategy must clearly be to maximize the average gain of the process. There are several ways to achieve this. Wolfe and Dantzig (1962) developed a linear programming strategy for this problem, but the algorithm used here is due to Howard (1960).

Howard's Algorithm

Howard's method is divided into two parts, which are applied sequentially and iteratively. Given an arbitrary starting decision as to the particular alternative q being adopted, the first operation, which is called the value-determination stage, produces the relative value v_j and the gain z. Then, with these values, the algorithm enters a policy-improvement stage, in which each alternative k is tested and the best selected. These new alternatives are then fed back into the value-determination operation, and the process continues until some convergence of decisions is reached.

First, the general equation for the decision process equation (11.33) is combined with the limiting equation (11.45) as

$$v_j + mz = g_j + \sum_k p_{jk} v_k(t + m - 1). \tag{11.46}$$

As the limiting value for $v_k(t)$ can also be determined from equation (11.45), equation (11.46) becomes

$$v_j + mz = g_j + \sum_k p_{jk}[(m-1)z + v_k] = g_j + \sum_k p_{jk} v_k + (m-1)z \sum_k p_{jk}. \tag{11.47}$$

Rearranging and simplifying equation (11.47) leads to

$$v_j + z = g_j + \sum_k p_{jk} v_k, \tag{11.48}$$

and this is a system of n equations in $n + 1$ unknowns. In its present form, it is not solvable, but because of the linearity of the limiting equations, it is the relative values of v_k that are important. Thus relative values for v_k and the gain z can be derived if one value for v_k is set equal to zero and the resulting equation system solved in the standard way.

The relative values from the modified equation system based on equation (11.48) can now be substituted into the optimization equation (11.34). Then

$$v_j(t + m) = \max_q \left\{ g_{jq} + \sum_k p_{jkq}[mz + v_k] \right\} = \max_q \left\{ g_{jq} + mz + \sum_k p_{jkq} v_k \right\}. \tag{11.49}$$

As the gain mz is constant for all k, then it is only necessary to select q from $g_j + \Sigma_k p_{jkq} v_k$, which can be computed for every alternative. The new reward vector

$[g_{kq}]$ and matrix $[p_{jkq}]$ are then fed back into equation (11.48), and equations (11.48) and (11.49) are reiterated until a stable set of alternatives occurs.

Where the number of alternatives is small, it is possible to maximize the gain of the process and thus choose the optimal design machine by solving equation (11.49) for all combinations of matrix $[p_{jkq}]$. However, for anything but a small number of alternatives and states, this approach cannot be used. For example, consider a process of twenty states, each with two alternatives. There are 2^{20} possible design machines, and any strategy designed to select one of these that is based on complete enumeration is clearly impossible.

Applications of Design Machines

The Alexander-Manheim Highway Location Problem

We have already applied the Markovian design machine in chapter 10 to a physical planning problem involving the search for new residential land in a small town. But we still need to compare variants of these methods, and here we will develop applications involving such comparisons. The design problem chosen was initiated by Alexander and Manheim (1962a) and it represents something of a classic in the sense that it is one of the best examples of the hierarchical approach to design expounded by various design theorists, but best developed by Alexander (1964) in his book *Notes on the Synthesis of Form*. The problem is one of locating a highway between Springfield and Northampton in western Massachusetts, and the approach to design used by Alexander and Manheim (1962a) and adopted here involves finding the best possible routes for the highway according to a set of conflicting physical factors. Twenty-six physical factors were chosen, and each factor was mapped in terms of its implied solution to the problem. Thus each factor showed areas that were acceptable or unacceptable locations for the highway. These twenty-six factors are presented in figure 11.3, which is taken from the original report. The gradation from black to white reflects the gradation from high to low potential land suitability for the location of the new highway.

Alexander and Manheim tackled the problem by first specifying all possible pairwise connections between the factors in binary terms. Factors were related to others if they were judged to strongly reinforce or conflict with those others. No formal measure of the strength of connection was developed in terms of the correlation analysis outlined in chapter 3, and thus the matrix of relations was derived intuitively. The matrix was then decomposed using a cluster-analysis method, and the resulting hierarchy was used as a sequence in which to synthesize or resolve the factors to a final solution. Alexander and Manheim were also at pains to point out that the process of synthesis was not just one of averaging, but one in which the designer strengthened the emerging solution according to his or her ideas about the

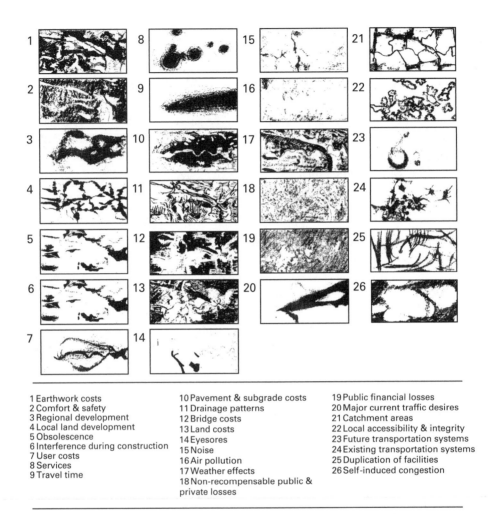

1 Earthwork costs
2 Comfort & safety
3 Regional development
4 Local land development
5 Obsolescence
6 Interference during construction
7 User costs
8 Services
9 Travel time

10 Pavement & subgrade costs
11 Drainage patterns
12 Bridge costs
13 Land costs
14 Eyesores
15 Noise
16 Air pollution
17 Weather effects
18 Non-recompensable public & private losses

19 Public financial losses
20 Major current traffic desires
21 Catchment areas
22 Local accessibility & integrity
23 Future transportation systems
24 Existing transportation systems
25 Duplication of facilities
26 Self-induced congestion

Figure 11.3
Factor maps for the Alexander-Manheim highway location problem.

Figure 11.4
The Alexander-Manheim hierarchical design solution.

anticipated form of the ultimate structure. Thus intuition was allowed to enter the process through the back door, so to speak, and this makes any formal comparison of solutions to this problem using different methods somewhat uncertain. Nevertheless, comparisons will be made with the solution synthesized by Alexander and Manheim, which is presented in figure 11.4.

Three different solutions to this highway location problem have been generated using the Markovian design machine. These Markovian solutions have all been computed using the result in equation (11.27), because their relationship matrices are symmetric. The Markovian equivalent of the Alexander-Manheim problem has been computed using the original binary relationships presented in the report, and this serves as a direct comparison. A second solution in which all factors are equally weighted is analogous to the sieve-mapping approaches of Keeble (1952) and McHarg (1969) outlined in the last chapter, and a third solution has been generated using a matrix of relations based on spatial conflict. It is possible to compute a spatial coefficient of association between each pair of factors, and the matrix of relations $[\hat{a}_{jk}]$ has been based directly on this association:

$$\hat{a}_{jk} = \frac{M_{jk}}{M_{jk} + U_{jk}}, \ 0 \le \hat{a}_{jk} \le 1. \tag{11.50}$$

Each factor has been coded on a regular grid of cells, and M_{jk} represents the number of matching cells between any two factors $A_j(t)$ and $A_k(t)$; U_{jk} are the non-matching cells. Clearly $[\hat{a}_{jk}]$ is symmetric, and the implicit theorem that the steady-state weights are proportional to the in-degrees or out-degrees holds.

In figure 11.5, the solutions generated by these three variations on the problem are presented, and a direct comparison with the Alexander-Manheim solution

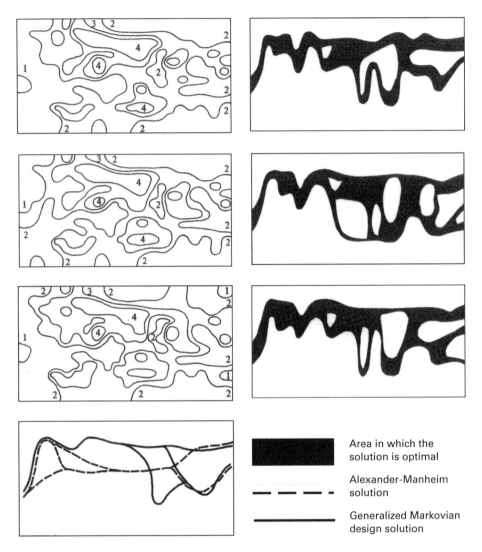

Area in which the
solution is optimal

Alexander-Manheim
solution

Generalized Markovian
design solution

Figure 11.5
Solutions generated by the design.

reveals some significant differences. In all three cases the solutions are very similar to one another, with the spatial association and Alexander-Manheim related solutions being somewhat firmer versions of the "sieve map" solution. In fact, this is an interesting result in itself, for it confirms the hunch that the differences between solutions to the same problem with different weighting structures are not as great as the differences in the weights themselves. In terms of the original solution in figure 11.4, the three solutions in figure 11.5 place the main line of the highway much farther to the west, whereas Alexander and Manheim's route is close to the Connecticut River. The Markovian solutions are similar to the original solution over the first quarter of the highway near Springfield, but it is difficult to see how Alexander and Manheim were able to select their route when this route is examined against each individual factor. However, this comparison is uncertain because of the intuitive bias mentioned by Alexander and Manheim and because of their obviously greater local knowledge of site conditions. Still, these comparisons are interesting and provide alternative ways of looking at a well-known problem.

In figure 11.6, the different weighting structures for the three Markovian problems are presented, and this reinforces the observation made above concerning the differences in the solution in terms of the differences in weights. Perhaps the most important aspect of these weights, however, is the comparison between factors. In

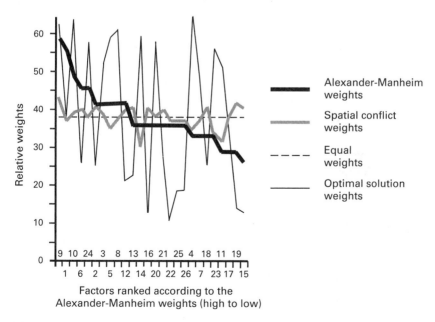

Figure 11.6
Weighting structures generated by the design machine.

a sense, the weights should reflect some intuitive, or at least explicable, notion concerning the importance of various factors. In the Alexander-Manheim problem, the most important factors are concerned with the construction of the road, whereas the least important factors relate to institutional constraints; whether or not this should be the optimal ordering in the mind of the designer is a matter for some debate, but the very fact that the Markovian design machine is able to generate such information serves to get such a debate started. This in itself is seen as an important advantage of the method.

Selection of an Optimal Design Machine

Perhaps the most speculative part of the theory presented here relates to the design of design machines. It is easy to see that different machines can exist and that some are more suitable than others, but it is not easy to devise relevant criteria on which the selection can be made. In this particular context an optimal machine has been synthesized from a range of possible alternative structures based on different combinations of relationships reflecting causal or statistical regularity. In the field of design, some debate has been centered on whether or not causal relationships or statistical relationships should form the basis on which design compromises are effected, yet this distinction between causal and statistical is only one way in which design machines might be judged for optimality. The fact that there appear to be so many ways in which optimality can be construed, and so few principles on which to evaluate this optimality, only serves to reinforce the speculative nature of this argument. As we have been at pains to point out throughout this book, our science is suggestive, not definitive, for problems of cities and city planning require multiple approaches involving multiple stakeholders. The purpose of our science is to inform the dialogue, not to generate "answers" or "solutions" per se, notwithstanding the fact that we represent the argument in these terms.

In figure 11.7, two graphs reflecting strictly causal or statistical relations are presented, together with somewhat arbitrary reward structures that are chosen intuitively. A third graph reflects the optimal structure selected using Howard's (1960) algorithm outlined above. One of the problems in implementing this method concerns the degree of connectivity of the chosen structure; clearly the optimal machine must be strongly connected for either Howard's method or the Markovian design solution procedure to have meaning. A check on the connectivity of the structures produced during the algorithm has been devised, and structures that do not meet the requirement of strong connectivity have been abandoned, leading to less optimal structures being chosen. In this context, it is likely that there are less arbitrary checking procedures, and this establishes an area for further technical research.

Another difficulty that arises, especially when dealing with design machines based on binary relationships, concerns the possibility that linear dependence might exist

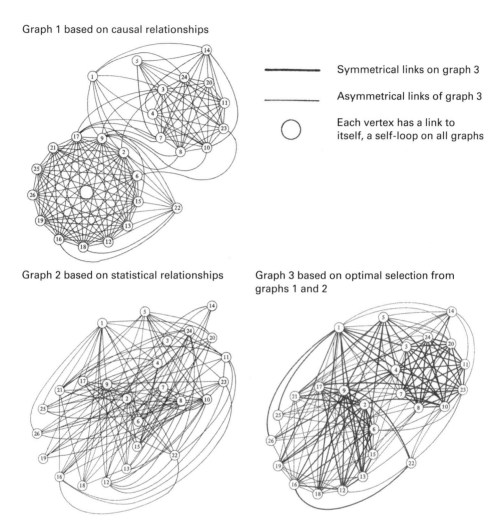

Graph 1 based on causal relationships

— Symmetrical links on graph 3

— Asymmetrical links of graph 3

◯ Each vertex has a link to itself, a self-loop on all graphs

Graph 2 based on statistical relationships

Graph 3 based on optimal selection from graphs 1 and 2

Figure 11.7
Selection of the optimal design machine from two alternative systems. As such graphs are immensely difficult to draw, an impression of the different structures has been obtained by constructing an ideal form for the first graph and drawing the second and third graphs according to this form.

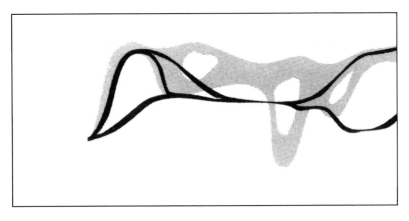

 Path of the Alexander-Manheim solution

Solution by the optimal design machine

Figure 11.8
The solution generated by the optimal design machine.

within the original machine or might be generated in the search for an optimal machine. Linear dependence within the $[p_{jkq}]$ matrix means that two or more factors have identical sets of relationships and, in terms of the design machine, are indistinguishable. This also has implications in terms of the wider design problem, and its importance depends upon the way in which the designer evaluates redundancy in the set of relationships. But from a purely technical point of view, it can be a problem, especially if it arises in the optimal search process, for it means that relative gains cannot be calculated. In fact, in this problem, where Howard's algorithm is started with the use of the statistical set of relationships, the issue does not arise, but it does if the process is started with the use of the Alexander-Manheim binary matrix, for this is linearly dependent; the implications of such dependency are likely to be important.

The weights characteristic of this optimal design machine have been computed with the use of the conventional algebra of equation (11.25), and the final solution is demonstrated in figure 11.8. This solution is similar to those in figure 11.5, and it also has similarities to the original solution in figure 11.4, thus demonstrating that the technique is robust in producing sensible and realistic solutions to design problems.

Design as Art and Science

The style of theory building advocated here, which is suggestive rather than definitive, deductive rather than inductive, is in direct contrast to other areas of planning and design, where there has been much more emphasis on the observation of the design process and on theory building on the basis of such data. Thus this approach could be criticized as being rather sterile, in that the emphasis on evolving a theory consistent with reality has been minimal. Yet no one approach is legion; it is essential to develop a perspective in which this kind of approach is judged on its contribution to an overall body of knowledge about design. Too often those interested in design swing from one extreme to another, rejecting all that has gone before rather than attempting to evolve a well-grounded set of principles and techniques. It is essential to link theory to practice and abstraction to reality, but the problems in doing so in this area are formidable and there are unlikely to be any easy answers. Both theoretical and empirical approaches are required, for there is an urgent need to establish a body of research in which design is regarded as a distinct activity that merges art with science, as Mumford (1928) implied in the quote at the heading of the last chapter. This will only become possible if we accept that analogies from areas such as the one proposed here might help to stretch the imagination and lead to better understanding. Alexander (1964) made this point half a century ago, and he is still making it (Alexander, 2012).

Although fitting or calibrating the design machine to a real situation has not been established here, there are ways in which this might be attempted. The interpretation that treats the machine as a social power structure in which decision making occurs seems promising, although data concerning such processes are extremely difficult to obtain and many significant characteristics may be impossible to observe. Some progress has, however, been made in social psychology using formalized group structures for problem solving and, although such studies may seem remote from the focus of this book, potential links need to be exploited. We outlined some of these in chapter 3, but we will develop them substantially in the next three chapters. Their value, we believe, lies in the raising of problems rather than in attempts at a solution, in providing models and analogs of the multiple interests in planning and design and how these might be reconciled through formal processes of conflict resolution. Many technical improvements to the theory could be easily incorporated in every facet, but more progress would probably come from exploring the practical implications of these concepts. In the social and design sciences, it is easy to speculate on theory, but it is immensely difficult to generate useful and lasting insights through good theory. The real challenge is to understand insights engendered in good design, and in the next chapter, the theory presented will be elaborated in an attempt to further clarify and bound this problem.

12

A Theory for Collective Action

I propose that we think of an organization as a *system of social exchange*, with a delimited set of events and a delimited set of persons as actors. The system is defined by the actors, the events, the structure of control of actors over events, and the structure of interest of actors in events.

—James S. Coleman, "Social Structure and a Theory of Action," in *Approaches to the Study of Social Structure* (1976, p. 86)

Our science of cities, which we began to elaborate in the first two parts of this book, focused our perspective on the city as an artifact, as a product rather than a process. But we very quickly reached the point where form gave way to function as urban processes began to dominate our models of how elements of the city relate to one another and how these elements provide the dynamics of change that drives the way cities evolve. The same change in focus has been apparent in the study of planning, in contrast to the study of cities, where there is now an increasing concern for processes. This change in emphasis is reflected in the new philosophies of planning and design that seek to see such processes as an integral part of a theory of social behavior in which such problems are tackled by society at large rather than by individual expert groups or professionals. This theme or paradigm is typical of post-rationalist planning theory based on planning as a process of community learning, originally advocated by theorists such as Friedmann (1973) and Rittel and Webber (1973), among others, and elevated through processes of argumentation and negotiation that now dominate contemporary planning theory (Forester, 2009; Healey, 2006). This change in approach is not hard to explain. The apparent failure of the systems approach to provide a tractable and well-defined theoretical base for planning, coupled with the deep mismatch between theory and practice, problems and solutions, have led to a retreat from the certainty of science. A view of planning that can easily accommodate changing value systems, the complexity of pluralism, and the notion of fallibility now seems much more appropriate. Thus the concept of planning as social learning and society-wide problem solving has become the accepted norm.

There are many ways of elaborating this new paradigm. For example, a dominant view is that planning should be based on a system of transaction and communication in small groups, somewhat similar in form to the theories of social exchange advocated by Blau (1964) and by Homans (1974). Bazjanac (1974), writing on Rittel's idea of design as an argumentative process, also emphasizes the role of communication and learning in decision making. He states, "Most decisions are in effect negotiated. This means that the process of arriving at better decisions is not a process of optimization in the operations research sense; it is rather a process of negotiation and compromise between parties" (p. 11). The model of design developed in the last two chapters quickly moved towards this perspective, treating design as a process of collective rather than individual action, or at least collective action where individuals come together to pool ideas and to effect rational compromises as opposed to individual action. The seeds of this approach lie in the idea of design as social problem solving. The simple algebra for design suggested in the last two chapters will propel this into a much richer framework of formal action, one that enables us to link actors reflecting the diversity of the context for design with the problems, policies, and plans they are motivated to solve.

This notion of design as collective action requires some explanation. It is very clear that the product of design—the plan, set of policies, or whatever its representation and form—is a public or collective good in the sense defined by Olson (1965). But in the past, the act of design has always been thought of as an individual action made by the expert or professional. Design, like many aspects of public government, has been the provision of collective goods through individual action. Newer theories of planning and design, however, suggest that traditionally, one of the field's failures has been this emphasis on expertise or individual action, and that a better product might result if design were treated as collective action. There is a great deal of appeal in this idea and already, Coleman (1966) has outlined the rudiments of such a theory. In essence, this chapter will elaborate the process by which a collectivity or small group might solve a problem of physical design, and this will relate directly to concepts of balance and equilibrium within the group. It will also draw on ideas about exchange and power, and the resources available to the group to accomplish the given task.

In the last two chapters, we developed a model of conflict resolution between a set of n factors, defined as A_{ik}, where we assumed or computed key relationships between each of these factors. In fact, as we need to embellish our notation quite extensively in this and subsequent chapters in order to distinguish between factors and actors, we will now use the subscripts i, j, k, ℓ to pertain to objects or agents such as actors, factors, plans, policies, and so on, and we will refer to a factor as pertaining to an actor i in zone or cell z as A_{zi}. Do not confuse these new definitions with the old in the two previous chapters, where the subscript i was used for location. In those chapters, we defined the set of relations as a communications network for routine transition and transformation of these factors using an averaging procedure. The weights were based on the number of connections that each factor has to all others, and we demonstrated that by using these connections, each factor would transform at each interaction to something closer to the ultimate average. We showed that this process of sequential averaging led to a steady-state compromise that was also mirrored in a Markov process, which provided an alternative and complementary interpretation of the averaging. We also demonstrated that in the case where the connection matrix was symmetric, the weight of each factor in the final solution was proportional to the in-degrees or out-degrees of the connection or adjacency matrix, a particularly simple result with respect to weighted averaging. Like the models of the last chapter, no empirical testing is possible for the model presented here, since our focus is largely normative. In effect, real processes of conflict resolution do not converge in the way we have assumed, and in this sense, the process we adopt is an idealization.

Models similar to this one have been explored by various authors in sociological and social psychological contexts. Perhaps the closest model is the one we noted in the two previous chapters suggested by French (1956), elaborated by Harary (1959), and generalized by de Groot (1974), but Rapoport (1949) appears to have been the first to speculate on the relevance of Markov chains for tracking changes in attitudes and other social characteristics. In a slightly different context, Anderson (1954) and Kreweras (1968) have explored the use of these types of models in analyzing changes in voting patterns. There is a long tradition from these early efforts that we noted in chapter 3, of which more recent work by Hegselmann and Krause (2002), DeMarco, Vayanos, and Zwiebel (2003), and Blondel, Hendrickx, Olshevsky, and Tsitsiklis (2005) is representative. The general consensus from these studies is that such models are useful first approximations to social processes of attitude change, but the Markovian assumption is particularly strong and is unlikely to hold empirically. It is in this sense that these models are statements of optimal processes and solutions.

In the models so far, we assumed that each actor is matched directly with one and only one factor. Despite this rather neat form given in previous chapters, such as in equations (11.19) to (11.24), there are several major limitations that must be resolved if the approach is to have anything more than theoretical interest. One of these limitations is this very rigid assumption that each factor is directly matched with one and only one actor or interest group; this is surely too unrealistic an assumption to continue with. It implicitly assumes that each actor has only one view of the problem, and although actors may have a dominant view, there is considerable variation that must be accounted for. Thus a major task of this chapter is to relax this assumption and to develop a model in which actors and factors are treated separately but consistently. Such an extension already exists in the work of Coleman (1966, 1972, 1973, 1998) where actors and factors are related through two separate sets of relationships—the interest each actor has in the issue or factor, and the control each individual has over the factor. These two sets of relationships might essentially be interpreted as bipartite graphs, in the manner outlined in chapter 3 and developed extensively for relationships between two sets of objects, nodes, and street segments in chapters 6 and 7. Here we have switched our focus to the design of cities rather than their representation, as our two sets of objects are now actors and factors. But as we will see, each set of relations is different from the other, whereas in our previous exposition we had but one set considered in two ways from a bipartite graph structure. Here we have two such graphs of the same order but that deal with different relations. This is Coleman's distinction, which provides a simple but effective way of extending the sequential averaging Markov model and relaxing the rigid assumption of correspondence between factors and actors. In the next section, we will extend our model accordingly.

Chains of Collective Action

Defining Interest and Control

The interest each actor i has in factor k is called x_{ik}, and this relates to the effect each factor has on each actor in the context of a design solution. Clearly, the number of factors and actors might differ. We assume here that there are n actors and m factors; thus the interest matrix $[x_{ik}]$ can be written in probability form as

$$[x_{ik}] = \begin{bmatrix} x_{11} & x_{12} & \dots & x_{1m} \\ x_{21} & x_{22} & \dots & x_{2m} \\ . & . & \dots & . \\ x_{n1} & x_{n2} & \dots & x_{nm} \end{bmatrix}, \sum_{k=1}^{m} x_{ik} = 1. \tag{12.1}$$

Each interest x_{ik} is defined as the proportion of the interest that actor i shows in factor k. In similar terms, each factor ℓ has a certain degree (proportion) of control exercised by an actor j in terms of the design solution. This control, called $c_{\ell j}$, can also be represented in probability matrix form as

$$[c_{\ell j}] = \begin{bmatrix} c_{11} & c_{12} & \dots & c_{1n} \\ c_{21} & c_{22} & \dots & c_{2n} \\ . & . & \dots & . \\ c_{m1} & c_{m2} & \dots & c_{mn} \end{bmatrix}, \sum_{j=1}^{n} c_{\ell j} = 1. \tag{12.2}$$

From these definitions of interest and control, two derived sets of relationships can be calculated: first, by mapping control into interests, and then by mapping interests into control. In terms of the first mapping, a new relationship between factor k and factor ℓ can be derived by finding the degree of control and the amount of interest in any factor in terms of a particular actor. Then, by summing over all actors,

$$q_{k\ell} = \sum_j c_{kj} x_{j\ell} \text{ where } \sum_\ell q_{k\ell} = 1, \tag{12.3}$$

which results from the multiplication of the two stochastic matrices $[c_{kj}]$ and $[x_{j\ell}]$. In equation (12.3), $q_{k\ell}$ gives the proportion of control multiplied by interest summed over all actors who have control in k and interest in ℓ. In a similar way, the second mapping is achieved by finding the product of the interest of actor i in any factor and the control of actor j in any factor and summing over all factors. Then

$$p_{ij} = \sum_k x_{ik} c_{kj} \text{ where } \sum_j p_{ij} = 1. \tag{12.4}$$

The two matrices defined in equations (12.3) and (12.4) are both stochastic; the first refers to relationships between events and the second to relationships between actors, and the idea that these might form the basis for an averaging of factors or attitudes by using the previous model immediately suggests itself. Furthermore, these two matrices are consistently and unambiguously related to one another, and thus their associated Markov chains must be related. We will now explore these two processes and then relate them in preparation for more substantive analysis in the next section.

Interpretation as a Markov Process
In terms of the set of relationships between factors $[q_{k\ell}]$, new factors at time $t + 1$ can be found from the averaging equation

$$A_{zk}(t+1) = \sum_\ell \sum_j c_{kj} x_{j\ell} A_{z\ell}(t) = \sum_\ell q_{k\ell} A_{z\ell}(t). \tag{12.5}$$

There is a rather nice interpretation of equation (12.5) in terms of the original matrices $[c_{kj}]$ and $[x_{j\ell}]$; first, each actor produces an attitude toward the solution by weighting each factor by his interests and summing over all factors, that is, $\Sigma_\ell x_{j\ell} A_{z\ell}(t)$. Then these attitudes are transformed into new factors by weighting each attitude by the degree of control vested in each factor and summing over attitudes.

The process continues in this way, and by recursion on equation (12.5), new factors are formed from

$$\begin{aligned} A_{zk}(t+m) &= \sum_\ell \sum_r \quad \dots \quad \sum_j c_{kj} x_{jo} \dots \quad c_{qr} x_{r\ell} A_{z\ell}(t) \\ &= \sum_\ell q_{k\ell}^{(m)} A_{z\ell}(t) \end{aligned} \tag{12.6}$$

Assuming that $[q_{k\ell}^{(m)}]$ is irreducible, the matrix converges to the steady-state matrix $[v_{k\ell}]$ and the fixed point vector $[V_\ell]$, that is

$$\lim_{m \to \infty} q_{k\ell}^{(m)} = v_{k\ell} = V_\ell. \tag{12.7}$$

The limit of equation (12.6) becomes

$$A_z = \lim_{m \to \infty} A_{zk}(t + m) = \sum_\ell V_\ell A_{z\ell}(t), \tag{12.8}$$

and the fixed-point vector can be calculated directly from

$$V_\ell = \sum_\ell V_\ell q_{k\ell}, \tag{12.9}$$

which bypasses the need for an iterative solution to the process.

The second set of relationships between actors given by $[p_{ij}]$ also sets up a chain of averaging in terms of the attitudes of actors in contrast to factors, the process being exactly analogous to that given above. It will be presented here for completeness and for purposes of definition, although nothing of note is really added. First, consider the set of original attitudes $\{\bar{A}_{zj}(t)\}$, which are transformed to new attitudes at time $t + 1$ by

$$\bar{A}_{zi}(t + 1) = \sum_j \sum_k x_{ik} c_{kj} \bar{A}_{zj}(t) = \sum_j p_{ij} \bar{A}_{zj}(t). \tag{12.10}$$

As in the first process, there is a physical interpretation of equation (12.10), which is the reverse of that for equation (12.5). Here the product of control and attitudes leads to an expected factor, which is in turn transformed to a new attitude when interests are applied. Recursion on equation (12.10) leads to

$$\begin{aligned} \bar{A}_{zi}(t + m) &= \sum_j \sum_q \quad \cdots \quad \sum_k x_{ik} c_{k\ell} \cdots \quad c_{pq} x_{qj} \bar{A}_{zj}(t) \\ &= \sum_j p_{ij}^{(m)} \bar{A}_{zj}(t) \end{aligned}. \tag{12.11}$$

For the irreducible transition probability matrix $[p_{ij}]$, equation (12.11) converges and the limit of $[p_{ij}^{(m)}]$ is

$$\lim_{m \to \infty} p_{ij}^{(m)} = R_{ij} = r_j. \tag{12.12}$$

By using the fixed-point vector $[r_j]$ given in equation (12.12), in the limit, equation (12.11) becomes

$$\bar{A}_z = \lim_{m \to \infty} \bar{A}_{zi}(t + m) = \sum_j r_j \bar{A}_{zj}(t), \tag{12.13}$$

and r_i can be calculated directly from the steady-state equation

$$r_i = \sum_j r_j p_{ij}. \tag{12.14}$$

These two Markov processes are clearly related, as a comparison of equations (12.6) and (12.11) indicates, and in the next section the two chains are linked in terms of steady-state equations, thus leading to some additional insights into these processes.

Analysis of the Steady State

By comparing equation (12.6) with equations (12.11) and (12.12), it is clear that in the limit, the steady-state matrix $[v_{k\ell}]$ can be written as

$$v_{k\ell} = \lim_{m \to \infty} \sum_i \sum_j c_{ki} p_{ij}^{(m)} x_{j\ell}. \tag{12.15}$$

Now, substituting $p_{ij}^{(m)}$ from equation (12.12) into equation (12.15) gives

$$v_{k\ell} = \sum_i \sum_j c_{ki} R_{ij} x_{j\ell}, \tag{12.16}$$

and, in an analogous way, it can be shown that

$$R_{ij} = \sum_k \sum_\ell x_{ik} v_{k\ell} c_{\ell j}. \tag{12.17}$$

Equations (12.16) and (12.17) provide direct links between the two Markov chains in the steady state, but these can be simplified further. Substituting the fixed-point vectors $[r_j]$ and $[V_\ell]$ for $[R_{ij}]$ and $[v_{k\ell}]$, respectively, in equation (12.16) leads to

$$V_\ell = \sum_i \sum_j c_{ki} r_j x_{j\ell} = \sum_i c_{ki} \sum_j r_j x_{j\ell}, \tag{12.18}$$

and because $\Sigma_i c_{ki} = 1$, equation (12.18) simplifies to

$$V_\ell = \sum_j r_j x_{j\ell}. \tag{12.19}$$

In the same way, an analogous equation for $[r_j]$ can be derived as

$$r_j = \sum_\ell V_\ell c_{\ell j}, \tag{12.20}$$

and equations (12.19) and (12.20) are now the fundamental equations linking the two processes.

There is an interpretation for these equations in the following terms: the weights each actor gives to each attitude in the system is the sum of the product of the weight of each factor with the degree of control over each factor. Alternatively, the weight of each factor is the sum of the product of the weights of each attitude with the degree of interest shown in each attitude. By using these ideas, it is clear that the weights reflect a state of balance between interests and control such that equations (12.19) and (12.20) form a simultaneous equilibrium. These concepts will be taken up later when the model is compared with Coleman's (1966) theory.

The relationship between factors and attitudes is also of interest, for this illustrates another way in which the two chains are closely associated. In the design problems examined in the last two chapters, it was customary to specify the solution in terms of factors by taking the interest of each actor and finding an average of the factors over all interests. Then, given the set of factors $\{A_{zk}(t)\}$, attitudes can be derived from

$$\bar{A}_{zi}(t) = \sum_k x_{ik} A_{zk}(t). \tag{12.21}$$

From equation (12.10), which tracks the changing attitudes through time, a new attitude is derived as

$$\bar{A}_{zi}(t+1) = \sum_j p_{ij} \bar{A}_{zj}(t) = \sum_\ell \sum_j \sum_k x_{ik} c_{kj} x_{j\ell} A_{z\ell}(t). \tag{12.22}$$

The last three terms on the right-hand side of equation (12.22), however, are given by equation (12.5), which calculates $A_{zk}(t+1)$. Thus equation (12.22) becomes

$$\bar{A}_{zi}(t+1) = \sum_k x_{ik} A_{zk}(t+1), \tag{12.23}$$

which is of the same form as equation (12.21). These arguments can, of course, be generalized to the limiting equations, and these are given as follows:

$$\lim_{m \to \infty} \bar{A}_{zi}(t+m) = \lim_{m \to \infty} \sum_j p_{ij}^{(m)} \bar{A}_{zj}(t)$$
$$= \lim_{m \to \infty} \sum_k \sum_\ell x_{ik} q_{k\ell}^{(m)} A_{z\ell}(t), \tag{12.24}$$

and

$$\lim_{m \to \infty} \bar{A}_{zi}(t+m) = \lim_{m \to \infty} \sum_k x_{ik} A_{zk}(t). \tag{12.25}$$

It has already been demonstrated in the two previous sections that $A_{zk}(t+m)$ converges to A_z and equation (12.25) confirms that $\bar{A}_z = A_z$ in the limit; thus the consensus in attitudes is the same as the consensus in factors. The same kind of convergence can be demonstrated for the associated chain. If the attitudes are specified initially and the factors are related to them through the control relationships by

$$A_{z\ell}(t) = \lim_{m \to \infty} \sum_j c_{\ell j} \bar{A}_{zj}(t), \tag{12.26}$$

then it can be shown that, in the limit

$$\lim_{m \to \infty} A_{z\ell}(t+m) = \lim_{m \to \infty} \sum_j c_{\ell j} \bar{A}_{zj}(t+m). \tag{12.27}$$

If equations (12.21) and (12.26) hold initially, this implies that $\bar{A}_{zi} = A_{zk} = A_z$, $\forall i, k$. In other words, the process is already in equilibrium and a consensus exists. This can easily be verified by substituting equation (12.21) into (12.26) and vice versa, and comparing the resultant forms with equations (12.25) and (12.27).

Coleman's Theory of Exchange

The Exchange Relations

The theory of collective action proposed by Coleman (1966, 1972, 1973, 1998) is based on the notion that each actor j ascribes a certain value V_ℓ to the degree of control $c_{\ell j}$ possessed over the factor ℓ. The actor is able and willing to influence events according to his or her interest in any factor $x_{j\ell}$ and according to available power or resources r_j. In this context, power is no longer a relational concept between actors but a measure of influence based on an actor's ability to act by using his or her resources. Coleman argues that in a system of perfect exchange, actors will move to an equilibrium in which the value of their control over a particular factor is equal to their resources, which are allocated to that factor in proportion to their interests. Such a situation can be brought about in several ways: by altering control, by altering interests, or by altering factor values or resources through social exchange. Formally, the following relationship must hold as

$$V_\ell c_{\ell j} = r_j x_{j\ell}, \tag{12.28}$$

and by summing equation (12.28), first over j and then over ℓ, value and power can be determined from

$$V_\ell = \sum_j V_\ell c_{\ell j} = \sum_j r_j x_{j\ell}, \tag{12.29}$$

and

$$r_j = \sum_\ell r_j x_{j\ell} = \sum_\ell V_\ell c_{\ell j}. \tag{12.30}$$

Note that Coleman's theory gives rise to equations that are identical to equations (12.19) and (12.20), which relate the two Markov chains in the steady state. Note also that V_ℓ and r_j can be interpreted as the steady-state weights as well as value and power. Coleman then suggests that one way in which actors might form a new control matrix $[c_{\ell j}^*]$ is by allocating their power in proportion to the ratio of their interests and value; that is,

$$c_{\ell j}^* = \frac{x_{j\ell}}{V_\ell} r_j = \frac{x_{j\ell}}{V_\ell} \sum_k V_k c_{kj}. \tag{12.31}$$

In the same way, a new interest matrix could be formed by

$$x_{j\ell}^* = \frac{c_{\ell j}}{r_j} V_\ell = \frac{c_{\ell j}}{r_j} \sum_i r_i x_{i\ell}, \tag{12.32}$$

although Coleman does not suggest this. If it is argued that new control and interest relations occur in equilibrium, then these equations must be solved simultaneously or probably iteratively.

The real focus here, however, is in the correspondence between equations (12.19) and (12.20) and equations (12.29) and (12.30). Clearly, Coleman's theory can be interpreted as a Markov process in which values and resources (power) are gradually changed as actors communicate with one another. In fact, from an initial set of values $\{V_k(t)\}$ and resources $\{r_i(t)\}$, new variables are derived at time $t + 1$ using the forward Markov equations

$$V_\ell(t+1) = \sum_k V_k(t) q_{k\ell}, \tag{12.33}$$

and

$$r_j(t+1) = \sum_i r_i(t) p_{ij}. \tag{12.34}$$

These equations converge in the limit to V_ℓ and r_j, respectively, but from the initial distributions of value and power, on any iteration $t + 1$, these quantities are not consistent with equations (12.19) and (12.20). That is, in general,

$$V_\ell(t+1) \neq \sum_j r_j(t+1) x_{j\ell}, \tag{12.35}$$

and

$$r_j(t+1) \neq \sum_\ell V_\ell(t+1) c_{j\ell}. \tag{12.36}$$

The question then arises as to how the difference between the actual values of control and resources and the perceived or expected values of these same variables, computed from the actual values, converge. From the initial distributions of actual values of control and power, perceived or expected values, called $V_\ell^*(t)$ and $r_j^*(t)$, respectively, can be calculated from

$$V_\ell^*(t) = \sum_j r_j(t)x_{j\ell}.$$

(12.37)

and

$$r_j^*(t) = \sum_\ell V_\ell(t)c_{j\ell}.$$

(12.38)

These perceived quantities occur in the associated Markov processes given by equations (12.33) and (12.34), for equations (12.37) and (12.38) are part of equations (12.34) and (12.33), respectively. Then

$$\begin{aligned} V_\ell(t+1) &= \sum_j r_j^*(t)x_{j\ell} \\ &= \sum_k \sum_j V_k(t)c_{kj}x_{j\ell} = \sum_k V_k(t)q_{k\ell}, \end{aligned}$$

(12.39)

and

$$\begin{aligned} r_j(t+1) &= \sum_\ell V_\ell^*(t)c_{\ell j} \\ &= \sum_i \sum_\ell r_i(t)x_{i\ell}c_{\ell j} = \sum_i r_i(t)p_{ij}. \end{aligned}$$

(12.40)

As in the expositions of these processes in earlier sections, there are physical interpretations of equations (12.39) and (12.40); for example, in the case of equation (12.40), a new distribution of resources is derived by finding the expected value of each factor, given the old level of resources and the interests, and then applying the available control to this expected value.

That the expected values converge to the actual values of r_j and V_ℓ in the limit is intuitively obvious from the previous argument, but to reinforce the point, it is worth demonstrating explicitly for, say, resources. The proof of convergence for one variable immediately leads to the other. The expected value of power $r_j^*(t+m)$ is then calculated from

$$r_j^*(t+m) = \sum_\ell V_\ell(t+m)\,c_{\ell j} = \sum_k \sum_\ell V_k(t)q_{k\ell}^{(m)},$$

(12.41)

and the actual value $r_j(t+m)$ is calculated from

$$r_j(t+m) = \sum_i r_i(t)p_{ij}^{(m)}.$$

(12.42)

Therefore, in the limit, it is only necessary to show that

$$\lim_{m\to\infty}\sum_k\sum_\ell V_k(t)q_{k\ell}^{(m)}c_{\ell j} = \lim_{m\to\infty}\sum_i r_i(t)p_{ij}^{(m)}. \tag{12.43}$$

In equation (12.43), $q_{k\ell}^{(m)}$ and $p_{ij}^{(m)}$ are replaced by their respective fixed-point vectors $[V_\ell]$ and $[r_j]$. Equation (12.43) then becomes

$$\sum_k\sum_\ell V_k(t)V_\ell c_{\ell j} = \sum_i r_i(t)r_j, \tag{12.44}$$

and equation (12.44) can be further simplified to lead directly to equation (12.20); this completes the proof that $\lim_{m\to\infty} r_j^*(t+m) = \lim_{m\to\infty} r_j(t+m)$.

Exchange as a Natural Markov Process

There is another interpretation of equations (12.29) and (12.30) in Coleman's theory, in terms of solution methods for these same equations, that leads quite naturally to the notion of a Markov process. Starting with a set of known vectors $\{r_i(t)\}$ and $\{V_k(t)\}$, new solutions for these variables at $t + 1$ can be calculated by using equations (12.37) and (12.38). In the next iteration, equations (12.39) and (12.40) can be used, and these solutions are then fed back into (12.37) and (12.38) and the procedure continued. This natural process of iterative solution has very interesting properties, for it defines a convergence for both variables in terms of their expected and actual values. Starting the solution with new values for V_ℓ^* and r_j^* at time $t + 1$ by

$$V_\ell^*(t+1) = \sum_j r_j(t)x_{j\ell}, \tag{12.45}$$

and

$$r_j^*(t+1) = \sum_\ell V_\ell(t)c_{\ell j}, \tag{12.46}$$

it is easy to show that when m is even,

$$V_\ell^*(t+m) = \sum_k V_k(t)q_{k\ell}^{(m)}, \tag{12.47}$$

and

$$r_j^*(t+m) = \sum_i r_i(t)p_{ij}^{(m)}. \tag{12.48}$$

And when m is odd,

$$V_\ell^*(t+m) = \sum_j\sum_i r_i(t)p_{ij}^{(m-1)}x_{j\ell}, \tag{12.49}$$

and

$$r_j^*(t+m) = \sum_\ell \sum_k V_\ell(t) q_{k\ell}^{(m-1)} c_{\ell j}. \tag{12.50}$$

Thus, when m is even the solutions characterize actual values calculated in the basic Markov processes given in equations (12.33) and (12.34), and when m is odd the expected values of these same variables are produced. Convergence of these processes is ensured by previous results. If, however, only one initial distribution is specified, then only one of the associated Markov chains arises in the course of solution, and only one expected value occurs. For example, if $r_j(t)$ is given, then $V_\ell(t+1)$ is calculated above, but $r_j^*(t+1)$ is given by

$$r_j^*(t+1) = \sum_\ell V_\ell^*(t+1) c_{\ell j}. \tag{12.51}$$

In such a case, the computed value for $r_j^*(t+m)$ is the actual value, whereas the value for $V_\ell^*(t+m)$ is the expected value. In fact, this latter solution procedure is clearly faster than the previous one, but both serve to demonstrate that the Markov model is completely consistent with Coleman's theory and is also a useful vehicle with which to interpret the way actors reach an equilibrium.

In the last chapter, we illustrated the idea that design machines could be "designed" in and of themselves to optimize some reward-penalty structure that related to the process whereby actors communicated information. It is a fairly straightforward matter to incorporate a choice of relationships that optimize some reward-penalty functions in Coleman's model, and the way we might do this would be to follow the treatment in chapter 11. We will not illustrate this here, for it is now necessary to extend the model to deal with a much larger array of issues pertaining to the objects and events comprising the plan-design problem. To this end, we will once again extend the model before we illustrate its use for a series of semi-real urban problems.

Extending the Theory

We can extend Coleman's theory by simply concatenating different sets of objects defining the system of interest and by noting that the relationship between any two sets of objects can always be specified in terms of the degree of interest and control one set of objects has over another. We can build chains of these relationships that link many sets of objects to one another, always noting we need to assume some closure of such chains so that equilibrium relationships can be uniquely associated with the degrees of interest and control each set of objects has with respect to another. Let us assume that actors have an interest in various problems, and control over these problems. In the same way, actors have an interest in and control over different policies. Thus, two patterns of interaction can be derived between actors—one through *problems*, the other through *policies*—and these patterns then determine the relationships between policies and problems. We need, of course, to connect problems and policies to the system of interest, the city. Problems and policies have so far been characterized as factors—mappable degrees of suitability for development—and thus we need to relate policies to problems through these factors. Problems obviously arise from the system of interest, just as policies are designed to change the system of interest. Problems are manifest through various factors, while policies are designed to alleviate problems by the manipulation of these factors. Thus within the theory, it is necessary to relate policies and problems to factors.

Accordingly, we define four sets of objects that characterize the system: actors, problems, policies designed to alleviate these same problems, and factors describing the system of interest used to specify problems and policies. Policies and problems are not coincident, for problems can be solved by combinations of policies, and thus any one policy affects a variety of problems. The lack of correspondence between policies and problems is further demonstrated by the ways in which the actors relate to these issues. The interest and control an actor has over a problem is in terms of how that actor might influence other actors in those respects. And the same goes for an actor's interest and control over policies, which will involve the degree of significance the actor thinks a policy has and how much control the actor has over this. Furthermore, interest and control are not coincident. As we have already seen earlier, this leads to the need for some conflict resolution so that a balance—an equilibrium—evolves between actors in terms of the value they endow through their interest and the power they exercise through their control.

The pattern of influence between actors through problems, and between actors through policies, can be directly predicted by relating interest and control. Furthermore, it is assumed that the importance of various factors in the specification of problems is also known, but that the importance of factors in forming policy bundles is something that is to be predicted by the theory. Associated with the various actors, problems, policies, and factors are their relative importance within the decision-making structure. It is the importance of these entities that represent the predictions sought from the theory. The following predictions can thus be made: first, the importance of problems; second, the significance of policies; and third, the value of factors in terms of the problems defined. Fourth, the importance of actors reflects the power actors have, first over problems, and then over policies, and in general, these distributions of power will not be the same. To summarize, then, the theory is able to predict the power of actors, the importance of problems, the significance of policies, and the value of factors through the patterns of influence formed by the interests and control of actors over problems and policies. The process required to make these predictions relates to the decision-making rules or "constitution" of the system, to be outlined below.

The constitution is based on the idea that the importance of any policy or problem to an actor is formed by combining the degree of interest shown by all the actors in the policy or problem with the power that the actors have over the policy or problem. If power does not correspond to interest, importance will be small, and vice versa. In the same way, the power of an actor is derived by combining the importance of policies or problems with the degree of control an actor has over these entities. If his or her distribution of control does not match the distribution of importance, the actor's power will be small, and vice versa. There is of course a strong element of mutual determination or simultaneity involved here that complicates the process. To predict power, one has to know the importance of, say, a policy, and vice versa. Thus the method for making these predictions must be simultaneous, or if this is not possible for technical reasons, the solution must be sequential but convergent. In one sense, the process of deriving importance from power and interest, and power from importance and control, can be regarded as a process of exchange in which actors exchange interest for control until some equilibrium occurs. In this sense, the actors act collectively, as we have already seen in Coleman's (1973) theory, sketched above for a simplified problem involving actors and factors.

One way of demonstrating the structure of this constitution is in functional terms. For the moment, assume that an initial distribution of power over problems is available. Then the first stage in the process is to derive the importance of the problems by combining power and interest. The means of combination is detailed below in the formal development, and we will follow the usual linear transformations already used in this and the previous two chapters. Functionally, the first four of these operations that form the looped sequence can be written as

$$[z] = f^1[r, X] \rightarrow [r^*] = f^2[z, C] \rightarrow [s] = f^3[r^*, W] \rightarrow [r] = f^4[s, G],$$

where z, r, r^*, s are, respectively, distributions of the importance of problems, the power by actors over problems, the power by actors over policies, and the significance of policies. X, C, W, and G are the interest of actors in problems, the control by actors over problems, the interest of actors in policies, and the control by actors over policies. It is easy to see that the way we have constructed this loop leads to an iteration with the presumption that a steady state will result as actors adjust their degrees of power and the values of policies and problems to changes. The value of factors v can be computed from $[v] = f^5[z, F]$, where F is the relevance of factors to problems (or, alternatively, from the significance of policies, as we show later).

The structure of this system is complicated, and is best visualized in the block diagram in figure 12.1. Each arrow-line represents one of the functional relations

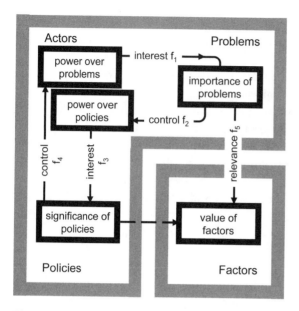

Figure 12.1
Structure of the decision-making system.

presented above. It is clear that aside from the direct relationships, various indirect relationships, for example between policies and problems and between policies and factors, can be derived. Of special importance here is the relationship given by the broken line, which shows the relevance of factors to policies. From this figure, it is immediately clear that relationships between actors and themselves (all other actors), which is the pattern of interaction based on the design network, are through policies and problems that represent the loops within the structure. There are a number of points that need to be made. The starting point is clearly arbitrary, as implied above, and one of the tasks of the more formal analysis that follows will be to explore this. Moreover, continued operation of the functional relations according to the sequence shown above will lead to changing distributions of power, the importance of problems, the value of factors, and the significance of policies. Whether or not the process is convergent or equilibrium-seeking will depend upon the way in which the functions f^* are made explicit, but to anticipate this point, the process can be seen as one in which actors try to restore a balance of power by modifying the overall importance of problems and significance of policies to reflect the given structure of interest and control. Thus the process can be likened to one of conflict resolution, in which a consensus is reached, at least implicitly, by seeking equilibrium or balance. In other words, for any pattern of interest and control in such a system, there should be unique distributions of power that reflect the functional operations specified previously.

In the previous two chapters, we argued that the plan-design problem could be seen as a way of balancing conflicting factors, which we argued could be interpreted as subsolutions to the problem. We have now extended this by introducing problems and policies that are phrased in terms of factors, but as we can show, if we were to collapse the extended theory, we can arrive at many of the methods given in the previous chapters. Consider the situation in which there is one actor, one policy, one problem, and several factors. The actor might be the planner, and in such a case, the policy and problem are by definition coincident. Thus, the theory collapses to reveal that the relevance of factors to a problem represents the set of weights or relevance of the factors in forming a policy. In this form, the factors can be seen as the components of the policy and their weights as being similar to the weights used in combining goals in the overlay, hierarchical, and sequential averaging methods introduced in chapters 10 and 11. This point will become clearer once the formal theory is developed and when the model is collapsed algebraically.

A Formal Theoretical Structure

Relationships to Coleman's Model

We will build as far as possible on the notation used earlier and in the previous chapters in part III of this book. But as we exhaust the usual notation quite rapidly, we will redefine certain terms and make fairly extensive use of vector-matrix notation, which is particularly useful in relating one set of entities to another. However, the theory will be first introduced using the more conventional notation so that we can relate it directly to Coleman's model, specified in its simplest form in equations (12.1) to (12.14).

Assume that the decision-making structure is composed of I or $L(I = L)$ actors, J problems, K factors, and M policies, and that the initial distribution of power of actors over problems $\{r_i(1)\}$ is given. Then the importance of problems given by $\{z_j(1)\}$ can be obtained by combining the power of actors with the degree of interest $\{X_{ij}\}$ each actor has in any problem. Then

$$z_j(1) = \sum_{i=1}^{I} r_i(1)X_{ij}, \quad j = 1, 2, ..., J, \sum_{j=1}^{J} X_{ij} = 1. \tag{12.52}$$

It is worth considering precisely what equation (12.52) actually predicts. By multiplying the power of any actor i by the degree of interest he or she shows in problem j, this gives the relative importance of the problem j to actor i. Then, by summing over all actors, the overall importance of problem j in the system is established. Having generated the importance of each problem, it is then possible, as a side relation, to predict the value of any factor by combining the importance of each problem with the relevance of each factor to the problem. Formally,

$$v_k(1) = \sum_{j=1}^{J} z_j(1)F_{jk}, \quad k = 1, 2, ..., K, \; \sum_{k=1}^{K} F_{jk} = 1. \tag{12.53}$$

From the importance of problems, a second distribution of power $\{r_\ell^*(1)\}$, which is the power of the actors used in determining the significance of policy, can be derived. By combining importance with the degree of control of each actor over the acceptance of any problem $\{C_{j\ell}\}$, this produces a new distribution of power

$$r_\ell^*(1) = \sum_{j=1}^{J} z_j(1)C_{j\ell}, \quad \ell = 1, 2, ..., L, \; \sum_{\ell=1}^{L} C_{j\ell} = 1. \tag{12.54}$$

At this point, it is worthwhile summarizing what has been achieved. Essentially, the theory has predicted the importance of problems by combining actor power with interest in problems, and then a new distribution of power by combining importance with control over problems. After the first operation, the value of factors is also predicted, and these operations involve working out the relationships shown by the arrow-lines f^1, f^2, f^5 in figure 12.1.

From the new distribution of power $\{r_\ell^*(1)\}$, it is possible to go through two similar operations, which involve predicting the significance of policies $\{s_m(1)\}$. First, combining power with the interest shown by each actor in the policy m, defined by $\{W_{\ell m}\}$, leads to

$$s_m(1) = \sum_{\ell=1}^{L} r_\ell^*(1)W_{\ell m}, \quad m = 1, 2, ..., M, \; \sum_{m=1}^{M} W_{\ell m} = 1. \tag{12.55}$$

A new distribution of power $\{r_i(2)\}$, which is to be used in predicting a further distribution of the importance of the problems, is then calculated by combining policy significance with the degree of control each actor has over each policy $\{G_{mi}\}$. Then

$$r_i(2) = \sum_{m=1}^{M} s_m(1)G_{mi}, \quad i = 1, 2, ..., I, \; \sum_{i=1}^{I} G_{mi} = 1. \tag{12.56}$$

Equations (12.55) and (12.56) represent the operations shown as f^3 and f^4 in figure 12.1, and it is clear at this point that one full cycle around the decision-making system has been made. We must now show whether the system reaches an equilibrium through continued operation of this sequence, and if so, we need to make the properties of the equilibrium clear. Before we do so, it is worth presenting the model in matrix terms, for this simplifies the ensuing analysis.

A Matrix Formulation

All the necessary terms have been introduced, but first note the following vector and matrix definitions: **r** is a *1xI* vector of the power of actors over problems, **z** is a $1xJ$ vector of the importance of problems, $\mathbf{v}$ is a $1xK$ vector of the value of factors, $\mathbf{r}^*$ is a $1xL$ vector of power of actors over policies, and $\mathbf{s}$ is a $1xM$ vector of the significance of policies. $\mathbf{X}$ is an IxJ matrix of interest of actors in problems, $\mathbf{F}$ is a JxK matrix of the relevance of factors to problems, $\mathbf{C}$ is a JxL matrix of the control of actors over problems, $\mathbf{W}$ is an LxM matrix of interest of actors in policies, and $\mathbf{G}$ is an MxI matrix of the control of actors over policies. Equations (12.52) to (12.56) can now be written more concisely. First, the importance of problems is calculated as

$$\mathbf{z}(1) = \mathbf{r}(1)\mathbf{X}, \tag{12.57}$$

and then the value of factors is given by

$$\mathbf{v}(1) = \mathbf{z}(1)\mathbf{F} = \mathbf{r}(1)\mathbf{X}\mathbf{F}. \tag{12.58}$$

Equation (12.58) shows that the value of factors can also be seen as a function of power by substituting for $\mathbf{z}(1)$ from equation (12.57). The new distribution of power is then calculated as

$$\mathbf{r}^*(1) = \mathbf{z}(1)\mathbf{C} = \mathbf{r}(1)\mathbf{X}\mathbf{C}. \tag{12.59}$$

Equation (12.59) is especially important, for by substituting from equation (12.57) for $\mathbf{z}(1)$, it is clear that the initial distribution of power over policies can be regarded as a function of power over problems. The new matrix $\mathbf{X}\mathbf{C}$ in equation (12.59) represents the relationships or *interactions* between actors through combination of their interest and control in problems. Thus in the sense alluded to above, this demonstrates that interaction is derivative to the system. We will define interaction between actors i and j through problems as $p_{i\ell}^*$. Then

$$p_{i\ell}^* = \sum_j X_{ij} C_{j\ell}, \; \sum_\ell p_{i\ell}^* = 1, \tag{12.60}$$

and equation (12.59) can now be rewritten as

$$\mathbf{r}^*(1) = \mathbf{r}(1)\mathbf{P}^*, \tag{12.61}$$

where $\mathbf{P}^*$ is an IxL (or IxI as $I = L$) matrix of interactions between actors through interest in and control over problems.

In a similar fashion, the significance of policies can be written as

$$\mathbf{s}(1) = \mathbf{r}^*(1)\mathbf{W}, \tag{12.62}$$

and the new distribution of power over problems is then generated by

$$\mathbf{r}(2) = \mathbf{s}(1)\mathbf{G} = \mathbf{r}^*(1)\mathbf{W}\mathbf{G}. \tag{12.63}$$

Using the same logic as above, define interactions between actors ℓ and i through policies as $p_{\ell i}'$. Then

$$p'_{\ell i} = \sum_m W_{\ell m} G_{mi}, \ \sum_i p'_{\ell i} = 1, \tag{12.64}$$

and equation (12.63) can be rewritten as

$$\mathbf{r}(2) = \mathbf{r}^*(1)\mathbf{P}', \tag{12.65}$$

where $\mathbf{P}'$ is an $L x I$ matrix of interaction between actors through interest in and control over policies. The whole process can in fact be seen in terms of the interaction matrices $\mathbf{P}^*$ and $\mathbf{P}'$. In equation (12.63), substitute for $\mathbf{r}^*(1)$ from equation (12.59). Then

$$\mathbf{r}(2) = \mathbf{r}(1)\mathbf{XCWG} = \mathbf{r}(1)\mathbf{P}^*\mathbf{P}'. \tag{12.66}$$

It is clear that the change in power (over problems and over policies) during one cycle of the model is related to the two types of interaction between actors. Define P_{ij} as the overall interaction between actors i and j through problems and policies. Then

$$p_{ij} = \sum p^*_{i\ell}p'_{\ell j} = \sum_m \sum_\ell \sum_k X_{ik} C_{k\ell} W_{\ell m} G_{mj}, \tag{12.67}$$

and equation (12.66) can then be written as

$$\mathbf{r}(2) = \mathbf{r}(1)\mathbf{P}, \tag{12.68}$$

where $\mathbf{P}$ is an $I x I = L x L$ matrix of interactions, which combines interest in and control over both policies and problems.

Analysis of the Model: Equilibrium Properties

Convergence to the Steady State

Equations (12.66) to (12.68) demonstrate the basic process of the model, and these equations can be stated more generally for any iteration or cycle t of the process as

$$\mathbf{r}(t+1) = \mathbf{r}(t)\mathbf{XCWG} = \mathbf{r}(t)\mathbf{P}. \tag{12.69}$$

Equation (12.69) is a first-order equation with a recursive structure, and by recursion on this equation, it can easily be demonstrated that power (over problems) on any iteration $t+1$ can be seen as a function of the initial vector of power $\mathbf{r}(1)$. Then

$$\mathbf{r}(t+1) = \mathbf{r}(1)(\mathbf{XCWG})^t = \mathbf{r}(1)\mathbf{P}^t. \tag{12.70}$$

In a similar way; each equation of the model can be presented as a function of the initial (exogenous) distribution of power over problems, and these equations are stated below:

$$\left.\begin{aligned}
z(t+1) &= r(1)(XCWG)^t X = r(1)P^t X \\
v(t+1) &= r(1)(XCWG)^t XF = r(1)P^t XF \\
r^*(t+1) &= r(1)(XCWG)^t XC = r(1)P^t XC \\
s(t+1) &= r(1)(XCWG)^t XCW = r(1)P^t XCW \\
r(t+2) &= r(1)(XCWG)^t XCWG = r(1)P^t P \\
&= r(1)(XCWG)^{t+1} = r(1)P^{t+1}
\end{aligned}\right\}. \tag{12.71}$$

The question is, of course: does this process converge to an equilibrium that is unchanging by repeated application of these operations? In a sense, we can anticipate this from the models we have already introduced above, and the answer will depend on the structure of the two interaction matrices formed through the interest and control of the actors.

First, let us examine the matrices P^* and P'. P^* is formed by multiplying the matrices X and C together. Now, X and C are both stochastic in the sense that they sum to 1 across their rows, and it is an elementary theorem of matrix algebra that the multiplication of two stochastic matrices leads to another stochastic matrix. Thus it is easy to show that

$$\sum_\ell p_{i\ell}^* = 1, \ \sum_i p_{\ell i}' = 1 \text{ and } \sum_j p_{ij} = 1. \tag{12.72}$$

In light of this knowledge, consider the recursive relation given in equation (12.65). This can be written out fully as

$$r_j(t+1) = \sum_i r_i(1) p_{ij}^{(t)}, \tag{12.73}$$

where $p_{ij}^{(t)}$ is the tth power by matrix multiplication of the matrix P. Examining equation (12.73), it is clear that this recursive form defines a finite Markov chain. As we have noted several times before, a well-known result from Markov chain theory states that the powering of a connected stochastic matrix such as P will converge toward a stable matrix, with each of its rows identical. In the last chapter, we developed this proof by Kemeny and Snell (1960). More formally then, and in the limit,

$$\pi_{ij} = \lim_{t \to \infty} p_{ij}^{(t)}, \tag{12.74}$$

and thus the process given by equation (12.73) converges to

$$\lim_{t \to \infty} r_j(T) = \sum_i r_i(1) \pi_{ij}. \tag{12.75}$$

The Kemeny-Snell convergence theorem introduced in chapter 11 demonstrates that each row of π_{ij} is identical, that is $\pi_{ij} = \pi_j$, $\forall i$. Then

$$\lim_{t \to \infty} r_j(T) = \sum_i r_i(1)\pi_j = \pi_j \sum_i r_i(1), \qquad (12.76)$$

and setting $\alpha = \Sigma_i r_i(1)$, the equilibrium distribution of power over problems, called r_j, is proportional to π_j, that is $r_j = \alpha\pi_j$. Note that if $r_i(1)$ is normalized to sum to 1, then $\alpha = 1$ and r_j is the steady-state or fixed-point vector of the stochastic matrix **P**.

In matrix terms, the convergence of the process can be written as follows,

$$\mathbf{r} = \lim_{t \to \infty} \mathbf{r}(1)(\mathbf{XCWG})^t = \mathbf{r}(1)\mathbf{\Pi}, \qquad (12.77)$$

where **Π** is the *IxI* limit matrix of **P**. The equilibrium relations of the process can be stated by examining the limit of equation (12.71). Substituting from equation (12.77),

$$\left. \begin{aligned}
\mathbf{z} &= \lim_{t \to \infty} \mathbf{r}(T) = \mathbf{rX} \\
\mathbf{v} &= \lim_{t \to \infty} \mathbf{v}(T) = \mathbf{zF} = \mathbf{rXF} \\
\mathbf{r}^* &= \lim_{t \to \infty} \mathbf{r}^*(T) = \mathbf{zC} = \mathbf{rCX} \\
\mathbf{s} &= \lim_{t \to \infty} \mathbf{s}(T) = \mathbf{r}^*\mathbf{W} = \mathbf{rXCW} \\
\mathbf{r} &= \lim_{t \to \infty} \mathbf{r}(T) = \mathbf{sG} = \mathbf{rXCWG}
\end{aligned} \right\}. \qquad (12.78)$$

Equation (12.78) is quite revealing, for it demonstrates the simultaneous nature of the equilibrium. The original relations between the various entities are preserved in the equilibrium, and this equation also demonstrates that the equilibrium distributions can be calculated simultaneously if required. For example, the steady-state equation for **r** in equation (12.78) can be written as

$$\mathbf{r} = \mathbf{rP}, \qquad (12.79)$$

which can be solved simultaneously. We noted in the last chapter that if the $I - 1$ equations are taken from matrix equation (12.79), and the Ith equation is taken as the normalization of power, which sums to unity, then the augmented set of non-homogeneous equations can be solved for the vector **r** using any conventional method (e.g., by the Gauss-Jordan or Cramer's method). All the other equilibrium distributions can then be immediately determined. Note also that the equilibrium distributions of power over policies and problems are related in the following way: $\mathbf{r}^* = \mathbf{rP}^*$ and $\mathbf{r}' = \mathbf{rP}'$, and these equations also reveal that the equilibrium could be sought by solving $\mathbf{r}^* = \mathbf{rP}^*\mathbf{P}'$, which is the complementary relation to equation (12.79).

Implicit Relationships in the Extended Model

A number of other relationships between the system's entities that are at present implicit can be formally examined using the relationships in equation (12.78). By substitution, the relationships between power over problems and the importance of problems is given by $\mathbf{r} = \mathbf{zCWG}$. Between the importance of problems, the power of policies, and the significance of policies, the derived relationships are $\mathbf{z} = \mathbf{r}^*\mathbf{WGX}$ and $\mathbf{z} = \mathbf{sGX}$. Between power over policies and the significance of policies, this is $\mathbf{r}^* = \mathbf{sGXC}$, and between the significance of policies and the importance of problems, this is $\mathbf{s} = \mathbf{zCW}$. In essence, these relations represent implicit arrow-lines that are absent from figure 12.1, but that can be derived from the indirect connections along the lines shown. However, there is one particularly important relationship between the value of factors and the significance of policies, which is useful in showing how policies can be conceived of as bundles of weighted factors. From the relation between factors and problems $\mathbf{v} = \mathbf{zF}$, we can substitute for $\mathbf{z}$ from $\mathbf{z} = \mathbf{sGX}$, and this leads to $\mathbf{v} = \mathbf{sGXF}$. The matrix $\mathbf{GXF}$ is stochastic and is defined as $\mathbf{T}$, that is,

$$T_{mk} = \sum_i \sum_j G_{mi} X_{ij} F_{jk}, \tag{12.80}$$

where each element of $\mathbf{T}$ gives the relevance or weight of factor k in policy m. Thus this matrix can be used to show how policies can actually be constructed in terms of the components or factors of the system of interest, and it provides a consistent relationship between problems, policies, actors, and factors. A block diagram showing the various matrix relations introduced above is given in figure 12.2, revealing how the functional logic of figure 12.1 can be translated into more formal and operational terms.

Two final points of analysis need to be established. The demonstration of convergence and equilibrium presented above was produced using an exogenous distribution of power over problems as the arbitrary starting point. However, it is necessary to trace the convergence from other starting points, and the question then is: would the same equilibrium be reached by starting with arbitrary distributions of power over policies $\mathbf{r}^*(1)$, or the significance of policies $\mathbf{s}(1)$, or the importance of problems $\mathbf{z}(1)$? The Markov property of the system ensures that the convergence is unique from any initial distribution, and it can easily be shown that this extends to any starting point. In one sense, this is already implicit in the equilibrium relations shown above in equation (12.78).

The same point can also be established in a somewhat oblique way. First substitute for $\mathbf{r}$ from the last equilibrium equation in (12.78) into the first. This gives

$$\mathbf{z} = (\mathbf{rXCWG})\mathbf{X} = \mathbf{rX}. \tag{12.81}$$

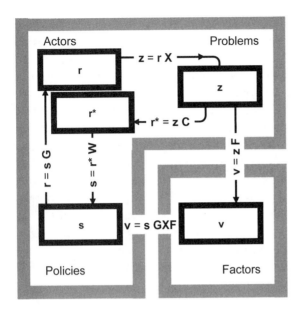

Figure 12.2
Formal relationships comprising the decision-making system.

This can be further rearranged as

$$z = zCWGX, \qquad (12.82)$$

and the matrix **CWGX** can be regarded as a matrix of interactions or a network linking the problems together. Moreover, this demonstrates that the above equation could be solved simultaneously for **z**. In exactly the same way, a matrix or network of interactions between policies can be derived. Manipulating **r** in equation (12.78) gives

$$s = (rXCWG)XCW = rXCW, \qquad (12.83)$$

and this equation can be simplified to

$$s = sGXCW, \qquad (12.84)$$

noting that **s** = **rXCW**. The matrix **GXCW** represents the derived interaction matrix, and this also demonstrates that equilibrium vector **s** can be calculated simultaneously. Besides implying that convergence is the same from any starting position, this discussion also identifies two further derived measures of interaction from the model concerning problems and policies, and these are likely to be useful in generating insights into the relationships between problems and policies.

Instability and Disequilibrium in System Structure

Implicit in the discussion so far has been the notion that the process defined on the proposed decision-making structure will generate some steady-state equilibrium. However, even a cursory review of the practice of decision making reveals that well-defined equilibrium states are rarely observed. It is much more likely for the process to be continually disturbed as it moves toward a theoretical equilibrium due to changes not at present accounted for by the theory. This and the next section propose to deal with such change in four ways. First, the idea of connectivity will be introduced, for when dealing with patterns of interaction, influence, or communication, the system is sometimes decomposable or separable to different degrees, and such partitions often account for disequilibrium. Second, the relationship between the system and its environment is important, because change in the relations or input from the environment to the system can generate instability. Third, changes in the structure of the system itself, due to constraints on channel capacity, filters on influence, and queuing will be considered. Finally, changes in the behavior of the actors—changes in the system's "constitution"—will be dealt with.

Consider a situation in which two actors are not connected to one another in the decision-making structure. If one actor's pattern of interest and control in either policies or problems is completely different from the other actor's, in the sense that the policies and problems considered by the one actor do not overlap with the other's, then the structure of decision making is separable. This separability cannot be determined immediately from an examination of the patterns of interest and control, for indirect effects also need to be taken into account. For example, the two actors' profiles may not overlap directly, but they may through an intermediate actor or set of actors, and thus the two actors could be indirectly connected. However, if the system is separable, then it would be possible to factor the matrix of overall interaction **P** into two or more separate matrices. But perhaps of more interest is the case where patterns of interaction reflect the dominance of cliques of actors or the unilateral effects of influence. In such cases, then it would be impossible for certain actors to influence other actors even indirectly, and this might be due to the dominance of a clique—actors would only influence others and remain uninfluenced. Or it might be due to the existence of two or more cliques, which may both influence actors not belonging to one of these cliques, or not influence one another's cliques.

These various cases, which reflect degrees of connectivity, can be easily dealt with using graph-theoretic interpretations of the overall interaction matrix **P** (Harary, Norman, and Cartwright, 1965). We used these measures in classifying the power structures associated with the elementary decision model in the previous chapters, but of most interest are variations in the connectivity of the networks, for this is key to the processes of convergence to an equilibrium. Clearly, the process will only converge if for some power t of the overall interaction matrix $\mathbf{P}$, each cell of this matrix is positive and greater than zero. This is the Markov equivalent of the requirement of strong connectivity for the matrix $\mathbf{P}$. In cases where the system is dominated by one clique of actors, the matrix $\mathbf{P}^t$ can be partitioned into an absorbing set and transient sets, which reveal that in the limit, the steady-state distributions from the matrix $\mathbf{P}$ are determined solely by states of the absorbing set. Convergence of a kind occurs in such a process, but in the case of two or more absorbing sets, or disconnected sets, no convergence is possible. Connectivity problems might occur in the original decision-making structure and if so, this probably suggests some problems of definition. It is possible that in thinking about how the structure of the system might change over time, connectivity might be some function of the connectivity or the structure at some previous point in time, and in this sense, the system may become disconnected.

The theoretical structure outlined previously has been based on a closed system in which the wider environment of that system was considered to be stable. In short, the equilibrium associated with the closed system is a function of the system itself, and never of the wider environment. Whether or not this rather strict form is appropriate will depend upon the context and the extent to which the real system is actually closed. The simplest possible way of envisaging a relationship from the environment to the system is one in which the state of the system is arbitrarily and randomly disturbed by an input from the environment. Consider the situation in which each actor's power is usually determined as a function of the process of decision making outlined above; occasionally, however, the power of an actor is suddenly disturbed by something outside the ambit of the system. In such a situation, the Markov property of the model would ensure that although the path of convergence of the process would be different, the final equilibrium would be unchanged. The process is organized around the notion that the equilibrium state is independent of the starting state, and thus changes from the outside environment simply constitute a new starting state for the process.

Arbitrary disturbances from outside the system can, however, produce a continuous disequilibrium. If the frequency of external disturbances is greater than the time taken to reach equilibrium, then the equilibrium state will never be achieved. However, the system will always be tending to equilibrium, and in the absence of further disturbances will attain an equilibrium. Of more speculative interest is the idea that the system structure might be influenced by the environment. If the structure changes, then a new pattern of convergence and equilibrium is implied, and the problem of connectivity referred to above might also become significant due to such change.

Capacity, Filters, and Queues in the Pattern of Influence

Many additional factors that change the process of convergence could be postulated here, but the real quest is to trace through the effect of changes, which can be handled analytically. Thus, it is worth focusing on those factors that can be introduced and explored using the algebra of the previous sections, but this should not suggest that significant factors that cannot be handled formally should be ignored. Indeed, in a later section and in the next two chapters, the notion of a simulation model of these processes of decision making will be broached. Extensions to the model to be incorporated here concern the patterns of interaction between actors in the system. Two concepts will be developed, the first based on the notion that actors are influenced by a buildup of tension or congestion in the system at regular intervals, the second by a lessening of tension, which is a cumulative phenomenon to be interpreted as a function of the system itself.

It is a fairly reasonable notion that decision-making structures begin to diverge from their given purpose when tension builds up within the system. For example, the speed at which actors have to process information often leads to a distortion of the information due to the buildup of a queue, and thus a loss of information or memory on the part of the actor. The capacity of the system might be too low, the means of transmission or reception too involved, the process of evaluating the importance or significance of entities too complex. One possible structure might involve the buildup of tension, which distorts the process on a regular basis. Consider a process in which actors process information in any single time period quite normally, but because of the effort of such processing, information in the next time period might be systematically distorted. In terms of the process proposed here, a measure of the distortion related to problems might be given by the *LxL* matrix $\mathbf{Q}$, which is applied to the combination of interest and control over problems. In a similar way, an *IxI* matrix $\mathbf{R}$ can be defined that is a measure of the distortion of policies. These matrices are applied regularly in every other cycle of the decision-making process, and they reflect filters to the transmission of interest and control. Take the process defined previously: in the first iteration or cycle, from the importance of problems, the power of actors over policies, and the significance of those policies, a new measure of the power of actors over problems can be defined by the sequence $\mathbf{z}(1) = \mathbf{r}(1)\mathbf{X}$, $\mathbf{r}^*(1) = \mathbf{r}(1)\mathbf{XC}$, $\mathbf{s}(1) = \mathbf{r}(1)\mathbf{XCW}$, and $\mathbf{r}(2) = \mathbf{r}(1)\mathbf{XCWG}$.

On the second iteration of the process, however, distortion occurs. The importance of problems is predicted as previously,

$$\mathbf{z}(2) = \mathbf{r}(1)(\mathbf{XCGW})\mathbf{X}, \tag{12.85}$$

but the new distribution of power of policies is distorted by the filter $\mathbf{Q}$:

$$\mathbf{r}^*(2) = \mathbf{r}(1)(\mathbf{XCGW})\mathbf{XCQ}. \tag{12.86}$$

The significance of policies is then predicted from

$$\mathbf{s}(2) = \mathbf{r}(1)(\mathbf{XCGW})\mathbf{XCQW}, \tag{12.87}$$

but the power of actors over problems is also distorted by the filter $\mathbf{R}$,

$$\mathbf{r}(3) = \mathbf{r}(1)(\mathbf{XCGW})\mathbf{XCQWGR}. \tag{12.88}$$

Examining the distribution of power over problems only, the general recurrence relations can be derived as follows: when the cycle (t) is odd,

$$\mathbf{r}(t) = \mathbf{r}(1)[(\mathbf{XCGW})(\mathbf{XCQWGR})]^{(t-1)/2}, \tag{12.89}$$

and when the cycle (t) is even,

$$\mathbf{r}(t) = \mathbf{r}(1)[(\mathbf{XCGW})(\mathbf{XCQWGR})]^{(t/2)-1}\mathbf{XCWG}. \tag{12.90}$$

Similar types of relation can be derived for the other distributions.

It is the behavior of the process in the limit that is of interest, however. Without loss of generality, it can be assumed that $\mathbf{Q}$ and $\mathbf{R}$ are both stochastic—in this sense, they can be considered as giving the proportion or degree to which each actor distorts the actual pattern of interest and control. It is clear that the matrix $(\mathbf{XCGW})$ $(\mathbf{XCQWGR})$ is also stochastic; this matrix will converge in the limit of t to a matrix $\mathbf{\Lambda}$. Then, in the limit for t, $\mathbf{r}^{odd} = \mathbf{r}(1)\mathbf{\Lambda}$ and $\mathbf{r}^{even} = \mathbf{r}(1)\mathbf{\Lambda P}$. These equations show that the steady state oscillates in a regular fashion due to the regular application of the filters. The pattern of oscillation is, however, quite revealing, for it implies that there is no learning process involved. Actors do not change their behavior due to the distortion. The actors act as if the distortion were a part of the structure of the system, never to be changed. This type of effect is particularly unrealistic, but it does serve to show again that for any kind of equilibrium to be maintained, the system's process must be a function of the system's structure, and not of some arbitrarily conceived instrument to be applied at regular intervals.

A much more realistic demonstration of the effect of distortion is accomplished if the distortion is made a function of the behavior of the system and the distortion itself at a previous time period. Imagine that the actors accept a measure of distortion at the beginning of the process, but that this distortion is modified when the actors relate it to the actual pattern of influence. In this sense, actors can be said to be learning. This process is demonstrated in relation to the distribution of power over problems, but can be easily extended to any of the other distributions. On the first cycle, $\mathbf{Q}(1)$ and $\mathbf{R}(1)$ represent the $L x L$ and $I x I$ distortion filters applied to the interaction through problems and policies. Then

$$\mathbf{r}(2) = \mathbf{r}(1)\mathbf{XCQ}(1)\mathbf{WGR}(1). \tag{12.91}$$

On the second iteration of the process, $\mathbf{Q}(2)$ and $\mathbf{R}(2)$ are based on the modified patterns of interaction produced by combining the original measures of distortion with the actual patterns of interaction. Formally,

$$\mathbf{Q}(2) = \mathbf{XCQ}(1), \tag{12.92}$$

and

$$\mathbf{R}(2) = \mathbf{WGR}(1). \tag{12.93}$$

A new distribution of power $\mathbf{r}(3)$ is calculated as

$$\begin{aligned}
\mathbf{r}(3) &= \mathbf{r}(2)\mathbf{XCQ}(2)\mathbf{WGR}(2) \\
&= \mathbf{r}(1)\mathbf{XCQ}(1)\mathbf{WGR}(1)(\mathbf{XC})^2\mathbf{Q}(1)(\mathbf{WG})^2\mathbf{R}(1)^{\boldsymbol{.}}
\end{aligned} \tag{12.94}$$

General recurrence relations up to time T for the process can now be stated:

$$\left.\begin{aligned}
\mathbf{r}(T+1) &= \mathbf{r}(1)\prod_{t=1}^{T}(\mathbf{XC})^t\mathbf{Q}(1)(\mathbf{WG})^t\mathbf{R}(1) \\
\mathbf{Q}(T+1) &= (\mathbf{XC})^T\mathbf{Q}(1) \quad \text{and} \quad \mathbf{R}(T+1) = (\mathbf{WG})^T\mathbf{R}(1)
\end{aligned}\right\}. \tag{12.95}$$

Equation (12.95) can be used to derive directly the appropriate equations for the other distributions by application of the five operations given earlier in equation (12.78).

Unlike the non-learning process described above, this does converge to a unique equilibrium. In equation (12.95), the matrices $(\mathbf{XC})^t$ and $(\mathbf{WG})^t$ are stochastic and converge in the limit. First set $\mathbf{z}_1 = \lim_{t\to\infty}(\mathbf{XC})^t$ and $\mathbf{z}_2 = \lim_{t\to\infty}(\mathbf{WG})^t$. Then we can show that the limiting sequence of the main equation in (12.95) can be written as

$$\mathbf{r}(T+1) = \mathbf{r}(1)\mathbf{XCQ}(1)\mathbf{WGR}(1)(\mathbf{XC})^2\mathbf{Q}(1)(\mathbf{WG})^2\mathbf{R}(1)\ldots\ldots\mathbf{z}_1\mathbf{Q}(1)\mathbf{z}_2\mathbf{Q}(2). \tag{12.96}$$

Equation (12.96) provides a unique equilibrium, but the limit distributions of power, importance, significance, and value cannot be calculated analytically; they must be calculated by iteration on the total process. The process is no longer Markovian, but convergence occurs because distortion of the pattern of influence comes closer and closer to the actual pattern of influence as the actors attempt to match actual and perceived interaction. Note that in this sense, distortion is regarded as a measure of perceived actor interaction. Note also that if the original distortion matrices are the identity matrices, then the limit of equation (12.96) is equivalent to the limit of the normal process equation (12.77) given above.

Finally, in this discussion of disequilibrium, the most dramatic effects which stop the system from converging are likely to be due to changes in the system's behavior—in the rules or constitution that govern the process of combining interests and control. Apart from external changes originating from the system's environment, changes of behavior within the system itself are likely to be due to some mismatch between actors' expectations and the actual situation. There are numerous possibilities, but the most important are noted below. First, when actors gradually realize their lack of power, they may form coalitions to share and thus extend their power. In situations where interest and control are negatively correlated for any actor, power will be small, and actors may thus attempt to match their interests more closely with their control, and vice versa. If their interests are too diverse, they may concentrate them, or they may speculate that by spreading their interests more widely, they are likely to choose one that will flourish. In the same way, actors with a great deal of power may "gang up" by forming coalitions and may diversify or concentrate interest and control. The precise form will depend on the way in which the various conditions are related. Coalitions may also break up due to organizational or performance problems, and coalitions may merge using the same rules as actors. In general, the system might be characterized by a strong element of bluff-counterbluff, tit-for-tat, and so on, due to the anticipation of consequences. What is obvious is that no analytic interpretation can be pursued directly. Convergence may occur, or it may not. The only way to explore such behavior is by simulation, and in the next section, a sketch of such a simulation model is attempted.

A Sketch for a Simulation of Decision-Making Processes

Changes in the behavior and structure of the decision-making process are most likely to be due to the degree to which each actor's goals are realized. In this theory, it is assumed that a reasonable indicator of how well an actor is doing is their measure of power. It is also assumed that each actor attempts to increase this power at every cycle of the process, but that the amount of power expected by each actor varies between actors. Moreover, these goals, measured by expectations, can also change during the process as actors lose or gain power. Thus, it is the actual amount of power at any cross-section in time, or the change in power over time, that leads to a measure of the actors' performance, and this measure enables the actors to decide whether or not to continue or change their behavior. There are two main ways in which behavior can change: first, actors can change the distribution of their interests or control in an effort to increase their power or "get a greater share of the action," and second, actors can aggregate themselves into larger groups or coalitions who have a greater chance of reaching individual goals by acting collectively. In the same way, coalitions can break up if they are unsuccessful or if intragroup tension increases beyond given thresholds.

The rules for changing behavior can be applied sequentially in the following fashion. At each stage of the simulation, when an action such as "form a coalition" or "change one's distribution of interest" is to be considered, there is a certain probability of its occurrence based on the set of actions already considered. We can simulate whether an action is taken using a random mechanism that involves the matching of a random number with the probability structure. Thus an actor who has just formed a coalition and is now considering whether to change their interest will have a different probability of action from an actor who is considering the same but is not in a coalition. There are several stages that must be considered by each actor. The probability structure for action will be different if the actor is already in a coalition, or not in one. Then the amount of power held, the change in power through previous time intervals, and the size of the coalition to which the actor belongs will all lead to different probabilities of forming a new coalition, breaking up a coalition, remaining the same, or any variation thereof. On the basis of these decisions, actors will consider whether to change their pattern of interests or control. Control may be substituted for interests, actors may diversify one and concentrate the other or concentrate both, and so on. Whatever the decision, the amount of change will be limited in some sense. It may also be limited by an element of inertia, which represents the structure of interest and control at previous time periods. Finally, the thresholds used to measure expected power might be altered according to new behavior, as the system creates its own goals by learning about its behavior through time.

This sequence of actions can be embedded into the model in two ways. As there are really two sets of actors, or in fact *one* set of actors who act in *two* capacities—first over problems and second over policies—then the process can be operated for problems, then policies, or vice versa. This will lead to sets of actors belonging to one coalition when dealing with problems, and another when dealing with policies. Therefore, actors will be attempting to maximize their power over problems and power over policies, and one interesting feature would be to assess how successful they are in one context compared to the other. In fact, it is impossible to anticipate in advance, by analysis, whether or not the system will converge. Moreover, the stability of the system and its sensitivity to different combinations of actions cannot be anticipated. It seems likely that the system will not converge in its present form, but one of the purposes of a simulation model is to set up a vehicle for exploring modifications to the theory. Once the simulation has been started, it is possible to add or delete various mechanisms from the model as they are found to contribute to the process or not. Essentially, this quest is to reveal pressure points in the system that indicate important structures. Many of the actions to be considered by actors during the process are likely to have only limited effect, and by using simulation as an experimental medium, such actions can be identified.

The theory that has been outlined so far represents a framework for understanding how decision-making systems actually work, not how they should work. Although the theory cannot be strictly categorized as being positive in emphasis, for it does contain certain normative notions, especially as to how actors combine interests with control, there is an implicit theme running through this presentation that the theory might be partly verified or falsified by comparison with a real situation. In one sense, this theory is only meant to represent the rudiments of an analytic approach to decision making, that is, how insights into real decision-processes can be gained using a system model that may have some modest empirical significance. This is a rather oblique way of using theory in social science, but given the inordinate observational and logical problems posed by prediction in social life, it may be the only way. Nevertheless, given the hypothetical semi-positive nature of the model, it is worthwhile to speculate on how a more explicitly normative dimension might be introduced.

It is somewhat ironic that this focus on decision making has been fostered by a desire to find better ways of optimizing some system of interest, for example a city, and that the problem appears to have been pushed one stage back—to find a method that will now optimize the decision-making system. In this sense, the problem could be pushed back indefinitely, as methods are sought to optimize the process used to optimize the process, and so on, like a set of Chinese boxes or Russian dolls. Yet the need to optimize the decision-making process does reflect a genuine change in outlook in planning and design from a product to process-oriented viewpoint. The quest is now to optimize the process, in recognition of the fact that it is exceedingly difficult to optimize the product, and that the product is intrinsically related to the process—that is, form follows function, which is one of the leitmotifs of our science of cities. In the elementary decision model introduced in the last chapter, an explicit optimizing routine was mapped onto its structure in which actors could choose to interact so that some criterion of benefit (reward-penalty criterion) was optimized. It is possible that the same kind of procedure could be mapped onto the analytic model presented above, but what is certain is that such a strict procedure could not be applied to the simulation model. As it is likely that the simulation model is more realistic, an attempt can be made to design procedures for optimizing the simulation. For example, a criterion function based on the requirement of convergence, the need to reduce inter-and intra-group hostility, and a realistic distribution of interest or control might be specified, and a trial-and-error process developed for optimizing such a criterion. In effect, such procedures relate to the design of institutional structures that provide effective organizations for decision making. Therefore, the theoretical structures proposed here seem quite consistent with the emerging paradigm in planning based on complexity, where the focus is shifting from the design of system structure to the design of system behavior.

To demonstrate the applicability and operationality of the theory, two case studies will be presented. Both are hypothetical in that they are not based on rigorous, empirically observable systems of interest or data sources and rather are based on stylized facts, but the focus here is upon the general way in which the theory can be used in applications that are close to real problems. At the same time, it is possible to demonstrate how the theory can handle aspatial and nonspatial problems, thus providing a rigorous framework for relating social and political processes to spatial forms and patterns.

Applications: I. Poverty in the Inner City

Our first case study examines the problem of poverty in the inner city. We believe that careful application of the theory will reveal how various problems of deprivation can be resolved by social policies based on differing mixes of spatial factors. In table 12.1, the actors, problems, policies, and factors relevant to this general concern are identified. Clearly, the appropriateness of the model will depend upon how well the list of entities reflects the real situation, and there is nothing within the theoretical structure proposed here that determines these questions of definition. Where such problems are highly ambiguous and uncertain, formal theories such as the one proposed here are of little use, but this reflects the more general problem of developing theory in subject areas that are intrinsically ill-defined.

The actors listed here relate to housing, education, social services, political parties, business interests, health, and transport. In no way is this list exhaustive, and at best, it is only a casual attempt at defining relevant actors—there are certainly others excluded, but at least the actors identified here all have some relevance to the poverty problem. The general problem itself is broken down into subproblems concerned with housing, income and expenditure, mental and physical health, and employment opportunities; again, the list is not exhaustive but suggestive. Policies are also framed in these terms, and the factors that link the spatial system to the decision-making system are all defined in terms of land uses and infrastructure. Thus, in essence, the decision-making environment captures social processes, while the system of interest is defined in terms of spatial form. What must be strongly emphasized here is that the system of interest need not be defined spatially in this way, if any other bias in comprehending the problem is sought. The theory is of sufficient generality to encompass any decision-planning problem, whether it be explicitly spatial in emphasis or manifestly nonspatial.

The analytic model given in equations (12.71) was applied by specifying an arbitrary vector of power over problems and solving the matrix equations in the given order. These equations were solved iteratively, and the convergence toward an equilibrium state was fast. A number of indices measuring the convergence can be defined in terms of the various distribution vectors; here a measure based on the

Table 12.1
Actors, problems, policies, and factors characterizing poverty in the inner city

Actors: Groups		Problems: Poverty and deprivation	
1	Private landlords	1	Housing facilities
2	Public landlords	2	Housing conditions
3	Education authority	3	Density: Overcrowding
4	Housing policy committee	4	Rent arrears
5	Social services department	5	Personal mobility (lack of)
6	Community action group	6	Undernourishment
7	Tenants' association	7	Physical health
8	Local political parties*	8	Mental health (coping with)
9	Local political leaders*	9	Crimes against property
10	National politicians*	10	Crime against persons
11	Government welfare agency	11	Job opportunities (lack of)
12	Business interest group	12	Work environment
13	Health authority		
14	Transport authority		

Policies: Local authority services		Factors: Land uses	
1	Housing: redevelopment	1	Transport infrastructure
2	Housing: rehabilitation	2	Public housing
3	Negative income tax	3	Private housing
4	Housing subsidies	4	Industrial areas
5	Rent control	5	Commercial centers
6	Special education	6	Education: schools
7	Transport facilities	7	Community centers
8	Fares-free public transport	8	Parks: Recreation areas
9	Health facilities		
10	Commercial facilities		

*Identifiable leading groups of politicians are implied here.

difference between power over problems at successive iterations t was used. This statistic is given by

$$\Gamma(t) = \sum_i |\, r_i(t) - r_i(t-1)\,|, \tag{12.97}$$

where $\Gamma(t) \to 0$ as $t \to \infty$. In fact, after the second iteration, the process had converged to within 2% of all steady-state values, and by the third iteration, an acceptable convergence had been attained. Convergence appears to be exceptionally fast in this model, probably due to the high degree of connectivity of actors through interests and control, but a finer specification of relationships could lead to a slower convergence, and thus it is impossible to generalize about the speed of convergence from this one case. Clearly, the size of the application in terms of number of problems, actors, policies, and factors affects the speed of convergence. The way the model is solved is based on equation (12.79), suitably augmented with a normalization constraint, as illustrated in equation (11.25) in chapter 11.

In exploring the application of this theory to a poverty problem, it is worth looking at the input data required to generate the output. Table 12.2 presents a list of the exogenous inputs, derived relationships from the inputs, and the equilibrium outputs. The inputs that describe the degree of interest and control each actor has

Table 12.2
Inputs, derived relationships, and outputs

Inputs		Matrix symbols
Data from observations		
1	Interest of actors in problems	X
2	Control of actors over problems	C
3	Interest of actors in policies	W
4	Control of actors over policies	G
5	Relevance of factors to problems	F
Arbitrary data to initiate the process: Initial conditions		
6	Power of actors over problems	r(1)
Derived relationships		
1	Actor interaction through problems	XC
2	Actor interaction through policies	WG
3	Actor interaction through problems and policies	XCWG
4	Problem interaction through actors and policies	CWGX
5	Policies interaction through actors and problems	GXCW
6	Relevance of factors to policies	GXF
7	Relationships between policies-problems, & vice versa	CW & GX
Outputs		
1	Power of actors over problems	r
2	Power of actors over policies	r*
3	Importance of problems	z
4	Significance of policies	s
5	Value of factors	v

in policies and problems has been specified quite hypothetically on the basis of a general knowledge of the problem culled from a variety of sources and experiences. This data set incorporates reasonable assumptions. Although a detailed description of these tables is unnecessary, some examples are in order. With regard to problems, it seems reasonable that, say, private landlords have greatest interest in problems of rent arrears, overcrowding, crimes against property, and crimes against the person. They have least interest in job opportunities, work environments, and mobility. When it comes to their control over problems, they have most control over housing facilities, housing conditions, and rent, and least control over employment, health, and criminal matters. All the actors can be viewed in these terms through problems and policies, but of specific interest is where the mismatch between interest and control occurs. It is difficult to generalize about such mismatch, but it appears that an actor's power will ultimately depend on how much support is realized through the interest of others, and how much of a monopoly the actor has over certain policies or problems, in terms of control. Indeed, it is the operation of the decision-making process that enables the distribution of power to be calculated; the relationship between the system's structure (interest and control) and behavior (rules for combining interest and control) leads to the evaluation of power, and thus the equilibrium distributions cannot be anticipated by reference solely to the system's structure.

A better set of measures of the system's structure and behavior is provided for by the derived relationships—the patterns of interaction or networks between actors, problems, and policies. Table 12.2 demonstrates how straightforward is the derivation of these patterns from the input data, and it is worth discussing briefly the significance of these relationships. First, examining the interaction between actors through problems—that is, the degree to which actors affect each other by their interest in and control over problems—it is clear that those concerned with housing, social services, and political interests are closely related. In one sense, this reflects reality, in that the housing problem and its solution in political and local municipality terms has long been regarded as the major way in which poverty can be alleviated in the inner city. The actors concerned with health and education also have significance, in that they interact with the housing problem through social services and political groups. In figure 12.3, a simplified pattern of interaction based on the strongest relationships between actors is displayed graphically; gross interaction, or the sum of interaction in both directions, is used to compute these and the following influence graphs. The complete pattern could be presented, but this is extremely complicated and time-consuming to construct, and in any event, the interest here is on the main directions of influence.

Turning to the interaction of actors through policies, a different pattern of influence is apparent. The housing policy group and the public housing agency dominate,

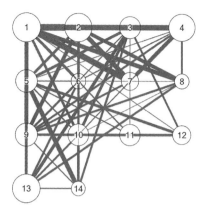

Figure 12.3
Interaction between actors through problems.

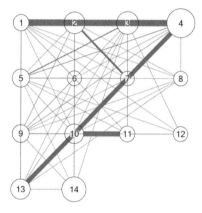

Figure 12.4
Interaction between actors through policies.

and local and national politicians also assume some importance. There is more self-interaction associated with an actor in terms of policies, as one might expect, but the influence of private landlords and the social services committee is slightly less in terms of policy formulation than in problem definition. This probably reflects a lack of power to directly implement public policy. The education agency has slightly more influence on other actors through policies than through problems, and this also reflects its power in policy formulation. The simplified pattern of influence is displayed in figure 12.4. Finally, the overall pattern of influence derived through policies and problems is of interest. As expected from the previous discussion, the

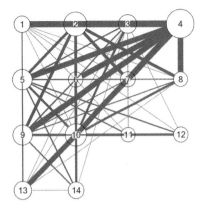

Figure 12.5
Interaction between actors through problems and policies.

housing policy committee is by far the most dominant actor. The public housing agency, the social services department, and local and national politicians also influence each other, and this is reflected in the simplified interaction pattern contained in figure 12.5. These patterns seem fairly realistic, for such poverty problems and policies always seem to be phrased in terms of housing. But it is likely that one of the goals of the planner or politician is to change the traditional pattern of influence so that certain policies or problems assume greater importance. In this sense, the theory could be used to show how this could be achieved by manipulating the pattern of influence; in short, designing organizational, institutional structures and networks would change the importance of problems, the significance of policies, and the power of actors.

The same kinds of interaction patterns can also be derived for the relationships between problems and the relationships between policies. The relationships between problems are consistent with the discussion above; housing and its various characteristics are the main problems, together with crime against property and persons, which are closely related to housing. This simplified pattern of interaction is shown in figure 12.6. With regard to relationships between policies, rehabilitation, subsidies, and rent control are closely related. Redevelopment and new commercial facilities are related, but transportation policies are not strongly related to any other factors. Again, the emphasis in policy interaction is in terms of housing, although special education and new health facilities are related through housing policy. The main relationships are illustrated in figure 12.7. In general, the five interaction patterns presented in figures 12.3 to 12.7 consistently show the influence of housing on poverty. To find out the importance of the various problems, policies, and actors, however, it is necessary to look at the equilibrium relationships derived from the decision model using these patterns of influence; these are described below.

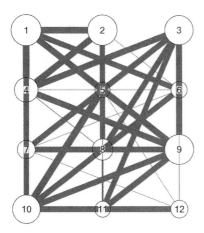

Figure 12.6
Interaction between problems.

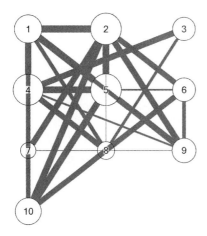

Figure 12.7
Interaction between policies.

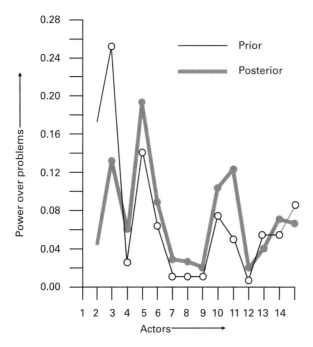

Figure 12.8
Prior and posterior distributions of power over problems.

The model is operated by specifying an initial distribution of power of actors over problems, which can be regarded as a prior distribution. A prior distribution, in which private and public landlords were assumed to have considerable power together with the housing policy committee, was defined, and in figure 12.8, a direct comparison of this distribution with the posterior distribution predicted by the model is made. This shows that the housing policy committee has more power than was assumed, and landlords less. This seems reasonable given the fact that the power defined here is a power to raise problems; that is, to establish the importance of problems. Figure 12.8 also shows that national politicians and the social services committee have more power than was assumed a priori. The power distribution over policies, however, is somewhat different. Landlords do have more power, as do local and national politicians. The housing policy committee still has considerable power, but it plays a less dominant role in exercising its power to formulate policy. The social services committee and the health authority also have a fair amount of power, and in figure 12.9(a), the equilibrium distributions of power over policies and over problems are contrasted. In terms of the equilibrium distribution, housing problems such as facilities, condition, and overcrowding together with problems of crime are most significant. Turning to policies, rehabilitation, housing subsidy,

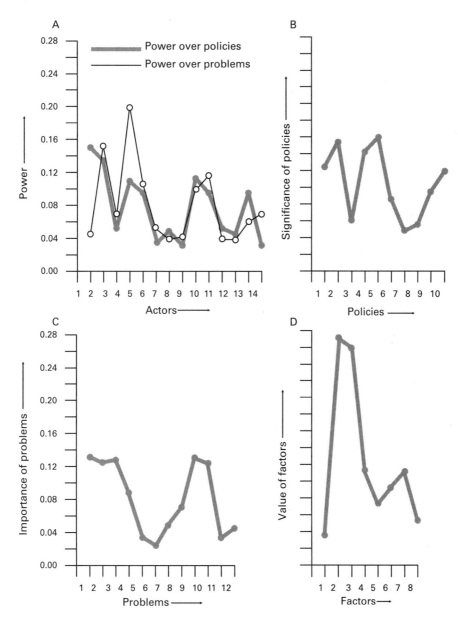

Figure 12.9
Predicted distributions from the decision-making model.

and rent control are judged to be slightly more important than redevelopment, and the location of new commercial facilities has some significance. The equilibrium distributions relating to policies and problems are shown in figures 12.9(b) and 12.9(c), respectively.

The value or importance of land uses, which comprise the physical system of interest, reveal how the policies and problems might be implemented spatially. Figure 12.9(d) illustrates the distribution of the value of these factors, demonstrating the importance of public and private residential land use in tackling the poverty problem. Of some importance, too, are the location of schools and community centers, but transport and recreational provision do not appear to have much importance in solving these kinds of problems. In short, this implies that problems of poverty are unlikely to be alleviated by changes in recreation and transport infrastructure; these seem quite reasonable conclusions, and further support our speculation that the theory predicts intuitively acceptable results. However, in no way does this imply that the theory has been verified. It is possible to derive the overlap between policies and problems from the model, but of more immediate interest is the composition of policies in terms of the importance of different factors and land uses. In figure 12.10, each graph shows the importance of each land use in

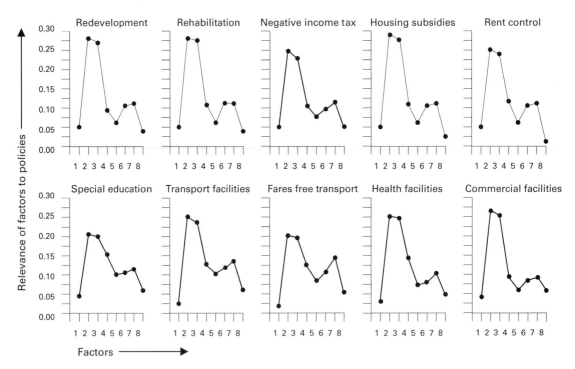

Figure 12.10
Weighting of factors in the composition of policies.

formulating any particular policy. In other words, policies are formulated as bundles or mixes of land uses, and these networks show the particular mix required for each policy. For example, the importance of residential land use in all policies is apparent, and community facilities, schools, and the location of industry all affect each policy significantly. However, figure 12.10 also reveals a certain weakness in the theory, for it is likely that such uniform mixes of factors for each policy are somewhat unrealistic. It is difficult, for example, to see how a policy of fares-free public transport could be mainly implemented by changing residential land use. This type of difficulty may be due to the specification of the data input, but it may also be due to the structure of the theory.

Applications: II. Resource Allocation in a Local Authority

The second application is again a semi-real case study that applies the theory to a "typical" resource allocation problem in a local authority with an approximate population of 500,000. In this problem, only actors, problems, and policies are defined, since resource allocation is essentially a nonspatial or aspatial problem, and thus the spatial dimension captured by the system of interest in the previous application is not relevant. In short, the decision-making environment is coincident with the system of interest, and the factor/land use dimension plays a dummy role in the operation of the decision model. Essentially, the equilibrium distributions of power of actors, the importance of problems, and the significance of policies give the proportions of some total resource to be allocated to the particular entity. For example, the resources controlled by each actor over problems and over policies represent the amount of resource each actor can command in raising problems and formulating policy. The resources allocated to each policy are proportional to the significance of each policy, and each problem area captures an amount of resources proportional to its importance. All these distributions of resources along these different dimensions are consistently related by the steady-state equilibrium relations. For example, each actor ℓ has a certain command or power over the total resource r_ℓ^* and distributes this resource to a policy m in proportion to that actor's interest $W_{\ell m}$. Thus the total amount of resources allocated to some policy is given as $s_m = \Sigma_\ell r_\ell^* W_{\ell m}$, and the same kind of interpretation can be made for all the other equilibrium distributions of resources.

The actors defined for this problem are based on the typical departmental structure of a local authority, together with two typical political groups, which we will define as left L and right R wing, thus reflecting the two-party system of local government in the United Kingdom. It is also assumed that the two parties are fairly evenly represented, and of some interest from the model's predictions is the relative power of each party. The problems defined are typical of those facing a medium-sized metropolitan area where land use planning problems are manifest in housing, transport, recreation, and industrial issues, together with problems of health, education, and crime. Policies, on the other hand, form a much-restricted set. Boaden (1971), for example, presents a series of statistics based on actual average resource allocations for certain local authority services in England and Wales. This set of services or policies is limited because of the availability of data, but we can speculate on how well a theory such as this one might lead to predictions about resource allocation similar to the actual data presented by Boaden (1971). The set of policies excludes transport and planning issues but contains a fairly fine disaggregation of social services—child welfare, library provision, and so on. Furthermore, by including a wide set of actors and problems, a partial set of policies can be defined quite realistically in terms of these wider sets of actors and problems. Table 12.3 lists the actors, problems, and policies involved in this application.

The model was operated in the same way as described above for the poverty problem. Data relating to interest and control were devised intuitively and are quite highly correlated. The distribution of interest was specified according to popular notions; for example, the L group, the Social Services Department, and the Planners are mainly interested in housing problems, whereas those groups with an interest in business matters, such as the R group and the Treasurer's Department, have their main interests in industrial and commercial problems. However, in terms of policies,

Table 12.3
Actors, problems, and policies characterizing resource allocation in a local authority

Actors: Groups		Policies: Local services	
1	Left-wing L party	1	Welfare
2	Right-wing R party	2	Health facilities and advice
3	Treasurer's department	3	Children: Care and health
4	Engineering department	4	Library provision
5	Planning department	5	Fire services
6	Education department	6	Police services
7	Housing department	7	Education
8	Social services department	8	House building and maintenance
9	Town clerk's department		

Problems: Urban issues			
1	Housing	4	Education
2	Transport	5	Crime
3	Industrial and commercial	6	Recreation and leisure

interest and control are not as highly correlated. In particular, the Planners and the Social Services Department have little control over but a high degree of interest in welfare, housing, and educational matters. The L group has wider interests than it has control, its control being mainly over housing. The R group have a strong interest and control over industrial matters and the police.

The equilibrium distributions from the model are interesting. In terms of power to raise problems, the Social Services Department, and to a lesser extent the Treasurer's Department, dominate. But with regard to the formulation of policy, the Treasurer's Department is most dominant. In terms of the relationship between the L and R groups, the L group has slightly more power in raising problems, but the R group has much more power over formulating policies to alleviate those same problems. These equilibrium distributions of power are also reflected in the derived interaction patterns between actors. In terms of problems, housing, education, crime, and health were predicted as being the most important, with transport and recreational provision least important. In one sense, the theory could be used to speculate on certain distributions of control and interest, which might lead to reversals in the typical pattern. For example, in areas where transport policy is all-important, it is likely that the interest of most actors in transport problems and policies is high but that control over other policies and problems is diffuse and low. This kind of speculation is akin to the search for decision-making structures that optimize the kind of criterion function referred to previously.

Finally, the predicted significance of policies can be contrasted with the actual significance (in terms of the resources allocation) presented by Boaden (1971). Graphs of predicted against observed significance are shown in figure 12.11, and it

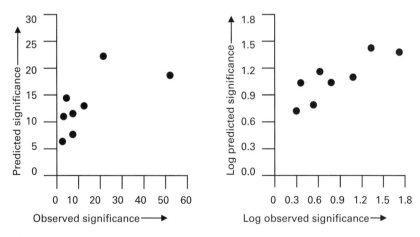

Figure 12.11
Observed and predicted allocations of resources.

is immediately apparent that the relationship is significant. In fact, the linear correlation between predictions and observations is 0.792. However, these predictions reveal another major problem posed by the theory, and this relates to questions of magnitude. It is particularly difficult with the theory in its present form to separate out the relative importance of an issue from more general questions about its absolute size importance. For example, it may be that an authority's role in providing services makes it obligatory to provide a complete educational system, but only a partial system to deal with industrial location. However, industrial location may have more significance in relative terms than education, but in absolute terms of resource allocation, education by its very nature may get more.

Part III of this book has switched the focus from a science of cities to a science of design and planning. One of the central themes here is how interactions and networks between the components defining cities are as important as the activities that generate them, if not more so. This was the basis for the models of cities in part II, and it is the basis of the models of planning in this and the previous two chapters in part III. In one sense, our models of cities are positive while our models of planning are normative, dealing not with how we make decisions and how we generate designs but with best or optimal ways of achieving decisions and creating designs. In fact, we have slowly begun to relax this focus on the normative in this chapter by suggesting how models might be built of the decision-making process, and there is a tradition that supports this originating from work such as that by Dearlove (1973) and Bolan and Nuttall (1975). In the next chapter, we will take our theory of collective action based on exchange further while still keeping the focus on decision making. It will take these ideas forward in terms of a real decision problem that is rooted in spatial development. This will steer our focus more firmly to the real than to the hypothetical. After this, we will conclude with ideas about how we can fashion these ideas into formal structures for committee decision making relevant to urban planning and design.

13

Urban Development as Exchange

The individuals in the market trade until each ... has ... ideal holdings, and then they leave the market. It is to be emphasized that it is only when equilibrium prices are attained that any trading takes place. The model is a dynamic model for arriving at a satisfactory price structure for goods, *not* for the trading of goods.

—James G. Kemeny and J. Laurie Snell, *Mathematical Models in the Social Sciences* (1962, p. 35)

At the heart of the theory of collective action outlined in the last chapter lies the idea of exchange: the notion that for a system to be in equilibrium, its agents must adjust their demands to the supplies controlled by other agents, or vice versa. In fact, the most likely condition in a perfect market is that both demand and supply adjust to the point where a balance occurs and all agents are satisfied that no further improvement in their conditions can take place. Price, of course, is the signature of this balance, and the process of adjustment that must take place for an equilibrium to occur is likely to be one of trial and error, but under a regime where convergence is assured. Of course, most markets are not perfect, but in economic theory, this rudimentary model of exchange is usually the starting point.

We have not so far exploited the notion of exchange in our theory, for we made clear that our concern is normative; that is, using these ideas to resolve conflicts between agents or actors by specifying a process of adjustment that is entirely rational and perhaps optimal, but probably artificial in practice. Yet the process of sequential averaging first introduced in chapter 11, where it was used as a basis for the simplest design methods, contains this notion that the system moves seamlessly to an equilibrium where all actors (or their representation as factors) achieve a consensus. This model as we developed it using Coleman's theory of collective action (1966, 1972, 1973, 1998) in the previous chapter clearly contains all the elements of demand and supply that lie at the basis of economic exchange, but this was still considered implicit in the way we used theory to explore hypothetical design and policy problems. The notion that this theory might constitute a working model of a real system, according to the perspective we developed for the various tools and methods used in urban simulation in part II, began to emerge at the end of the last chapter. Of course, the theory of economic exchange is widely used to develop models of real economic systems, notwithstanding the numerous critiques of this approach. Because these ideas are controversial in practice, we have stressed the use of these ideas in normative terms, as optimal rather than empirical or positive models of a reality, although real-world extensions to these kinds of applications are tempting.

In this chapter, we will digress from our focus on planning and design and explore how these ideas can be used to examine questions of conflict between agents with an interest and control over the land market. In doing so, we will show how these ideas might underpin the way agents reach an equilibrium where land is actually exchanged. Although the predominant emphasis in building models of city systems has been on the macro-structure of land use and economic activity, problems of urban development are usually conceived by physical planners in somewhat different terms. The gap between our current perceptions of urban problems and available analytic technique is considerable, and consequently there have been moves among model-builders to explore ways of "relaxing" the harder features of their theory in an effort to embrace less well-defined issues. Many of the tools introduced in part II push toward finer spatial scales at the level of buildings and geometry, especially where an emphasis on morphology and the complexity sciences have forced new, more individualistic bottom-up approaches to modeling that incorporate ideas about agents and cells, agent-based models, and cellular automata. In this chapter, we will connect up with ideas in earlier chapters, particularly regarding the simulation and modeling explored in part II, and we will develop a model of urban development that treats actors or agents more literally as agents of change in the urban system in contrast to more top-down agencies, which have dominated approaches to design methods in the previous chapters (see Gabbour and Cartwright, 1974).

To introduce these ideas, the process of change in land ownership will first be considered as a process of economic exchange, alluding to classic ideas in economic theory. This implies a fairly conventional theory of general economic equilibrium, but it is proposed here to use Coleman's (1973) particularly flexible form of social exchange theory, which we elaborated in some detail in the previous chapter. The model is first stated as a formal equilibrium system, and its structural characteristics will then be examined through an interpretation of the exchange processes, which give rise to the imputed equilibrium. We then digress to examine an application to land development in north Central London, where key conflicts between agents have dominated the development process for the last thirty or more years. Our focus is actually on a historical urban conflict based on changing land ownership and property speculation in the Euston area of lower Camden, known locally as Tolmers

Square. The model to be built relates to the crisis over property speculation in Tolmers Square, which came to a head in 1973. A rich source of information is available concerning the chronology of events in this area between 1960 and 1976, from the brilliant impressionistic account by Wates in his book *The Battle for Tolmers Square* (1976). These events are briefly outlined, and an account of how the problem is structured before the model is developed is then given. Four versions of the model are developed and their predictions in terms of changes in land owner-ship are evaluated.

Exchange Processes and Urban Land: An Equilibrium Model

The well-known model of economic exchange is predicated on the assumption that individual actors increase their overall utilities or levels of satisfaction by trading commodities they have control over for other commodities they desire but that are held by different actors. This theory assumes the existence of a general economic equilibrium in which the process of trade or exchange is essential for market clear-ing and for an efficient allocation of resources. The model can be easily adapted to the land market, and indeed has been by Lowry (1967), who uses it as a basis for comparison of several traditional urban models. The model developed here will be stated in two ways: as a set of formal accounting relations, and then as a set of input-output equations.

We will present the model as we have already developed it in the last chapter, but first in descriptive rather than formal terms so that we can adapt it to processes of land exchange. Any theory involving changes in land ownership and occupancy at the site, plot, or building scale must implicitly or explicitly involve the idea of exchange. Indeed, the market for land represents one of the most well-defined of all economic markets in that its major mechanism is based on explicit, almost literal exchange; that is, the market process is based on how buyers and sellers are tied together and how land is exchanged for other land or cash through the finance market. In some cases the goods in question are actually exchanged for one another without cash ever entering the picture.

The land market, however, is invariably much more complicated in that such direct exchanges are rarely possible. More common is the case in which land is exchanged for cash, where actors may own part rather than the whole of a site and the very process of exchange alters and determines the value of land itself. Moreover, because there are many actors involved, and because such exchanges cannot be tied together in the simple way described above, there are often long "chains" or "net-works" of buyers and sellers that make the market slow and difficult to clear. Here, however, we will be involved in a land-exchange process that is especially character-ized by actors who are in direct conflict with each other. In situations where some actors own land and wish to retain their ownership, but where there are other actors who do not have such control but desperately want to own the land in question, severe conflict can emerge. The situation may well be resolved by massive changes in the value of land, particularly if those who do not control the land, but wish to do so, have access to more resources (through financial or political means) than those who presently own the land.

This is the type of situation that characterizes, for example, the invasion of one land use over another, especially where more "profitable" land uses are expanding into areas of less-than-profitable use; the expansion of offices around central areas of major cities, invading the poor, inner areas of such towns is the classic example. This is the type of situation the Tolmers Square model attempts to simulate. Modestly well-off owner-occupiers and poorer tenants of small plots, confronted by "rich" property speculators wishing to own those plots and with the resources to "tempt" occupiers to sell or leave, and in addition confounded by local agencies—typically local governments wishing to own the land to comprehensively redevelop it, with the powers to do so in principle, but the inability to do so in practice: this is the typical picture. Clearly, here is a situation with all the ingredients for a severe conflict of interests and ideologies. Our primary concern, however, is how the rudiments of such a process of exchange might be modeled, and before we adapt Coleman's theory to this context, we will talk the reader through how the model works.

The essence of exchange is the immediate conflict between what people actually own and what they desire to own. Following Coleman (1973), we will call actual ownership *control* and desired ownership *interest*. We will also assume that each actor initially has a budget—a level of resources fixed by what they control at present, but that is altered as the actors exchange their resources in the quest to bring their actual ownership closer to their desired ownership. We will call the actual absolute allocation of resources to any land parcel or site k that each agent or actor j controls $\hat{C}_{kj}$, and the desired allocation that is the agent i's interest in each site k, $\hat{X}_{ik}$. As in chapter 12, we will assume that there are m sites and n actors or agents. If we add the resources invested by one actor over all sites, this gives us the total investment of an actor in sites, while if we add the resources for individual sites over all actors, then this gives us the investment in each site. The matrices $\hat{C}$ and $\hat{X}$ are essentially flow matrices measured in cash terms between actors and the sites they have control and interest over. The essence of the exchange problem is to initiate a process of "averaging," or "*tâtonnement*," as Leon Walras, one of the founders of economic equilibrium theory in economics, called it 150 years or more ago (Morishima, 1977). This ensures that the differences $\hat{X} - \hat{C}^T$ between these disappear in the equilibrium; that is, where interest is equal to control, or where desired control is equal to actual control. One rather nice visual illustration of this quest for balance is to plot the matrices $\hat{C}^T$ and $\hat{X}$ as bipartite graphs, which immediately shows the differences between the actual and desired flows of resources. Note that $\hat{C}$ and $\hat{X}$ are transposes (T) of one another. These measure the way in which interests and control by actors in sites are combined in the process of reaching a balance. There are, of course, different ways of defining this balance by altering interest or control, or both, although in our exposition here, we will assume that resources associated with each actor do not change, whereas the resources allocated to sites, hence the ownership of sites, will change. Over long periods of time, however, actor resources may change, but here, any such change involves the actors in question losing interest in the problem and thus leaving the system.

The model does not actually involve a literal process of exchanging land until the last stage of the simulation. Instead, the model works as follows. On the basis of the mismatch between interest $\hat{X}$ and control $\hat{C}^T$, we compute an explicit degree of mismatch between any pair of actors i and j, say, and argue that it is necessary to begin to exchange resources (power) in proportion to this strength (degree) of interaction (mismatch). In essence, we will be exchanging resources using the inter-actions between pairs of actors as a kind of communications network or power structure. If we operate this process continually, the amount of resource "captured" by any actor will converge to a stable value, and an equilibrium will emerge. When this happens, these resources can be used to work out how much exchange is necessary to satisfy the pattern of interests. This is very much an approximation to the exchange process, but it is consistent. In fact, there are more realistic ways of explaining how the model works, but to increase realism, we need to resort to the slightly more formal representations of the last two chapters; hence our emphasis on viewing land exchange as the final outcome of a process of trading resources according to a fixed power structure.

The first thing we have to do is to work out how much interaction is necessary between actors according to their mismatch of interest and control. Before we do this, we must note that the actual and desired allocations of cash to sites is not strictly relevant in making comparisons, because these amounts will change as the amount of resources change. What is important is the pattern of allocation. For actors, what we wish to know is the proportion of their interest they have in each site, while in terms of these sites, we need to know the proportion of control each actor has over a site. Let us first begin with a set of resources—cash, say—which each actor has. We can then work out how much cash each actor invests in sites k by expressing the degree of interest each actor has in a site as a proportion of the actor's interest in each site. If we then multiply this by the cash each actor has, this gives us the amount each actor would like to invest in each site. If we then sum these amounts for actors over each site, we determine how much cash is desired to be invested in each site. This, in general, will be different from the amount of cash that is actually invested in each site because this pertains to the degree of control or ownership actors have in the various sites.

With this new desired pattern of investment in sites, we can express the actual control each actor has in each site as a proportion of the control an actor has in all sites, and if we then multiply this control that the actor has by the desired amount of investment in each site, this gives us a second round of cash that the actor would control if the investment in sites were to be expressed solely in terms of the interest they had in the various sites. Thus we have determined a new allocation of resources for each actor with respect to sites. This in general will be different from the cash the actors have invested in sites the first time around. It represents a new allocation based on merging—or averaging interest with control, and desires with actual ownership—and it implies a process of compromise. If the process continues in this fashion, initial differences between resources and the value of sites will begin to disappear, and if every actor is linked with every other (which we assume is the *de facto* situation), then convergence or equilibrium will occur. Control and interest, however, will not change although the resources available to actors investing in sites and the value of the sites will do so. The last stage is to either ensure that control is changed—that is, ownership is changed to meet the desired investment in sites, which is the interest—or the interest is changed to match the initial control or ownership. We presume the former since this relates to land development, but some intermediate position where both ownership and interest change is eminently possible.

An Outline of the Land Development Model

The Tolmers Square model is largely the same as Coleman's (1966, 1972, 1973, 1998), which we outlined in chapter 12 in equations (12.1) to (12.14). Now, however, we give the model a distinct substantive interpretation as a "market" for land in which cash is the medium of exchange. The measurement of actual ownership, or control, and desired ownership, or interest, is in money, and the total amount of money in the system can now be defined as M. Then we can define the total resources of each actor associated with control and interest as

$$r_j^c = \sum_k \hat{C}_{kj} \text{ and } r_i^d = \sum_k \hat{X}_{ik}, \tag{13.1}$$

and the total value of the sites as

$$v_j^c = \sum_j \hat{C}_{kj} \text{ and } v_k^d = \sum_i \hat{X}_{ik}. \tag{13.2}$$

The total money M associated with these investments is given by the general conservation equations

$$\sum_j r_j^d = \sum_i r_i^d = \sum_k v_k^c = \sum_k v_k^d = \sum_k \sum_j \hat{C}_{kj} = \sum_i \sum_k \hat{X}_{ik} = M. \tag{13.3}$$

An equilibrium defined in terms of an equal exchange is defined in terms of the coincidence of actual and desired ownership—interest and control—where the two flow matrices are identical, that is

$$\hat{C}_{kj} = \hat{X}_{ik}, \quad i = j, \tag{13.4}$$

from which equal resource and value vectors are defined as

$$\left.\begin{array}{l} r_i = r_i^c = r_i^d \\ v_k = v_k^c = v_k^d \end{array}\right\}. \tag{13.5}$$

Exchange will therefore only occur if there is a mismatch between actual and desired resource allocations, and in such a case $\hat{C}_{kj}$ and $\hat{X}_{ik}$ will adjust so that eventually equations (13.4) and (13.5) are satisfied. This was the process we described informally above, and which can be likened to the process of *tâtonnement*. In essence, what we do is balance the resources and values, keeping the initial control and interest fixed, and once the convergence has taken place, we then adjust the control, which implies a new pattern of site investment consistent with the ultimate values and resources. We do not assume that interest changes, for this remains stable, but in the model variants that follow, changes in interest and control are also invoked by letting agents enter or leave the system.

To demonstrate the model, we follow the logic developed in the previous chapter, but now note that we define the relative control and interest relationships in terms of our known patterns of actual and desired ownerships. In this, the underlying patterns of interest and control are assumed stable, and we define them in relative terms as follows: relative interest X_{ik} is

$$X_{ik} = \frac{\hat{X}_{ik}}{r_i^d} = \frac{\hat{X}_{ik}}{\sum_z \hat{X}_{iz}}, \quad \sum_k X_{ik} = 1, \tag{13.6}$$

and relative control is

$$C_{kj} = \frac{\hat{C}_{kj}}{v_k^d} = \frac{\hat{C}_{kj}}{\sum_z \hat{C}_{kz}}, \quad \sum_j \hat{C}_{kj} = 1. \tag{13.7}$$

The model can now be formulated and solved as the two-stage process outlined in descriptive terms in the previous section. The logic follows quite closely equations (12.37) and (12.38), which can be represented in terms of the relationships between actor and actors and site and sites in equations (12.39) and (12.40). As our notation slightly differs here, we will repeat this sequence of equations (so that readers do not have to turn back to chapter 12). We now write the iterative two-step process as

$$v_k(t+1) = \sum_i r_i(t) \, X_{ik} \quad \Rightarrow \quad r_j(t+1) = \sum_k v_k(t+1) \, C_{kj}, \tag{13.8}$$

where we start with resources $r_i(t = 1)$ and effect the continued substitutions implied by the sequence in equation (13.8) until we reach some limit of convergence defined as

$$\left.\begin{aligned} v_k &= \sum_i r_i X_{ik} \\ r_j &= \sum_k v_k C_{kj} \end{aligned}\right\}. \tag{13.9}$$

Equation (13.9) can be solved simultaneously for v_k and r_j. However in general at this stage, the resource distributions associated with actual and desired control (interest) are not equal; that is, $v_k C_{kj} \neq r_i X_{ik}$, $i = j$. Thus the next stage of the model consists in adjusting C_{kj} to $\overline{C}_{kj}$ to ensure that $v_k \overline{C}_{kj} = r_i X_{ik}$, $i = j$. From these definitions, it is clear that

$$\overline{C}_{kj} = \frac{\hat{X}_{ik}}{v_k} = \frac{r_i X_{ik}}{v_k} \, , \quad i = j, \tag{13.10}$$

where $\overline{C}_{kj}$ is now the final pattern of ownership or control.

A number of points should be noted. The model clearly represents a caricature of an exchange process, although its equilibrium form appears consistent with a variety of such processes (Coleman, 1977). In this form, it is close to the virtual process of *tâtonnement* outlined originally by Walras, in which change in ownership of goods occurs after the process of iterating prices for demand and supply has worked itself out. The units in which value and resources and their patterning are measured clearly depend on $\hat{C}$ and $\hat{X}$, and this is arbitrarily fixed to the total investments M in equation (13.3). The distributions of value and resources predicted from the model are relative to the application, and as such, they do not correspond very closely to land values or actual resource income levels in this context, since these are determined by a host of factors not included in this model.

Structural Aggregation and Model Dynamics

In chapter 12, we aggregated Coleman's model in various ways, building networks between actors through their interest and control in events or sites and between sites through the control and interest in them by actors. In short, we produce networks by concatenating the two bipartite sets of relations—interest and control—in opposite or dual ways. This was a technique of network analysis, which we widely exploited in chapter 6 in exploring links between road segments through their intersections, and vice versa when we examined street networks in cities. We will repeat these relations by noting that the two steady-state equations in (13.9) are intrinsically linked with respect to how resources relate to values and vice versa. Resources can thus be expressed as

$$r_j = \sum_k v_k C_{kj} = \sum_i r_i \sum_k X_{ik} C_{kj}, \tag{13.11}$$

where

$$P_{ij} = \sum_k X_{ik} C_{kj}, \quad \sum_j P_{ij} = 1. \tag{13.12}$$

This enables the steady-state relation to be written as

$$r_j = \sum_i r_i P_{ij}. \tag{13.13}$$

Equation (13.13) is equivalent to the steady-state equation of a discrete-time, discrete-state Markov process. It is, in effect, French's (1956) model as interpreted by Harary (1959), which we used for sequential averaging in chapter 10, where in its averaging form it first appeared as equation (10.40). All other results we have related to this earlier model apply, of course, and we have demonstrated several times that this steady state can be computed iteratively or by solving the related set of linear simultaneous equations as in (11.25).

Equation (13.13) is the reduced form of the first equation in (13.9), and it can also be interpreted as the outcome of an averaging process. Assume that actors are initially endowed with a distribution of resources $r_i(t = 1)$. According to equation (13.13), actors will then exchange resources $r_i(t = 1)$ using the weighted interaction network implied by $\{P_{ij}\}$. At the next time $t + 1$, the new distribution of resources $r_j(t + 1)$ can be computed as

$$r_j(t + 1) = \sum_i r_i(t) P_{ij}, \tag{13.14}$$

and with suitable restrictions on $\{P_{ij}\}$, equation (13.14) converges to equation (13.13) as $t \to \infty$. $\{P_{ij}\}$ can be regarded as a social power structure or network (Taylor and Coleman, 1979), although it is actually useful to interpret this as the technical coefficient or flow matrix of a closed input-output model (Gale, 1960). In equilibrium, the flow of resource from actor i to j is given by $T_{ij} = r_i P_{ij}$, and the difference $T_{ij} - T_{ji}$ measures the degree of actor dominance in any bilateral trade.

We refer to this type of interaction—over sites to generate actor interactions—as the primal process of exchange. A dual process also exists. Substituting for r_j in the first equation in (13.13) gives the following reduced form,

$$v_\ell = \sum_j r_j X_{j\ell} = \sum_k v_k \sum_j C_{kj} X_{j\ell},$$ (13.15)

where

$$Q_{k\ell} = \sum_j C_{kj} X_{jk}, \quad \sum_\ell Q_{k\ell} = 1.$$ (13.16)

This enables the steady-state relation to be written as

$$v_\ell = \sum_k v_k Q_{k\ell}.$$ (13.17)

The relevant flow network given by $S_{k\ell}$ shows the transfer of investment or value between sites and is defined as $S_{k\ell} = v_k Q_{k\ell}$, and the difference $Q_{k\ell} - Q_{\ell k}$ measures the degree to which the initial site investment dominates the ultimate one.

In some senses, this primal-dual interpretation of the model is its most attractive feature. For example, for an appropriate definition of agents as activities and events as sites or zones, the primal process can be considered akin to input-output analysis, and the dual to spatial interaction modeling. Furthermore, as we implied in chapter 3 in part I, there is considerable potential for developing the analysis of structure in such models using Atkin's (1981) Q-analysis, which is performed on relationships defined over sets, such as those involved here. The aggregate flow matrices $[T_{ij}]$ and $[S_{k\ell}]$ only refer to the first stage in the equilibrium model using fixed interest and initial control. Using final control $\bar{C}_{kj}$, final flow matrices $[\bar{T}_{ij}]$ and $[\bar{S}_{k\ell}]$ can be predicted. It can easily be shown that these flow matrices are symmetric about their main diagonals, defining an equal exchange in the sense of Berger and Snell (1957). Thus, in equilibrium, such equal exchanges imply reversibility of the related Markov processes (Kelly, 1979), and there are connections here to the design of optimal committee decisions using ideas of symmetry (Theil, 1976). We will explore some of these issues in the next chapter, but as the statement of the model is now complete, it is appropriate to introduce the application.

The Battle for Tolmers Square

Tolmers Square is an area of north Central London bounded by Euston Centre on the west, Euston Station on the east, and University College London on the south. To the north, the area merges into a mixture of urban blight and gentrified housing characteristic of some parts of inner areas of British cities, and much of the London borough of Camden, of which Tolmers Square is a part. The area itself, however, has a certain historic style in many of its buildings, and the best early nineteenth-century housing in the area is now listed as heritage development. Back in the early 1970s, when the controversy this model addresses was ongoing, the area was quite run-down, with a mixture of owner-occupied and rented housing and shops, small industrial premises, and institutional uses. In some senses, the area was an anachronism even then in that it is so close to the commercial heart of Central London that it might be expected to have been redeveloped and commercialized many years earlier. In fact, this represents the problem. Property developers were not slow to see the area's potential, and by 1974, land values in the area were anything upward of £600,000 per acre. Moreover, since the late 1950s, local authority involvement had led to indecision and lost opportunities, thus encouraging the onslaught of planning blight with the consequent fragmentation of the local community and increasing dereliction of property.

The saga is complicated by the affairs of one large property company called Stock Conversion. This company, through the "shrewdness" of its main director, Joe Levy, had, like many such companies in the 1960s, built up enormous assets based on the profits of property speculation (CIS, 1973). Stock Conversion had already developed a similar area of land to the west of Tolmers Square—Euston Centre—which had yielded immense profits. In fact, as early as 1962, Stock Conversion was buying properties in the Tolmers Square area and by 1974, owned well over half the area of interest here (Wates, 1973). A map of the area and the pattern of land ownership in 1974 is shown in figure 13.1. The series of planning schemes prepared by the London Borough of Camden and its predecessor local authority, St. Pancras Borough Council, came to nothing, largely because of the excessive cost of compulsory purchase. By 1972, Camden had decided that a scheme of its own was impossible, and thus they negotiated with Stock Conversion to agree a joint development that involved a much larger proportion of offices than had hitherto been envisaged. Up to this point, there had been little opposition to Camden's plans. Most of the residents considered the local authority to have little interest in any scheme for the area and there was a general feeling that "nothing would ever happen." However, once knowledge of the proposed deal between Camden and Stock Conversion became known, opposition was mobilized, opposition largely external to the area itself. This mobilization took the form of a proposal that a new company called Claudius would develop the area and "hand back" any development profit to the local authority. The proposal had impeccable backing and quickly convinced the local authority to abandon the "Levy Deal," as it had become known.

Events then moved swiftly. A local pressure group, the Tolmers Village Association or TVA (Wates, 1975) was founded, and squatters moved into many of the empty derelict properties in the area. But most important was the arrival on the scene of a new group of influential actors, none of whom had been active prior to

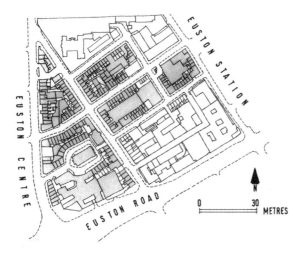

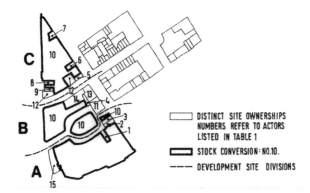

Figure 13.1
Built form and land ownership in Tolmers Square.

1973 and many of whom had never heard of Tolmers Square before that date. The eventual outcome of this pressure was that in 1975, Camden finally decided to buy out Stock Conversion's interest and undertake a more equitable and socially relevant development for the area (Hall, 1975). Other issues of more global import were also seen as affecting the outcome. In particular, the election of a left-wing Labour government in 1974 and the prospect of a tax on land profit through the Community Land Act made property speculation less attractive and perhaps forced Stock Conversion into acquiescing to Camden's proposals. Nevertheless, it is clear from Wates' (1976) account that the new actors emerging after 1973 had a profound effect upon the outcome, and probably reversed the redevelopment by corporate interests to commercial land uses that had earlier seemed inevitable.

To return to our primary subject, the intended goal is to simulate this change in Camden's policy toward their ownership of land as a switch in control within the exchange model. However, the model will represent only the crudest of caricatures of these events. Wates (1976) identifies at least 132 actors and 62 sites in his account, and to attempt to build a model at this level of detail would be far too extensive. We can illustrate the logic of this approach with a smaller, simplified representation of the conflict. Thus, in the interests of manageability and interpretability, our attempt at modeling the conflict and its resolution has been restricted to the western group of sites known locally as the "Levy Triangle." These are indicated in figure 13.1, where the actors owning plots are identified by number and listed later in Table 13.1.

Structuring the Problem

The most obvious way of modeling this process of land exchange is to identify all the actors involved over the time period of interest, whether or not any of these actors are passive or active at different points during this time. Final control could then be predicted with fixed interest, or if it was essential to change interest periodically, the equilibrium model could be embedded into a recursive temporal structure that would enable this. Thus the actors have first been divided into two sets: those who were active in bargaining over land prior to 1973, and these together with an additional set of actors who became active due to the events of 1973. The model will be run as follows: the patterns of final control and related exchange networks will be predicted using the subset of actors active prior to 1973, and this prediction will provide an outcome to the process without the effects of the 1973 crisis. The model will then be run with the full set of actors post-1973 with these predictions incorporating the 1973 crisis.

There are 15 actors who own or occupy land in 1973 and 4 additional actors— Camden, the TVA, Claudius, and the Camden Labour Party, whose "interest" was

Table 13.1
Actors and events

Agents/actors			Events/sites		
			1	A	Tolmers Square South
			2	B	Tolmers Square North
			3	C	The island site
Owners/occupants of residential sites			*Owners/occupants of residential sites*		
1	Tolmers Square South		*4*		
2	North Gower Street		*5*		
3	North Gower Street		*6*		
4	Tolmers Square North		*7*		
5	North Gower Street		*8*		
6	North Gower Street		*9*		
7	North Gower Street		*10*		
8	Hampstead Road		*11*		
9	Hampstead Road		*12*		
Owners/occupants of commercial sites			*Owners/occupants of commercial sites*		
10	Stock conversion (developer)		*13*		
11	Cecil House (club)		*14*		
12	Victor Laurence (shop)		*15*		
13	TGWU (trade union office)		*16*		
14	London CHA (club-office)		*17*		
Local governments					
15	Greater London Council				
16	London Borough of Camden				
Pressure groups					
17	Housing subsidies				
18	Rent control				
19	Special education				

awakened by the 1973 crisis. Events in the model essentially involve the 3 sites shown in figure 13.1. The Levy Triangle has been divided into Tolmers Square South (A), Tolmers Square North (B), and the "Island" site (C), and the 15 actors in the pre-1973 situation all have some control over these 3 sites, which is also indicated in Table 13.1. However, the most important characteristic of this set of events is the necessity for a very large "rest of the world" sector. In a model such as this, which is attempting to simulate the pressures posed by developers on longer-standing owners to sell their land and the pressure of government and other groups on developers to sell up, it is necessary to incorporate a set of alternative locations, or "pseudo-locations." Such events enable a type of migration of actors into and out of the system to take place. Consequently, associated with the first 14 actors, there are 14 pseudo-locations in the rest of the world sector, which these actors could control if they relinquished their control in land in Tolmers Square. These pseudo-locations are listed in Table 13.1 alongside the actors to which they pertain. Note that actor 15, the Greater London Council, does not have a pseudo-location because it is assumed this organization would never have any interest in abjugating its interests or control in an area for which it was responsible. The same point can be made for the 4 additional actors. The following problems to be modeled can now be identified. There is the pre-1973 model, based on the first 15 actors and all events, and the post-1973 model involving all actors and events. However, since the effect of the rest of the world sector in such a model was unknown, it was decided to consider each of these two versions, with and without this sector, thus giving four model versions in total. These models numbered 1 to 4 in the accompanying empirical analysis are listed as:

- pre-1973 with the rest of the world sectors: actors $i = 1, 2, \ldots , 15$; events (sites) $k = 1, 2, \ldots , 17$
- post-1973 with the rest of the world sectors: actors $i = 1, 2, \ldots , 19$; events (sites) $k = 1, 2, \ldots , 17$
- pre-1973 without the rest of the world sectors: actors $i = 1, 2, \ldots , 15$; events (sites) $k = 1, 2, 3$
- post-1973 without the rest of the world sectors: $i = 1, 2, \ldots , 19$; events (sites) $k = 1, 2, 3$

These four problems are displayed in figure 13.2, where the sparsity of their associated interest and control matrices $\hat{X}$ and $\hat{C}$ is indicated. These sparsity indices are based on a direct count of positive entries divided by the total possible (positive) entries, and as such represent a measure of structure rather than density. As expected, the control matrices are sparser than the interest matrices, which one suspects is typical of this sort of problem.

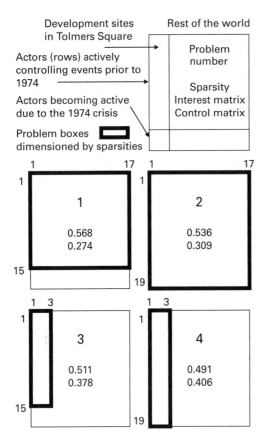

Figure 13.2
Dimensions and sparsity of the four models.

Applications and Predictions

The measurement of interest and control is inevitably problematic. The most straightforward is the 15 x 3 bottom left-hand submatrix of control, which reflects actual land ownership in the Levy Triangle. No data was available in anything more than a qualitative form to guide the specification of the rest of the control matrix or total pattern of interests. Accordingly, it was decided to attempt to elicit the strength of these relationships from a panel of informed experts using the attitudinal pairwise scaling technique from Saaty (1980) that we briefly noted in chapter 10. The panel was first briefed on the Tolmers Square case study and shown a film of these events. They were then asked to make a large number of paired comparisons, reflecting comparisons of control and of interest from which the strength of interest of any actor over events and of control by any actor in an event could be computed.

In fact, although the exercise was salutary, it was aborted due to the relative ignorance of the panel concerning the problem, and the data elicited were used merely to inform the author's own usage of this method to achieve the same. The resulting pattern of control was then "pruned" to reflect the obvious lack of control in certain instances, and the resulting matrix was scaled to give $\hat{C}$ in land value terms. The interest matrix $\hat{X}$ was much more uniform, and this was therefore sharpened to reflect greater differentiation of interest. Finally, a check on the resulting patterns was made using the qualitative discussion provided by Wates (1976).

For each of the four versions of the model, there are four baseline models that can be formed if it is assumed the system is already in equilibrium. When the system is in equilibrium, equation (13.4) holds, and if it is assumed that the initial pattern of control in value terms $\hat{C}$ is in equilibrium, then $\hat{C}^T = \hat{X}$. This is equivalent to $v_k \overline{C}_{kj} = r_i X_{ik}, i = j$, and it follows that resources and values can be found immediately from the known flow matrix $\hat{C}$. Then

$$r_j = \sum_k \overline{C}_{kj} \text{ and } v_k = \sum_j \overline{C}_{kj}. \tag{13.18}$$

Such a model is in equal exchange, and its flow matrices $[T_{ij}]$ and $[S_{k\ell}]$ are symmetric. Models of this type in fact will be referred to as "symmetric" in the following sections. The resource and value distributions predicted using equation (13.18) will be plotted below for each of the four models and compared to the actual predictions from equation (13.9) in an effort to assess how close these predictions are to the existing observed pattern of ownership.

The predictions from each of the four models will be examined, first in terms of resources and values, then in terms of exchange between actors measured from $T_{ij} = r_i P_{ij}$, and finally in terms of the patterns of landownership, into which final control predictions can be translated. The resource levels are plotted in figure 13.3 for the symmetric and actual cases. In fact, the symmetric are close to the actual values and thus represent a good first approximation to resource levels, as demonstrated in other experiments with this type of model (Batty, 1981; Michener, Cohen, and Sorensen, 1977). Taking out the rest of the world sector increases the dominance of the "large actors"—Stock Conversion, Camden, and Claudius—and it is of interest to note that Camden and Claudius exert considerable power in the post-1973 models, considerably more than Stock Conversion. This seems a reasonably accurate portrayal of the situation. In figure 13.4, the distributions of equilibrium values are plotted. For models 1 and 2, the addition of new actors reduces the value of the three sites and increases the importance of ensuring that Stock Conversion is bought out. This is a reasonable prediction in that the additional actors were mobilized precisely for this purpose—indeed, this is the essence of "The Battle for Tolmers Square." In the case of the first model 1, the symmetric case gives much greater weight to the Island site and to getting Stock Conversion out of the area, while in

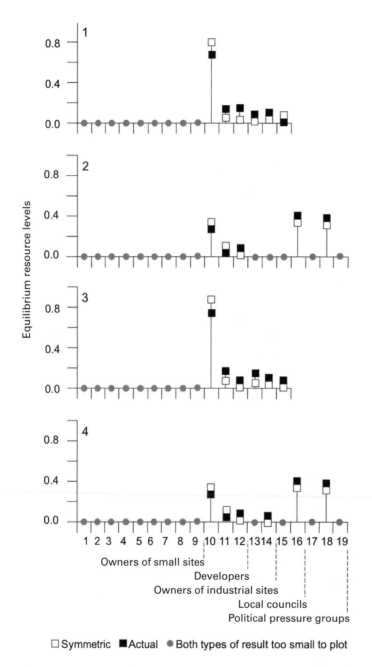

Figure 13.3
Equilibrium distribution of resources.

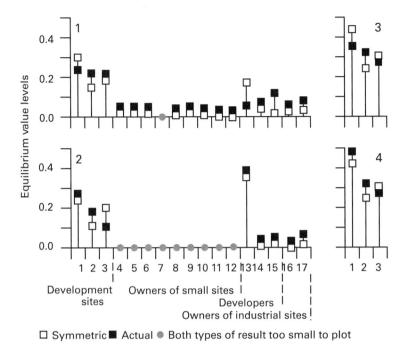

Figure 13.4
Equilibrium distribution of investment values.

the case of model 2, the symmetric results are close to the actual. For models 3 and 4 without the rest of the world sectors, all the value is concentrated on the three sites, with the Island site still the most important.

It is also worth looking at the exchanges between actors involved in reaching the equilibrium using the Markov averaging mechanism. Figure 13.5 shows a plot of $\{T_{ij} + T_{ji}\}$ for each model where we set $M = 100$ units. These networks show the percentage of resources exchanged and thus identify the critical actors in this process. In model 1, it is quite clear that Stock Conversion dominates in a situation where there is very little exchange between actors: 48% of the exchange is internal to Stock Conversion itself. For model 2, the pattern changes radically. Three actors—Stock Conversion, Camden, and Claudius—engage in about 70% of the exchange, which of course is notional exchange, or more influence than actual exchange. When the rest of the world sectors are deleted, these patterns of exchange intensify. In case of the pre-1973 model (model 3), 63% of trade becomes internal to Stock Conversion.

In model 4, the three key actors who are involved in 70% of the exchange in model 2 are now involved in 86% of the exchange. These patterns of influence are certainly consistent with those described by Wates (1976) and provide further confidence in this approach.

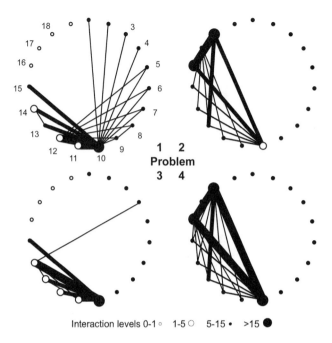

Interaction levels 0-1 ○   1-5 ◯   5-15 •   >15 ⬤

Figure 13.5
Equilibrium exchange between actors.

Finally, the four predicted patterns of land ownership are presented in figure 13.6 alongside the initial land holding in 1973. In these diagrams, land ownership has been plotted using a general rule of thumb that distinct site areas nearest to existing ownerships are taken over in the event of that ownership expanding, while adjacent sites are lost in the event of that ownership declining. The predictions for the pre-1973 models are perhaps the most acceptable, in that these represent outcomes without the 1973 crisis and can be seen in terms of existing rather than hypothetical ownership patterns. In model 1, owners and occupiers disappear at the expense of the property developers, while without the rest of the world sector in model 3, there is little change from the pattern of initial control: this indicates the importance of the rest of the world sector. In the case of models 2 and 4, the predictions are more ambiguous. For model 2, the main interest is that Claudius loses its control over the Island site while increasing its land holdings in Tolmers Square South. In model 4, the exclusion of the rest of the world sector leads to less change from the pattern of initial control in that owner-occupiers remain but the developers, Camden and Claudius, behave in a manner similar to that predicted by model 2.

The predictions from these models at least appear consistent with the actual chronicle of events in Tolmers Square. However, there remains the perennial problem

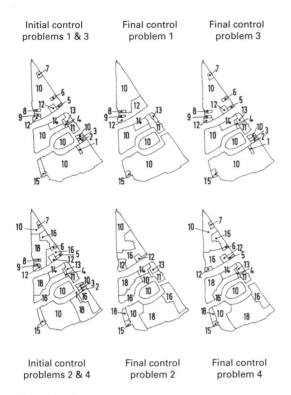

Initial control   Final control   Final control
problems 1 & 3   problem 1   problem 3

Initial control   Final control   Final control
problems 2 & 4   problem 2   problem 4

Figure 13.6
Equilibrium predictions of land ownership patterns.

of assessing to what extent the initial structuring of the model and specification of data makes such predictions tautological. In an effort to further explore these issues, the idea of the "symmetric" baseline equilibrium model seems useful. It is also necessary to develop the process of eliciting data for this model more thoroughly and to examine the structure of the rest of the world sector in such models. Furthermore, it is important to continue evolving the mathematical structure of the model to examine explicit dynamics and to explore changes in interest as well as control. Two additional speculations are worth making. Since it is necessary to explore the dimensional structure of actors and events in the model, it may be possible to begin to disaggregate those spatial events to the level of the configuration and geometry of the built form, thus extending the model to a domain where it can be used quite consciously in the design of the built form itself. Moreover, as models such as these must be continuously evolved, they are highly suited to the present focus on computation where models need to be developed interactively, as we sketched in chapter 9. Such developments are being attempted and open up exciting possibilities for interaction with a wider community of model users and participants.

Predicting Networks

Although it is easy to see how relationships between two sets of objects that differ—in this case actors involved in land and the land parcels or sites they control or have an interest in—can be represented by bipartite graphs, the real essence of this representation is to enable us to predict the form of graphs, interaction networks, or patterns that can be derived from these structures. The majority of work in social analysis has to date focused on observing networks rather than predicting their form, but consistent use of the bipartite structures gives us a handle on ways in which we might generate, and hence predict, interaction patterns. In chapters 6 and 7, we first illustrated these relations for almost trivial kinds of networks—street networks where we represented the street segments as the first set of objects and the junctions or locations where they intersected as the second set. The networks we were able to derive between streets and other streets, or junctions and other junctions, were observable, since all this involved was counting common properties. In this last part of the book, and in this and the previous chapter, we have introduced models that enable us to predict more substantive properties of networks. In all the chapters in part III, we have shown how we might predict patterns of communication between actors that might be observable in pure examples of conflict resolution by successive averaging. It is likely, however, that actually observing such networks in practice would more likely reveal the fact that such processes are far from pure, with averaging taking place in much noisier, volatile environments.

In this chapter, we have come closer to networks we might actually observe. The idea that flows of capital might underlie the communications networks used to engage in exchange is a very obvious one, and the networks in the form of probabilities $[P_{ij}]$ and $[Q_{k\ell}]$ or flows $[T_{ij}]$ and $[S_{k\ell}]$ are possibly observable from actual transactions. From such data, we might be able to more finely tune our models. In short, if we think of the model as predicting these interactions, then we can change the model to reflect more realistic processes of conflict resolution and exchange. In chapter 12, we sought to broaden our approach to deal with strings of interactions matrices between actors, problems, and policies. Here we have reverted to a focus on land and location as one of the key sets of objects defining conflict resolution. In this sense, these kinds of models are very much in the spirit of this book, which treats location as a consequence of interaction. Our concern is not so much in predicting and manipulating location as it is with predicting and manipulating interaction patterns, from which location inevitably emerges as a natural consequence of such analysis. We will provide another perspective in the next chapter.

14

Plan Design as Committee Decision Making

Groups, collectivities, etc., can and should be studied as systems of interaction, and there seems little doubt that systems-theoretical concepts can be applied fruitfully in the social sciences. Systems theory has only superficially penetrated the vocabulary of the social sciences . . .

—Anthony Giddens, *New Rules of Sociological Method: A Positive Critique of Interpretative Sociologies* (1976, p. 121)

One very clear reason why Giddens' (1976) observations on the relevance of using systems theory for representing and understanding group interaction patterns and consequently networks resonate with our ideas here is that such patterns tend to be well-defined. Traffic, migration, trade, even information flows across networks can, in principle, be measured with some accuracy. Networks exist as physical artifacts or wireless transmission, at least with respect to the movement of physical materials, populations, and digital information, and these are entirely detectable given appropriate access. Many social networks are harder to observe, for they may only be implicit in social interaction, but the idea of systems partitioned into a hierarchy of subsystems and their entities still provides a useful analog to thinking about how human agents influence one another in purposive situations. This paradigm was somewhat less controversial in the science developed in part II, where our focus was on the city in more passive terms, but it still holds strong in this part of the book. Here interactions and flows between populations in terms of the way they are described by actors or agents lie at the core of the various methods we are proposing, and these are useful in informing the way we go about solving urban problems, designing city plans, and creating spatial policies.

In this last substantive chapter, we pull together the themes of this part of the book. We begin again with the idea of design based on resolving conflicts between different factors that define subsolutions to the planning problem, and quickly move to the notion that key factors affecting a design problem needed to be interpreted as ideas about subsolutions associated with key actors. Resolution of problems of this kind are predicated on the idea of determining a set of weights, and ways of doing this involve examining the structure of relationships between factors (or actors) with the generation of weights being dependent on the strength and pattern of these relationships. This leads to the notion that optimal solutions can be based on consensus, and the processes defined on these networks are formulated in terms of sequential averaging. In chapters 11 and 12, we took this further and illustrated how any actor might have a variable interest and control over any factor. In this approach, we explode the design problem as one in which its solution could be seen in terms of the power (or weight) of actors in determining a solution to the value that factors had in determining such outcomes. In this chapter, we will twist the ideas back again to solutions based on sequential averaging, beginning with a more formal optimization problem, and turning back full circle to our extended Markov model of consensus based on Coleman's (1973) theory of collective action.

Our typical problem of physical plan design has involved choosing a spatial configuration of land uses that reduces or resolves conflict between those goals that such a plan is required to meet. In practice, the plan is chosen according to a technical process in which the planner selects and measures the appropriate goals, decides their importance, and reaches some compromise by evaluating the goals against one another. The process is subject to certain basic constraints on what can or cannot be achieved, and it is usually set in a wider political context that has a direct influence on the goals adopted and the importance ascribed to them. We have not explored this kind of rational planning process in any depth here, largely because it now tends to represent more a metaphor for how planning might be pursued in any entirely technical context, and the last fifty years has led to the idea that any plan is so complex that such a technically inspired process is purely a guide to action. Nevertheless, optimization, which such a method often features, is still a starting point in exploring how more formally structured human agencies might generate such plans collectively.

The spatial emphasis to such plans requires that they be characterized in terms of the amount and location of land use activities, and the goals are often specified in terms of spatial cost-benefit measures. Many attempts have been made to structure this process formally. Optimization models, such as those based on linear programming (Schlager, 1965; Anderssen and Ive, 1982) have been developed, but we argued in chapter 10 that such methods did not allow qualitative considerations, and we moved the debate to qualitative design methods that give the designer more control over the process. For example, simple sieve mapping or more complicated weighted analogs were developed in chapter 10, thus raising the basic idea that planning solutions were weighted averages of some set of relevant factors in which conflict between various parts of the planning problem needed to be resolved (Steinitz, Parker, and Jordan, 1976). Indeed, the basic logic of geodesign depends on these ideas (Steinitz, 2012). Two problems, however, continue to characterize these techniques. The first is the assumption that all goals can be collapsed into a linear objective function, which as Hopkins (1977) notes, assumes that the goals are independent of one another. The second relates to the choice of weights for combining these goals into a composite objective. Notwithstanding the attempts by theorists of design methods (Alexander and Manheim, 1962a; Alexander, 1964) to bypass this problem by arguing that the way goals are combined implicitly determines their weights, most weighting schemes are determined exogenously to the problem, often arbitrarily, and have been subject to much dispute. In fact, in chapter 10, we argued that such weights should be endogenized in that they should emerge naturally from our understanding and manipulation of the problem.

These methods are all conceived from the viewpoint of the individual designer or decision maker—in this case the planner—and only take account of the broader political milieu in which planning exists insofar as the planner adopts a "representative" set of goals for the plan. It is in fact generally accepted that planning embodies group and public decision making or at least that plans are modified and tempered once they are considered in the political process. Indeed, it is possible to see plan design as a rather structured process of group decision making in that planning committees and teams are formally engaged in the production and choice of plans. This suggests that formal approaches to plan design should embrace group or committee decision making, building on ideas that we introduced in the previous two chapters. But existing approaches to broaden the optimization paradigm have not generally extended this far (Voogd, 1983). It is the purpose of this chapter to begin a tentative exploration of some formal ideas necessary to cast such decision-making models in formal processes that might be operated in the context of small-group decision making. We developed these ideas informally in the last two chapters, but here we will make the link directly, comparing a formalized model with the theory of collective action elaborated previously.

Modeling Decision Processes

The clear distinction between individual and collective approaches to decision making and design is overlain by a deeper methodological distinction involving questions of positive and normative theory. Planning as design, from the viewpoint of the individual designer, has been conceived mainly as a normative affair, and its methodology has been heavily influenced by technical methods of problem solving and optimization. Indeed, the most articulate and persuasive statement of this philosophy is that of Harris (1978), who argues strongly that planning should be defined and studied as a normative activity informed by positive theory (Harris, 1983). Indeed, the development of planning support systems that we alluded to in chapter 9, originally articulated by Harris (1989) and Harris and Batty (1993), builds on these notions of normative planning theory. In contrast, the study of planning as a social phenomenon and as a political activity has been largely conceived in positive terms, as the study of institutional and less-formal decision making (Forester, 1989; Healey, 2007), although even here, normative approaches to planning have found some favor (Goldstein, 1984).

These two broad approaches to planning processes are not necessarily in conflict. Los (1981) argues that planning as design can be seen as complementary to planning as social change, and indeed that the theory of design exists within the theory of planning, thus reconciling individual with collective action and positive with normative theory. The approach developed here is consistent with Los's (1981) characterization, in that the models proposed contain elements relating both to design and to planning and, at the same time, indicate a shift from design based on individual, normative activity to the study of design in planning based on collective decision making. Nevertheless, the distinction is not as clear-cut as Los's argument might suggest. The two models suggested here, for example, reflect optimizing and nonoptimizing behavior in different ways, although not in the broader paradigm of planning as optimization. The models are clearly in the conventional tradition of those based on replicating known phenomena, rather than on providing normative strategies for action.

The models of formal decision making proposed represent highly stylized versions of collective decision making, and to use Choukroun's (1984) terminology, are obvious metaphors for the exploration of real processes. Decision processes are largely unobservable. Thus the concern here is with testing these models in terms of the relevance of their predictions versus observed outcomes of such processes, rather than with the form of such processes per se. No formal evaluation of the appropriateness of the models as theories of decision making is offered, apart from the most casual observations of the seeming relevance of the assumptions made and the way the models are articulated in formal terms. The kinds of processes for which these models are relevant are fairly easy to define. Highly structured processes involving formal decision making, with no scope for "non-decision making," versus the comparative independence of decision makers in such contexts where there is scope for collaboration and collusion but where actors must in the last analysis be independent—these are the types of processes of interest. Such processes are those observed in formal committee structures, where preferences may well be influenced by the wider environment, but where preferences are relatively stable during the time when decisions are actually being made. This stability of preferences also suggests that these models may be useful in exploring decision making in highly structured teams or groups where preferences are similar but where there is a problem of organizing information.

By considering plan design as committee decision making, three issues that affect the traditional design-based approaches alluded to above are resolved. First, the idea of optimization as a collective (group) rather than just an individual characteristic must be defined, and this raises immediately the notion of conflict between individuals. Second, the process of group decision making must be related to the process of optimization, and third, the way in which groups or committees take account of or establish the importance (weight) of goals is made apparent. In this context, it is sometimes possible to link these approaches to more formal theories of utility and economic rationality. Here we will explore two models, which both assume a collective goal as the weighted linear sum of individual goals, and which both determine weights of goals by reference to the impact individual decisions have on one another. Since the optimal decision is based on a weighted linear sum of individual objectives, this serves to link these models to more conventional plan-design methods introduced in chapter 10. The first model is based on the traditional utility-maximizing approach, first generalized to a group or committee context by van den Bogaard and Verslius (1962), and then further justified by Theil (1963, 1964). The second model is not strictly an optimization model per se, in that decisions are reached through the resolution of conflict by exchange. This is the model from Coleman (1973) used extensively in chapters 12 and 13, and it is a disaggregate form of the more general Markov sequential averaging model lying at the heart of all the methods in this part of the book. All of these are appropriate to a variety of problem-solving contexts as an analog for reaching consensus (French, 1956; de Groot, 1974; Hegselmann and Krause, 2002; Jackson, 2010).

As we now know, Coleman's model produces weights as a consequence of an exchange process, and these are consistent with certain conditions equivalent to the weights produced in the Bogaard-Verslius committee model. We will in fact explore both models and trace their equivalences, but we will develop the Bogaard-Verslius model first, before we then illustrate how this classic economic optimization model, albeit developed in a political budgetary context, can in fact be interpreted as part of a wider theory of collective action. It has been operationalized for a particular case of resource allocation by committee observed in a small town in northwest Germany by Roskamp and McMeekin (1970). This application will be described and reworked using the same data in the Coleman model, thus tracing the equivalences between the two models in something more than theoretical terms. Although we will emphasize equivalences, perhaps somewhat inconclusively, there will also be some focus on the way in which weights are derived in both models. Most design models are particularly weak on procedures for generating weights, and thus the models can be viewed as procedures in their own right for weight generation. The example chosen is not strictly one of physical land use plan design, but it is sufficiently close to illustrate how such a plan-design problem might be tackled using these methods. As an introduction to such decision making, we will first present the Bogaard-Verslius model before contrasting this with Coleman's model, and then reworking Roskamp and McMeekin's example as a method for how we might tackle a problem of land-use plan design using these ideas.

The Optimal Committee Decision Problem

Van den Bogaard and Verslius (1962) set up their model as one in which a set of n actors i, j form a committee to discuss or agree a plan for allocating resources to a set of m components, elements, or events k, ℓ that constitute the plan. Now, a plan can be defined for each individual actor i as the row vector

$$\mathbf{a}_i = [A_{i1}, A_{i2}, ..., A_{im}], \tag{14.1}$$

where A_{ik} is the resource allocated by actor i to event k and the nxm matrix $\mathbf{A} = [A_{ik}]$ is the set of individual plans. The superscript T will be used to denote transposition, and thus the row vector $\mathbf{a}_i$ can be transposed to the column vector $\mathbf{a}_i^T$.

Each actor is assumed to have a set of preferences for the level of resources to be allocated, which can be represented by a quadratic utility function $U_i(\mathbf{a})$ of the form

$$
\begin{aligned}
U_i(\mathbf{a}) &= \sum_k \gamma_{ik} A_{ik} + \frac{1}{2}\sum_k \sum_\ell A_{ik}\gamma_{ik\ell}A_{i\ell} \\
&= \boldsymbol{\gamma}_i \mathbf{a}_i^T + \frac{1}{2}\mathbf{a}_i \boldsymbol{\Lambda}_i \mathbf{a}_i^T
\end{aligned}
\tag{14.2}
$$

where $\boldsymbol{\gamma}_i$ is a $1xm$ row vector of parameters γ_{ik}, which ensure that the individual's utility increases as he or she is able to allocate more resources to events, and $\boldsymbol{\Lambda}_i$ is an mxm negative-definite matrix, which reflects the diminishing marginal rate of substitution between events. The theory of quadratic preference functions is very well worked out (see Theil, 1964), and it is easy to show that the optimal allocation of resources to plans for actor i is called $\tilde{\mathbf{a}}_i$ when $U_i(\mathbf{a})$ is maximized. Then

$$\tilde{\mathbf{a}}_i^T = -\boldsymbol{\Lambda}_i^{-1}\boldsymbol{\gamma}_i^T, \tag{14.3}$$

and substituting equation (14.3) into equation (14.2) leads to the utility maximum for each individual:

$$U(\tilde{\mathbf{a}}) = -\frac{1}{2}\boldsymbol{\gamma}_i \boldsymbol{\Lambda}_i^{-1}\boldsymbol{\gamma}_i^T. \tag{14.4}$$

To resolve the difficulties over providing a suitable scale for utility, van den Bogaard and Verslius (1962) define an individual loss function $L_i(\mathbf{a})$, which when minimized is equivalent to the maximization of equation (14.2). Then

$$L_i(\mathbf{a}) = U_i(\tilde{\mathbf{a}}) - U_i(\mathbf{a})$$
$$= -\frac{1}{2}(\mathbf{a}_i - \tilde{\mathbf{a}}_i)\mathbf{\Lambda}_i^{-1}(\mathbf{a}_i - \tilde{\mathbf{a}}_i)^{T}. \tag{14.5}$$

A social welfare function (in this case, a social loss function) for the committee is formed as a weighted linear sum of the individual loss functions in equation (14.5). Harsanyi's (1976) justification for such a linear form of social utility is invoked, and this would seem acceptable for the case of a committee where each actor was assumed to allocate resources independently of the others. Nevertheless, a set of weights $\{w_i\}$ associated with each loss is required. Without loss of generality, if we assume $\Sigma_i w_i = 1$, then the social loss function $L(\mathbf{a})$ can be written as

$$L(\mathbf{a}) = \sum_i w_i L_i(\mathbf{a})$$
$$= -\frac{1}{2}\sum_i w_i(\mathbf{a}_i - \tilde{\mathbf{a}}_i)\mathbf{\Lambda}_i^{-1}(\mathbf{a}_i - \tilde{\mathbf{a}}_i)^{T}. \tag{14.6}$$

The optimal budget plan $\mathbf{a}$ for the committee is then chosen by maximizing equation (14.6), which gives

$$\mathbf{a}^T = \left(\sum_i w_i \mathbf{\Lambda}_i\right)^{-1} \sum_i w_i \mathbf{\Lambda}_i \tilde{\mathbf{a}}_i^T, \tag{14.7}$$

which simplifies to the weighted mean

$$\mathbf{a}^T = \sum_i w_i \tilde{\mathbf{a}}_i^T \tag{14.8}$$

when all the preferences are equal, that is when $\mathbf{\Lambda}_i = \mathbf{\Lambda}_j, \forall i, j$. This much of the model is well established in the theory of quadratic preference structures (Theil, 1964). Its innovative nature relates to the way weights $\{w_i\}$ are chosen in this regard.

The "Symmetry of Loss" Criterion

If actor i receives utility $U_i(\tilde{\mathbf{a}})$ from the implementation of his or her own optimal plan, it is clear the actor can compute the loss inflicted by the implementation of actor j's optimal plan $\tilde{\mathbf{a}}_j$ using equation (14.5), but noting that the non-optimal plan $\mathbf{a}_j$ is now $\tilde{\mathbf{a}}_j$. Then

$$L_{ij}(\mathbf{a}) = U_i(\tilde{\mathbf{a}}_i) - U_i(\mathbf{a}_j)$$
$$= -\frac{1}{2}(\tilde{\mathbf{a}}_j - \tilde{\mathbf{a}}_i)\mathbf{\Lambda}_i^{-1}(\tilde{\mathbf{a}}_j - \tilde{\mathbf{a}}_i)^{T}, \tag{14.9}$$

where $L_{ij}(\mathbf{a})$ is the loss incurred by i from j's plan being implemented. From equation (14.9), the matrix $\mathbf{L}(\mathbf{a})$ formed from the set of best individual plans $\{\tilde{\mathbf{a}}_i\} = \tilde{\mathbf{A}}$ gives a summary of the losses and gains due to these plans being implemented.

As actor i's loss incurred by j is $L_{ij}(\mathbf{a})$ and actor j's loss incurred by i is $L_{ji}(\mathbf{a})$, it would seem "fair" to ensure that these losses were equal in some sense; that is, that the optimal loss structure be constrained so that

$$\tilde{L}_{ij}(\mathbf{a}) = \tilde{L}_{ji}(\mathbf{a}), \; i \neq j. \tag{14.10}$$

The losses in equation (14.9) will not, in general, meet the condition of symmetry in equation (14.10), and thus one strategy for choosing the weights $\{w_i\}$ would be to determine these so that when applied to the raw losses in equation (14.9), the resulting loss matrix would be as near symmetric as possible. Formally, weights should be chosen to ensure that

$$w_i L_{ij}(\mathbf{a}) = w_j L_{ji}(\mathbf{a}), \; i \neq j. \tag{14.11}$$

Theil (1963) provides an axiomatic justification for the symmetrization of the loss matrix.

A special case emerges if the individual preference weights Λ_i are the same, that is, if $\Lambda_i = \Lambda_j$, $\forall i, j$. In this case, it is quite clear that $\mathbf{L}(\mathbf{a})$ computed from equation (14.9) is already symmetric, that is $\mathbf{L}(\mathbf{a}) = \mathbf{L}(\mathbf{a})^T$. In this case, equations (14.10) and (14.11) imply that the weights should all be equal, or $w_i = w_j$, $\forall i, j$. If we assume the usual normalization, this means $w_i = 1/n$, $\forall i$ and that equation (14.7) simplifies to

$$\mathbf{a}^T = \frac{1}{n} \sum_i \tilde{\mathbf{a}}_i^T. \tag{14.12}$$

In other words, the optimal budget plan is the arithmetic mean of the individual budget plans when the preference structure of each individual is the same.

The more usual case, however, will exist where preferences are different and a non-uniform distribution of weights is required to approach equation (14.11). In general, it will not be possible to choose n weights to meet the $n(n-1)$ conditions of equation (14.11), so an approximation to symmetry must be accepted. Van den Bogaard and Verslius (1962) suggest that marginal symmetry of the weighted-loss matrix would provide a reasonable compromise. Then weights would be chosen so that

$$\sum_i w_i L_{ij} = \sum_j w_j L_{ji}, \; i \neq j. \tag{14.13}$$

The n conditions in equation (14.13) can in fact be met uniquely, with the assumptions of the non-negativity and full independence of the matrix $\mathbf{L}(\mathbf{a})$. Van den Bogaard and Verslius (1962) prove this assertion and provide a method for estimating the vector of weights **w**. Various improvements to the procedure have also been tested by McMeekin (1974).

The model was first made operational by van den Bogaard and Verslius (1962) for a three-member-committee decision problem involving policymaking for the Dutch economy in 1957. They embedded the model in a more comprehensive econometric model of the Dutch economy and demonstrated how the approach could be used to determine the optimal level of five policy instruments. The only known application of the model with any attempt at testing, however, is provided by Roskamp and McMeekin's (1970) example. Here the authors attempted to predict an agreed budget share for eight items of municipal expenditure for a small town in northwest Germany in the 1960s. They had available budget shares from 1966, 1967, and 1968, and they had surveyed the preferences of each of the eight committee members with respect to desired budgets in 1967. They calibrated the model to the 1967 preference structures and assessed how close were the predictions of the model in terms of the budget share in 1968. There are limitations to this test, but the data provided are useful and will be reworked for Coleman's model later. Discussion of this example will be postponed until then.

Decision Making as Exchange: Markovian Dialogues

A major limitation of the Bogaard-Verslius model relates to the fact that it says little about the process of decision making. If any process is implied, it is assumed that losses are "calculated" by the committee members themselves, and then that the chairman, say, calculates the weights and determines the optimal budget plan. In fact, it is more likely that, given a formal committee, weights are chosen by the chairman in the light of discussion or dialogue rather than by calculation. In this section, a procedure based on such a dialogue will be explored as a prelude to a richer model of decision making to be presented below.

If it is assumed that losses incurred between committee members are determined according to equation (14.9), one way of proceeding would be for members to attempt to choose weights that reflect the overall importance of losses inflicted on themselves or by themselves. Two processes of averaging these overall losses, either losses on or by themselves, can be specified. First, consider the process that involves assessing the importance of the loss inflicted on an actor by all others. The relative importance P_{ij} of any loss to i incurred by j is

$$P_{ij} = \frac{L_{ij}}{\sum\limits_{z} L_{iz}}, \quad \sum\limits_{j} P_{ij} = 1. \tag{14.14}$$

Now assume that the actors begin with an initial distribution of weights reflecting their overall losses $\{Z_i(1)\}$. Each actor j then works out the importance of the loss he inflicts on another actor i as $Z_i(1)P_{ij}$ and then modifies the importance he gives to his own overall loss as

$$Z_j(2) = \sum_i Z_i(1)P_{ij}. \qquad (14.15)$$

In short, the actor is working out the overall loss inflicted on others, then deciding that the importance of that loss should be set equal to this quantity. There is an element of "fairness" in such a scheme.

The first-order process described by equation (14.15) is Markovian, and if $\{P_{ij}\}$ is non-negative and irreducible, iteration on equation (14.15) will lead to $\{Z_j(t)\}$, converging to $\{Z_j\}$ in the limit of t. This limit is described by the steady-state equation

$$Z_j = \sum_i Z_i P_{ij}. \qquad (14.16)$$

Clearly, equation (14.16) does not imply symmetry, but it does imply that j equates his own loss with the importance of the loss (to others) inflicted on all others. The process of discussion associated with this scheme is called a "Markovian dialogue," and in this context it was first noted by Theil (1976).

An analogous process exists if actor i equates his or her total loss with the loss inflicted on him or herself by all the other actors in the following way. This is perhaps a less "natural" process than that already sketched, but it is useful, as will become apparent below. The importance of the loss inflicted on i by j, measured relative to j's total inflicted loss, is

$$\hat{P}_{ij} = \frac{L_{ij}}{\sum_i L_{ij}}, \quad \sum_i \hat{P}_{ij} = 1. \qquad (14.17)$$

Starting with a distribution of weights $\{\hat{Z}_i(t=1)\}$, we begin a process of iteration using

$$\hat{Z}_i(t+1) = \sum_j \hat{Z}_j(t)\hat{P}_{ij}. \qquad (14.18)$$

In the limit, this process leads to the steady-state equation

$$\hat{Z}_i = \sum_j \hat{Z}_i \hat{P}_{ij}, \qquad (14.19)$$

which also forms a Markovian dialogue. The processes implied in equations (14.16) and (14.19) are similar to the averaging processes first developed by French (1956),

Harary (1959), and Abelson (1979) but were recently used as the basis for many models of opinion pooling (Zollman, 2012). In chapters 12 and 13, we also referred to these processes as duals of one another. Moreover, the deterministic model suggested by de Groot (1974) for the pooling of subjective probabilities and its stochastic equivalent (Kelly, 1981) are close in spirit to the ideas introduced here.

The two steady-state distribution vectors $\mathbf{Z}$ and $\hat{\mathbf{Z}}$ will not of course be the same, but it is of interest to consider cases when averaging in one direction over the losses is equivalent to averaging in the other direction in terms of the equality of $\mathbf{Z}$ and $\hat{\mathbf{Z}}$. Assume then that the two steady-state equations are equal,

$$\sum_i Z_i(t)P_{ij} = \sum_i \hat{Z}_i(t)\hat{P}_{ji}, \quad i \neq j, \tag{14.20}$$

which can also be written as

$$\sum_i Z_i \frac{L_{ij}}{\sum_j L_{ij}} = \sum_i \hat{Z}_i \frac{L_{ij}}{\sum_j L_{ji}}. \tag{14.21}$$

Equation (14.21) is a form of the Bogaard-Verslius marginal symmetry condition in equation (14.13). In fact, $Z_i = \hat{Z}_i$, $\forall i$ if the loss matrix is symmetric, but it is of interest to note that if the loss matrix is only marginally symmetric, that is, if

$$\sum_j L_{ij} = \sum_j L_{ji}, \tag{14.22}$$

then it is easy to see from equation (14.21) that $Z_i = \hat{Z}_i$, $\forall i$. This is obvious from equation (14.13), and it extends the equal weighting result for symmetric loss matrices to marginally symmetric cases. Finally, the weights in equation (14.13) can now be considered as being related to the ratio $Z_i/\Sigma_j L_{ij}$ in the Markovian dialogue.

A Disaggregate Model of Social Exchange

The dialogue model reflects a rather artificial process in that the type of averaging involved is solely motivated by the idealistic criterion of "fairness." However, it is possible to construct a model in which the optimal budget plans themselves are averaged according to an exchange scheme that reflects the intrinsic conflict over the realization of each individual's optimal plan. The model is a disaggregate Markovian dialogue model that is built around ideas from social exchange theory. In the form developed here, the model is similar to that developed by Coleman (1973), which we used for our theory of collective action in chapter 12 and which we applied to the imperfect market of land exchange in the last chapter.

The optimal set of individual plans $\tilde{\mathbf{A}} = [\tilde{A}_{ik}]$ has already been defined. In this model, we will define a "desired" distribution of resources as $\mathbf{A} = [A_{ik}]$, which might be considered as being optimal in the sense of the utility maximization introduced earlier, but only in certain circumstances. A second distribution, $\mathbf{B} = [B_{ik}]$, can also be introduced. $\mathbf{B}$ represents the set of "actual" plans each individual is able to realize, and as such reflects the degree of control each actor has over the allocation of resources. Clearly, the conflict between actors is now related to the difference between their desired and actual resource allocations, as well as to the differences between different desires and between different realizations. It is also necessary to define

$$\alpha_i = \sum_k A_{ik}, \tag{14.23}$$

where α_i is the desired level of resources necessary for actor i to realize his or her desired allocation, and

$$\beta_k = \sum_i B_{ik}, \tag{14.24}$$

where β_k is the actual value of each event or element in the plan, as reflected by the amount of resources the actors are able to allocate to event k.

Imagine a process in which the actors begin with a given distribution of resources $\{r_i(1)\}$. We will assume this distribution is different from $\{\alpha_i\}$ and that if the actors wish to indulge their desires, it is necessary to allocate these resources in proportion to their desired allocation as

$$r_i(1)\frac{A_{ik}}{\alpha_i}. \tag{14.25}$$

The value $v_k(1)$ of the resources actually allocated to each event or plan k by this process can be found from

$$v_k(1) = \sum_i r_i(1)\frac{A_{ik}}{\alpha_i}. \tag{14.26}$$

Equation (14.26) implies that $v_k(1)$ of the total resources in the system have been allocated to event k. But associated with event k is a degree of control (ownership, say) that each actor has. Thus, if $v_k(1)$ is the resource value of the event, and actor i controls B_{ik}/β_k of that event, then the event provides $v_k(1)B_{ik}/\beta_k$ resources to actor i. The total resources can then be found from

$$r_i(2) = \sum_k v_k(1)\frac{B_{ik}}{\beta_k}. \tag{14.27}$$

In general, $r_i(2) \neq r_i(1)$, $\forall i$, and thus the resources actually realized from this exchange are different from those the actors began with. Reiteration of the process is implied, and, if equations (14.26) and (14.27) are solved sequentially, the resource and value distributions will converge to an equilibrium or steady state, assuming that **A** and **B** are irreducible and nonnegative. The process is in fact Markovian, as can be easily demonstrated.

To show this, first define the proportions in equations (14.26) and (14.27) as

$$X_{ik} = \frac{A_{ik}}{\alpha_i}, \quad C_{j\ell} = \frac{B_{j\ell}}{\beta_\ell}, \quad \sum_k X_{ik} = 1, \text{ and } \sum_j C_{j\ell} = 1, \tag{14.28}$$

which in matrix terms are

$$\mathbf{X} = \boldsymbol{\alpha}^{-1}\mathbf{A} \text{ and } \mathbf{C} = \mathbf{B}\boldsymbol{\beta}^{-1}. \tag{14.29}$$

Note that **X** and **C** are *nxm* stochastic matrices and $\boldsymbol{\alpha}^{-1}$ and $\boldsymbol{\beta}^{-1}$ are *nxn* and *mxm* diagonal matrices with α_i^{-1} and β_k^{-1} on the main diagonals. These differ only in dimension from $\hat{\mathbf{X}}$ and $\hat{\mathbf{C}}$ in chapter 13. In fact, it is a little clearer if we define the matrices **A** and **B**, hence **X** and **C**, as having the same dimensions *nxm*, and this is reflected in the greater use of the transpose operator in the sequel.

Using these definitions and noting that **r** and **v** are *1xn* and *1xm* row vectors associated with resources and values, respectively, we can now write the steady-state forms of equations (14.26) and (14.27) as

$$\mathbf{v} = \mathbf{r}\mathbf{X} = \mathbf{r}\boldsymbol{\alpha}^{-1}\mathbf{A} \tag{14.30}$$

and

$$\mathbf{r} = \mathbf{v}\mathbf{C}^T = \mathbf{v}\boldsymbol{\beta}^{-1}\mathbf{A}. \tag{14.31}$$

Substituting equation (14.31) into equation (14.30) and vice versa leads to the aggregate steady-state forms

$$\mathbf{v} = \mathbf{v}\mathbf{C}^T\mathbf{X} = \mathbf{v}\boldsymbol{\beta}^{-1}\mathbf{B}^T\boldsymbol{\alpha}^{-1}\mathbf{A} \tag{14.32}$$

and

$$\mathbf{r} = \mathbf{r}\mathbf{X}\mathbf{C}^T = \mathbf{r}\boldsymbol{\alpha}^{-1}\mathbf{A}\boldsymbol{\beta}^{-1}\mathbf{B}^T, \tag{14.33}$$

which are both immediately recognizable as steady-state Markov forms.

Equations (14.32) and (14.33) can be regarded as duals of one another in the economic sense, as portrayed in general equilibrium theory (Gale, 1960). It is easy to see that these equations are the limit equations of first-order Markov processes because $\mathbf{X}\mathbf{C}^T$ and $\mathbf{C}^T\mathbf{X}$ are stochastic matrices. Thus, for suitably defined matrices, **r** and **v** exist, are unique, and can be scaled to sum to 1 or to the total resources in the system. Coleman (1973) develops the model for use in a variety of political contexts, and of interest is the possibility of interpreting the steady-state resource distribution **r** as a measure of power. There are few empirical applications of the model that have led to testable propositions, but in a plan-making context, the model we developed to simulate land development in the central city outlined in the previous chapter shows what is possible. In general, as we have developed these ideas here, the focus is very definitely on normative decision making and on structures that are idealized rather than on real decision-making processes.

It is important to be clear about the measures **r** and **v**. Here it seems appropriate to relate these to the magnitude of **A**, **B**, or both, but in Markov chain theory, these are probability vectors. Let us write equation (14.30) in elemental form as

$$v_k = \sum_i \frac{r_i}{\alpha_i} A_{ik}, \tag{14.34}$$

which resembles an averaging equation in which the weights are given by r_i/α_i. In fact, assume that the weights $\{w_i\}$, used to obtain the optimal budget, are actually proportional to $\{r_i/\alpha_i\}$; then

$$\sum_i w_i = K \sum_i \frac{r_i}{\alpha_i} = 1, \tag{14.35}$$

where K is the constant of proportionality. Using these weights, we write a weighted linear average that follows the form of equation (14.7) based on equal individual preferences as

$$a_k = \sum_i w_i A_{ik}. \tag{14.36}$$

Using equations (14.34) and (14.35), we can rearrange equation (14.36) to give

$$a_k = \sum_i \frac{r_i}{\alpha_i} A_{ik} \Big/ \sum_i \frac{r_i}{\alpha_i} = v_k \Big/ \sum_i \frac{r_i}{\alpha_i} = K v_k. \tag{14.37}$$

Thus the value $\{v_k\}$ produced in equilibrium is proportional to the optimal budget allocation. The same sort of normalization can be achieved in relation to equation (14.31), which enables resources to be scaled to reflect the weighted distribution of resources allocated to events. In fact, it is easy to show that the Markov process implied by equations (14.26) and (14.27), and represented in equilibrium terms by equations (14.30) and (14.31), can be written in terms of optimal (equilibrium) plans as

$$a_k = \sum_i \frac{a_i}{\alpha_i} A_{ik} \Big/ \sum_i \frac{a_i}{\alpha_i} \tag{14.38}$$

and

$$a_i = \sum_k \frac{a_k}{\beta_k} B_{ik} \Big/ \sum_k \frac{a_k}{\beta_k}. \tag{14.39}$$

Examining these equations, we see that the exchange process can be considered as an averaging process in which new, optimal but nonequilibrium budgets are determined through successive iteration until an equilibrium criterion is achieved. There is an obvious correspondence to the Bogaard-Verslius model through the linear equilibrium forms involving the optimal budget allocations.

Symmetric Structure and Equal Exchange

Coleman's model, however, has a much stronger emphasis on the symmetry of relationships between actors and the formation of weights than does the Bogaard-Verslius model. First, define the exchange matrix, the amount of resources actors exchange with one another to enable the system to be stable, as

$$r_i P_{ij} = r_i \sum_k \frac{A_{ik}}{\alpha_i} \frac{B_{jk}}{\beta_k}. \tag{14.40}$$

The obvious analog to the "symmetry of loss" criterion given in equation (14.11) is the "symmetry of exchange" derived from equation (14.40) as

$$r_i P_{ij} = r_j P_{ji}. \tag{14.41}$$

In fact, equation (14.41), which implies that the resources actor i gives to j are the same as those j gives to i, is the equation for equal exchange explored by Berger and Snell (1957), and it is also the condition for reversibility of the Markov chain (Kelly, 1979). To relate equation (14.41) to equation (14.11), first assume that the weights $\{w_i\}$ are given as $\{r_i/\alpha_i\}$. Then in the Bogaard-Verslius model, these weights are equal when loss is symmetric. If in equation (14.41)

$$\frac{r_i}{\alpha_i} = \frac{r_j}{\alpha_j}, \quad i = j, \tag{14.42}$$

this implies that the term $\mathbf{A}\boldsymbol{\beta}^{-1}\mathbf{B}^T$ is symmetric. It is of interest to explore the conditions under which such symmetry might exist in the Coleman model.

One obvious symmetry exists when $\mathbf{A} = \mathbf{B}$, that is, when desired allocations are equal to controlled or actual allocations. Using this definition, we can write equation (14.33) as

$$\mathbf{r} = \mathbf{r}\boldsymbol{\alpha}^{-1}\mathbf{A}\boldsymbol{\beta}^{-1}\mathbf{B}^T = \mathbf{r}\boldsymbol{\alpha}^{-1}\mathbf{A}\boldsymbol{\beta}^{-1}\mathbf{A}^T. \tag{14.43}$$

It is easy to show that $\mathbf{r} = \boldsymbol{\alpha}$, if this type of symmetry or equilibrium exists (Batty, 1981). As $\mathbf{r}$ is unique, simply substitute $\boldsymbol{\alpha}$ into equation (14.43) and rearrange as

$$\mathbf{r} = \mathbf{1}\mathbf{A}^T = \boldsymbol{\alpha}, \tag{14.44}$$

where **1** is a 1x*n* unit row vector. Equation (14.44) is clearly the definition of **α** as given at the beginning of the previous section. With **A = B**, symmetry also exists in the dual relationship. Then equation (14.32) becomes

$$\mathbf{v} = \mathbf{v}\boldsymbol{\beta}^{-1}\mathbf{B}^T\boldsymbol{\alpha}^{-1}\mathbf{A} = \mathbf{v}\boldsymbol{\beta}^{-1}\mathbf{A}^T\boldsymbol{\alpha}^{-1}\mathbf{A} = \mathbf{1A} = \boldsymbol{\beta}. \tag{14.45}$$

The weights $\{r_i/\alpha_i\}$ and $\{v_k/\beta_k\}$ are clearly equal with respect to their distribution over *n* and *m*. With these results in equations (14.35) to (14.37), it is easy to show that

$$a_j = \frac{1}{n}\sum_k B_{jk} \text{ and } a_i = \frac{1}{n}\sum_k A_{ik}. \tag{14.46}$$

By defining the weights as proportional to $\mathbf{r\alpha}^{-1}$ and $\mathbf{v\beta}^{-1}$, which determine the dual exchange processes, symmetry in exchange implies that no conflict in the realization of optimal plans exists; exchange is fair in that what actor *i* gives to *j*, *j* gives back to *i*; and that the original resource allocation schedules **A** and **B** are the same. Moreover, the weights emerge "naturally" from the group decision-making process and are given a substantive interpretation as the ratio of optimal to original resource allocations. Coleman's model is thus richer than the Bogaard-Verslius model, but its results will differ as the criterion relating to symmetry is different. In fact, loss could be considered equivalent to exchange in principle, although in practice these are defined differently in the two models. In Coleman's model, it is also useful to examine how near to symmetry the equilibrium is.

The Coleman model contains one final prediction, which involves symmetry with unequal weights. It is argued that resources and values are determined according to the degree of conflict present in the system (that is, from equations [14.32] and [14.33]) at a first stage, and then this conflict is resolved through actors altering their actual schedule of resource allocations in line with the modified desired schedules. Then control or ownership is altered to $\tilde{C}_{ik}$ to ensure

$$v_k\tilde{C}_{ik} = r_iX_{ik} = \frac{r_i}{\alpha_i}A_{ik}, \tag{14.47}$$

which implies that

$$\tilde{C}_{ik} = \frac{r_iX_{ik}}{v_k} = \frac{r_i}{\alpha_i}\frac{A_{ik}}{v_k}. \tag{14.48}$$

Substitution of $\tilde{C}_{ik}$ from equation (14.48) back into the resource-value equilibrium equations (14.32) and (14.33) generates the following structures, which contain symmetric exchange relationships,

$$\mathbf{v} = \mathbf{v}\mathbf{V}^{-1}\mathbf{A}^T\boldsymbol{\alpha}^{-1}\mathbf{R}\boldsymbol{\alpha}^{-1}\mathbf{A} \tag{14.49}$$

and

$$\mathbf{r} = \mathbf{r}\alpha^{-1}\mathbf{A}\mathbf{V}^{-1}\mathbf{A}^T\alpha^{-1}\mathbf{R}, \tag{14.50}$$

where $\mathbf{V}$ and $\mathbf{R}$ are diagonal matrices formed from $\mathbf{v}$ and $\mathbf{r}$, respectively. In this model, it is assumed the committee actually adjusts its preferences, and perhaps its organization, to reflect this change in control.

The Roskamp-McMeekin Optimal Budget Problem

Roskamp and McMeekin (1970) have applied van den Bogaard and Verslius's (1962) model to the problem of predicting an optimal budget allocation for eight items of municipal expenditure determined by eight members of a municipal budget committee. The eight items of expenditure in the order taken by the model were general administration, public safety (police), education, culture, social services, health care, construction and housing, and local economic initiatives. The eight committee members empowered to decide the resources allocated to each of these budget items were local politicians, by occupation: retired police officer, public employee, police officer, factory worker, union secretary, two businessmen, and a baker. This is also the order used in the model. The model was developed for this budget problem in a small town (5,000 inhabitants) with the pseudonym "Frisia" located in northwest Germany.

In 1967, Roskamp and McMeekin interviewed each member of the committee to elicit their attitudes toward the best level of resource allocation. From these surveys, preference functions and then optimal budget allocations for each member were computed. The Bogaard-Verslius loss function was used, and two methods were developed for choosing weights according to the symmetry relationships in equation (14.13) (McMeekin, 1974). The weights and preferences were used to compute the optimal committee budget as given in equation (14.7). This budget was then compared with the observed allocation in 1968. The fit between the predicted and observed levels of resource allocation was close, but Roskamp and McMeekin (1970) did not explore the extent to which one might expect committees to act optimally, nor did they examine the process through which the committee collectively determined the budget. As we will note below, the example is not an ideal one for empirical testing, but it is useful as a demonstration device.

A strict comparison of the Bogaard-Verslius model with Coleman's model applied to the same problem is not intended here. But as the example provides one of the only sources of data on committee decision making, we have used it in applying the Coleman model, and we will make a brief comparison with Roskamp and McMeekin's (1970) results in passing. In the 1967 survey, each actor was asked to indicate his "desired" budget allocation, which we will take as the *nxm* matrix $\mathbf{A}$. Actors were also asked to indicate the minimum level of resources they considered necessary for each budget item, and the intensity of their preferences was recorded on a 0–2 point scale. Nowhere in the survey was there any question relating to the actors' ability to control the level of resources, but it can be considered that the level of preferences represents a reasonable proxy to this in that we have assumed that the greater this intensity, the greater control they will exert on the item in committee discussion. The *nxm* preference matrix has been set as **B**. **A** is measured in terms of budget level in thousands of DM, and **B** is measured on the 0–2 point preference intensity scale. These data are contained in Roskamp and McMeekin's (1970) article and in McMeekin's (1974) thesis. Clearly, **A** and **B** can be scaled to equivalent levels without loss of generality.

To aid comparison, all variables and results have been scaled in the following presentation to sum to 100, and thus their distributions represent percentages. Thus in the following section

$$\sum_i w_i = \sum_i r_i = \sum_i \alpha_i = \sum_k v_k = \sum_k \beta_k = \sum_k a_k = 100. \tag{14.51}$$

The Coleman model has been solved in equations (14.30) to (14.33), and the weights computed from equations (14.34) to (14.37). Values for **w**, **r**, α, $v\beta^{-1}$, **v**, and β are shown in Table 14.1. Also shown there is the weight vector associated with the symmetry-of-loss criterion given by solving equation (14.13) and presented by Roskamp and McMeekin (1970). If the weights from the Roskamp-McMeekin model are compared with those from Coleman's model (columns 1 and 2 in Table 14.1), there is quite a strong inverse relationship, which seems to imply that, if the symmetry of loss weights reflect the degree to which weak actors need to be compensated to ensure fairness, then the Coleman model weights would appear to represent the degree of self-interest, or perhaps power. In fact, **w** from Coleman's model is also highly correlated with **r**, traditionally assumed by Coleman (1973) to represent power. This suggests that the symmetry criterion in the Coleman model is based on simply adjusting an initial distribution α (which usually will reflect existing power) to enable a balance to be struck. There is a clear need to explore the two models further in this regard and perhaps to devise a model that incorporates both optimality and exchange considerations.

The real interest, however, in these models is on the group budget allocation, whether it be viewed as an optimal budget or simply as a negotiated one reflecting existing power relationships, as in Coleman's model. A particularly important point must be made at the outset. If equation (14.36) is examined, it is clear that the value of any allocation $\{a_k\}$ must lie somewhere on the range from $\min_i A_{ik}$ to $\max_i A_{ik}$. If this range is rather narrow, substantially different weights will not yield very different allocations. The same point is involved if the distribution of $\{A_{ik}\}$ values over i for a given k is skewed toward one end of the range. Thus to indicate where the

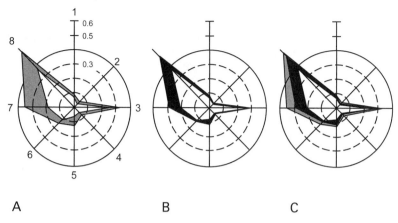

Figure 14.1
Range of desired and observed optimal budget allocations. (a) Range of $\{A_{ik}\}$, the desired allocation, graphed on k-axes (1–8). (b) Actual allocation. (c) Superposition of (a) and (b).

solution might lie, we have graphed the range of $\{A_{ik}\}$ in percentage terms in figure 14.1(a), where the range on each of the eight axes $\{k = 1, 2, \ldots, 8\}$ is from $\min_i X_{ik}$ to $\max_i X_{ik}$. Figure 14.1(a) shows that the solution from the model cannot be dramatically different from the original desired allocation, which the model takes as input data.

The actual budget allocations observed in 1966, 1967, and 1968 can also be presented in the same manner. Minimum and maximum values of $\{a_k(t)\}$ where $\{t = 1966, 1967, 1968\}$ are graphed in figure 14.1(b), and we have also overlapped figures 14.1(a) and (b) in (c) to indicate the closeness of the desired to actual allocations. No matter what weights are chosen, the model will predict a distribution of resources "close" to the observed value range. In short, there is not much conflict between actors in this problem, and as such, it does not represent a particularly good test case for the model. Nevertheless, the optimal budget $\{a_k\}$ has been computed from equation (14.36) using the weights $\{w_i\}$ in Table 14.1. We have also computed $\{a_k\}$ with equal weights under the assumption of symmetry $\mathbf{A} = \mathbf{B}$, and these are compared with the Roskamp-McMeekin budget predictions in figures 14.2(a), (b), and (c). These budgets have also been computed with $\mathbf{C} = \mathbf{I}$ and $\mathbf{C} = [1/n]$, and even these dramatically different patterns of control do not lead to different predictions. There is hardly any difference between any of these predictions, and thus we conclude that a stronger test of the model must involve desired allocations dramatically different from those that are eventually observed, hence predicted.

Table 14.1
Weight distributions associated with the Roskamp-McMeekin and Coleman models[†]

Actor i	Roskamp-McMeekin weights	Coleman weights						
		$w_i = \dfrac{r_i}{\alpha_i}$	r_i	α_i	$\dfrac{v_k}{\beta_k}$	v_k	β_k
1	8.4 (7)	18.1 (1)	16.9 (1)	13.7 (5)	12.1 (4)	7.8 (4)	9.3 (8)
2	6.6 (8)	16.8 (2)	15.8 (2)	12.2 (7)	3.8 (5)	2.9 (5)	11.0 (6)
3	14.5 (2)	7.9 (7)	9.9 (7))	16.3 (1)	21.8 (3)	22.3 (3)	14.7 (3)
4	31.4 (1)	3.2 (8)	3.6 (8)	14.8 (3)	0.2 (8)	0.2 (8))	12.1 (4)
5	10.0 (4)	16.4 (3)	13.9 (5)	11.0 (8)	0.6 (7)	0.4 (7)	10.1 (7)
6	8.6 (6)	14.6 (4)	14.4 (3)	12.9 (6)	0.7 (6))	0.6 (6)	11.5 (5)
7	9.6 (5)	13.3 (5)	14.4 (3))	14.1 (4)	21.9 (2)	23.9 (2)	15.7 (1)
8	10.9 (3)	9.7 (6)	11.1 (6)	14.9 (2)	38.8 (1)	41.8 (1)	15.5 (2)

[†]Percentages are rounded to a single decimal point; rank order is given in parentheses.

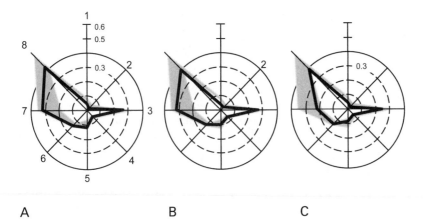

A   B   C

Figure 14.2
Predicted optimal budget allocations. (a) From Coleman's model in equation (14.36) in the text. (b) From Coleman's model using equal weights in equation (14.36). (c) From the Roskamp-McMeekin model using equation (14.7).

It is, however, possible to explore the appropriateness of the Coleman model a little further through the degree to which the sequential averaging process leads toward symmetry or equal exchange. A measure of deviation from symmetry for any matrix, $[Z_{ij}]$, can be computed as

$$\Omega = \left(\sum_{i \geq j} \frac{|Z_{ij} - Z_{ji}|}{Z_{ij}} \right) \bigg/ \frac{n(n-1)}{2}. \tag{14.52}$$

Ω is calculated in percentage terms and is equal to 0 when $\mathbf{Z} = \mathbf{Z}^T$. In the Roskamp-McMeekin example, the raw loss matrix $\mathbf{L}$ deviates by about 57% from symmetry, and this is reduced to 20% when the weights are calculated and applied. For the Coleman model, the original exchange matrix $\mathbf{A}\boldsymbol{\beta}^{-1}\mathbf{B}^T$ deviates from symmetry by 126%, and this is reduced to 5% when the weights $\mathbf{r}\boldsymbol{\alpha}^{-1}$ are computed and applied. In this narrow sense, the Coleman model would appear to be a better compromising or averaging device.

Plan Design and Real Decision-Making Processes

The two models presented here are equivalent in a very special technical sense. Both have an equivalent structure in that their weights or the importance of committee members are equal when what one committee member gives a level of resource to another that he or she receives back from the other. This process of equal exchange may imply fairness, as in the Bogaard-Verslius model, but it may also imply self-interest, as in Coleman's model. The other appealing equivalence relates to the weighted form of group budget, which is generated through optimization in the Bogaard-Verslius model, but exists as an equilibrium condition in the Coleman model. The link between these is suggestive, and although the example here is inconclusive, it would appear that the two models are simulating rather opposite averaging processes. One clear implication of this approach involves exploring the extent to which notions of power as conceived in Coleman's model might be used in influencing the optimization in the Bogaard-Verslius model. For example, it would be possible to determine weights using Coleman's model and then embed these into the group welfare function optimized in the Bogaard-Verslius model. It may even be possible to embed some form of iterative quadratic optimizing into Coleman's model and to determine both the weights and the optimal budget allocation as an equilibrium state.

In all the models introduced in part III of this book, we have been at pains to ensure that definite outcomes are produced by their execution and implementation. In network models, it is easy to see how processes would not converge to unique outcomes or would not converge at all. If networks can be decomposed into mutually exclusive subnetworks, this implies that the system or the problem-solving processes can so be decomposed into separate systems and separate problems. The classification of graphs from disconnected to strongly and completely connected introduced in part I as well as in chapters 10 and 11 makes this clear. If graphs are anything less than strongly connected, processes operating on their links, such as opinion pooling and sequential averaging, will not converge, or if they do converge it will be to extreme states. Different processes, however, might be defined by developing more elaborate models rather than exploring different network structures. For example, it would be straightforward to develop models in which averaging was based on second- or higher-order processes where averages were formed from weighted combinations of events, not only at the previous time period but at ones further in the past. In short, stronger inertia might be built into attitudes this way, and different actors might be classified into different groups with different degrees of resistance or receptivity to change. A second strategy would be to introduce new attitudes externally, that is, to assume that these processes were informed by new information periodically or at each time period, thus distorting the movement to equilibrium. These kinds of processes are similar to dynamic input-output models where change is introduced over and over again in response to changed events. Of course, if such change is removed or becomes identical at each time period, then convergence of attitudes will take place.

In this chapter, we have explored a more realistic model of decision making than any we have introduced so far, and by illustrating an application have shown how it might be adjusted to a real context. However, our progress in this part of the book to full-fledged decision models has been somewhat muted since we are clear that the models of these chapters are designed to choose best solutions, which are not necessarily capable of being reproduced by real decision making. In this sense, our models are nearer design than they are decision, and they are presented in the spirit of informing designers and decision makers regarding new ways of thinking about design and planning in structural terms. Yet the applications are suggestive, and good test examples are required to make empirical progress. Problems of physical or land use plan design are quite relevant here, for such planning usually takes place as formal negotiation, in teams or planning committees.

It is still necessary to be careful about the choice of examples. A high degree of controversy resolvable in a highly structured way is a prerequisite to a strong empirical test. Moreover, it should be possible to observe at least externally the process of decision. In the Roskamp-McMeekin example, it is impossible to discriminate between models on the basis of the predicted budgets because the initial data are so strongly correlated to these outcomes. Nevertheless, a closer observation of how the budget was determined would have revealed, at least in part, the process of negotiation and decision making. Such observation is essential if the problem of equifinality is to be broached and resolved. The problem of optimality should also be tackled head-on. Coleman's model is appealing in that its process implies a kind of local optimality that is not guaranteed to lead to any collective optimality. This seems to accord well with emerging approaches to decision making which build on Simon's (1983) long-standing ideas about satisficing. Coleman's model presents a framework for improving these ideas, but of critical significance is the fact that it is equivalent to the Bogaard-Verslius model at several points. This equivalence has been exploited here and provides an important pointer to applications of these tools for generating optimal designs through planning processes.

Conclusions: A Future Science

If we put all these principles together, we discover that there are too many. They are inconsistent with each other.

—Richard P. Feynman, *The Character of the Physical Law* (1965, p. 155)

The canvas that we have attempted to lay out is a mere beginning. As we have argued throughout, it would be presumptuous to think of this effort as the only science of cities, for the city and its planning admits many viewpoints. As we have begun to recognize its complexity, we have become ever more aware that both its understanding and design need to be based on a synthesis of many different and often conflicting perspectives. Indeed, the models in part III were all based on the idea that city plans are vehicles that resolve conflict, continuing to generate a consensus about wherein lie the best urban futures. A major distinction between the models in part II and those in part III lies in the difference between positive and normative perspectives on the world. Positive thinking, which relates loosely to "positivism," involves an approach that focuses on understanding "what is," while normative thinking concentrates on "what should be." The first accepts the world as it is, while the second explores how it should be; this difference is sometimes referred to as idealism in contrast to realism. In fact, things are not so clear-cut because the positive and normative combine in diverse ways, and as we began to illustrate in part III, although our models of design are normative in intent, they imply a degree of realism in that such processes are potentially observable and the kinds of normative activity they imply do appear to exist in limited situations. Our models suggest ways in which actual decision processes might be improved, embodying norms that pertain to the design of future cities.

From one perspective, the science of cities and of design can be reconciled in terms of the evolutionary models first introduced in chapter 8. These lie at the basis of complexity theory. They suggest ways in which agents, actors, or activities that comprise the city system learn about each other and their environment with respect to decisions concerning the best places to locate activities and the best ways in which to interact with one another to achieve more sustainable environments. At the heart of such models there is learning, and it is this that unites the real and the ideal, positive and normative, in that depending on one's perspective, the real can be interpreted as a process of reaching for the ideal, while design can be seen as a process in which ideals and norms are compromised in the move toward a consensus. In fact, learning is intrinsic to evolution, and many contemporary views of design stress how opinions and attitudes mutate as they evolve to some overall criterion of fitness that represents the quest to achieve certain norms and goals (Steadman, 2008). Indeed, these norms themselves may mutate in and of themselves, and in this sense, the future is a moving target.

What we aspire to in terms of how this new science might develop are models that span both the domains we have developed here: that is, an understanding of how locations and interaction lie at the basis of how cities form and evolve, and the design of these same forms as a product of the collective action of all those who have a stake in such futures. Our models of land exchange go some way to advancing this perspective, but there are models of a more theoretical variety that suggest such a synthesis. The essence of design decision making in cities is unanimity with respect to some pattern of location that realizes a set of goals, and the core of such processes involve moving to a consensus or resolving conflict by some form of compromise based on opinion pooling, repeated averaging, and Bayesian learning, all of which imply some form of convergence to an equilibrium. The French-Harary models first introduced in chapter 10 define the rudiments of such processes, and over the last fifty years, these models have been elaborated in the social network tradition in many ways (see Jackson, 2011). In fact, these elaborations have come from several different directions. In particular, statistical learning has been reconciled with these processes, first by de Groot (1974) and then by others from Chatterjee and Seneta (1977) onward, but this tradition has hardly been linked to social networks. In a different domain, those working with computation in social networks focus more on how convergence can be embedded into digital processes. Blondel et al. (2005) have developed variants of these outside the social networks domain, but essentially these are all based on the convergent opinion polling model first articulated by French (1956). The focus, however, has been more on how these processes are distorted on the way to an equilibrium, and in this sense, imply much greater affinity with the positive models presented in part II.

There is, however, a model that is very close in spirit to cellular agent-based models of how cities grow, introduced in chapter 8 and in the opinion pooling models of later chapters. This is based, essentially, on particle dynamics. Vicsek et al. (1995) have introduced a model that essentially averages "particles" with respect to what is happening to the particles in their "neighborhoods," which is equivalent to thinking of how opinions might be averaged with respect to those agents to whom the agents in question are directly connected. They write: "The only rule of the model is at each time step a given particle with a constant absolute velocity assumes the average direction of motion of the particles in its neighborhood of radius r with some random perturbation added" (p. 1,226). In short, the random perturbation keeps the set of particles from converging on one another, but the direction of convergence is soon established. In a spatial context, the model produces flocking, with the random component fixing the density of the flock. In a system where every particle is connected either directly or indirectly to every other particle, convergence is assured when the random component is zero.

This model has strong similarities to the logic of models such as the diffusion-limited aggregation model introduced in chapters 1 and 8, but if we think of the particles as agents or actors and the space as the space in which opinions are defined, then this essentially is a generalization of the French-Harary model. It illustrates that the models in part III are not that far away from some of those in part II, and it implies an obvious strategy for enriching our science. With the tools provided here, it is eminently possible to generalize the French-Harary model to a spatial domain where the elements of land or space that define the factors associated with actors are dealt with individually. Indeed, it may even be possible to consider elements of the system to be movable in the sense of spatial interactions and to fashion a general model of motion in a city as some integration of the Vicsek-French-Harary approach. These are perhaps wild musings, but we consider that this is one very clear direction in which this science should be developed further.

There are, of course, other many other directions and as Feynman (1965/1992) implies of physics, many if not most of these do not fit together as a tight jigsaw. More aggregative approaches to understanding cities, such as those based on land use–transportation models, do not fit so well with more disaggregate models involving individuals and agents, which a comparison of chapters 8 and 9 reveals. Macro-economic approaches in contrast to micro approaches, imply different perspectives on the city system, while many characteristic processes have clear spatial implication, which cannot be readily measured and modeled. In short, many important aspects of cities are hard to explore using the methods and tools introduced here. Real and ideal, and positive and normative models, which comprise the essential distinction between the main two parts of this book, do still conflict with one another in some senses, although as we have implied in this conclusion there is hope for some sort of integration. Questions of spatial scale also change the emphasis, while dynamics has not been much exploited here and still remains one of the major issues in forging an applicable science (Batty, 2005). In addition, the wider environment in which policy takes place has barely been broached here, and this clearly affects the way the models and tools developed here might be adopted and adapted to real situations. Nevertheless, this cornucopia of possibilities provide a rich environment in which to progress this science, and it is for others to take up the mantle and demonstrate whether these approaches lead to the better understanding and design our cities sorely need.

References

Abelson, R. P. 1979. Social Clusters and Opinion Clusters. In *Perspectives on Social Network Research*, ed. P. W. Holland and S. Leindhardt, 239–256. New York: Academic Press.

Acevedo, W., T. W. Foresman, and J. T. Buchanan. 1997. Origins and Philosophy of Building a Temporal Database to Examine Human Transformation Processes. Washington DC: US Geological Survey. http://landcover.usgs.gov/LCI/urban/umap/pubs/asprs_wma.php. Accessed 18/10/2012.

Adamic, L. A. 2002. Zipf, Power-laws, and Pareto—A Ranking Tutorial. Information Dynamics Lab. Palo Alto, CA: HP Labs. http://www.hpl.hp.com/research/idl/papers/ranking/ranking.html. Accessed 4/1/2010.

Alexander, C. 1964. *Notes on the Synthesis of Form*. Cambridge, MA: Harvard University Press.

Alexander, C. 1965. A City is Not a Tree, *Architectural Forum*, 122 (April), 58–61 and 122 (May), 58–62.

Alexander, C. 2003. *New Concepts in Complexity Theory: Arising from Studies in the Field of Architecture*. http://www.katarxis3.com. Accessed 21/10/2012.

Alexander, C. 2012. Harmony-Seeking Computations: A Science of Non-Classical Dynamics Based on the Progressive Evolution of the Larger Whole. http://www.livingneighborhoods.org/library/harmony-seeking-computations-v29.pdf. Accessed 20/9/2012.

Alexander, C., and M. Manheim. 1962a. The Use of Diagrams in Highway Route Location: An Experiment. Research Report RR-R62–3. Department of Civil Engineering. Cambridge, MA: MIT.

Alexander, C., and M. L. Manheim. 1962b. HIDECS 2: A Computer Program for the Hierarchical Decomposition of a Set with an Associated Graph. Publication No. 160. Civil Engineering Systems Laboratory. Cambridge, MA: MIT.

Alonso, W. 1964. *Location and Land Use*. Cambridge, MA: Harvard University Press.

Alonso, W. 1976. A Theory of Movements: Introduction. Working Paper No. 266. Institute of Urban and Regional Development. Berkeley, CA: University of California.

Alonso, W. 1978. A Theory of Movements. In *Human Settlement Systems: International Perspectives on Structure, Change and Public Policy*, ed. N. M. Hansen, 197–211. Cambridge, MA: Ballinger.

Anderson, C. 2006. *The Long Tail: How Endless Choice is Creating Unlimited Demand*. New York: Random House.

Anderson, T. W. 1954. Probability Models for Analyzing Time Changes in Attitudes. In *Mathematical Thinking in the Social Sciences*, ed. P. F. Lazarsfeld, 17–66. Glencoe, NY: The Free Press.

Anderssen, R. S., and J. R. Ive. 1982. Exploiting Structure in Linear-Programming Formulations for Land-Use Planning. *Environment and Planning B* 9:331–339.

Ashby, W. R. 1956. *An Introduction to Cybernetics*. London: Chapman and Hall.

Atkin, R. H. 1974. *Mathematical Structure in Human Affairs*. London: Heinemann Educational Publishers.

Atkin, R. H. 1981. *Multidimensional Man*. Harmondsworth, UK: Penguin Books.

Atkins, P. W. 1994. *The 2nd Law: Energy, Chaos, and Form*. New York: W. H. Freeman.

Auerbach, F. 1913. Das Gesetz der Belvolkerungskoncentration. *Petermanns Geographische Mitteilungen* 59:74–76.

Austwick, M. Z., O. O'Brien, E. Strano, and M. Viana. 2013. Spatial Networks and Clusters in Bicycle Sharing Systems. London: Centre for Advanced Spatial Analysis, UCL.

Bailey, N. J. T. 1964. *Elements of Stochastic Processes*. New York: John Wiley.

Bairoch, P. 1988. *Cities and Economic Development: From the Dawn of History to the Present*. Chicago: University of Chicago Press.

Bankes, S. 1993. Exploratory Modeling for Policy Analysis. *Operations Research* 41: 435–449.

Barabasi, A. L. 2002. *Linked: The New Science of Networks*. New York: Perseus Publishing.

Barabasi, A. L. 2005. The Origin of Bursts and Heavy Tails in Human Dynamics. *Nature* 435:207–211.

Barabasi, A. L., and R. Albert. 1999. Emergence of Scaling in Random Networks. *Science* 286:509–512.

Barthelemy, M. 2010. Spatial Networks. *Physics Reports* 499 (1–3):1–86.

Batty, M. 1971. Modelling Cities as Dynamic Systems. *Nature* 231:425–428.

Batty, M. 1974a. Social Power in Plan Generation. *Town Planning Review* 45:291–310.

Batty, M. 1974b. Spatial Entropy. *Geographical Analysis* 6:1–31.

Batty, M. 1976. *Urban Modeling: Algorithms, Calibrations, Predictions*. Cambridge, UK: Cambridge University Press.

Batty, M. 1981. Symmetry and Reversibility in Social Exchange. *Journal of Mathematical Sociology* 9:1–41.

Batty, M. 1983. Linear Urban Models. *Papers of the Regional Science Association*. 53:5–25.

Batty, M. 1984. Plan Design and Committee Decision-Making. *Environment and Planning B* 11:279–295.

Batty, M. 1985. Fractals: Geometry between Dimensions. *New Scientist* 105 (1450):31–35.

Batty, M. 1992. Urban Modeling in Computer-Graphic and Geographic Information System Environments. *Environment and Planning B* 19:663–685.

Batty, M. 2005. *Cities and Complexity: Understanding Cities with Cellular Automata, Agent-Based Models, and Fractals*. Cambridge, MA: MIT Press.

Batty, M. 2006a. Rank Clocks. *Nature* 444:592–596.

Batty, M. 2006b. Hierarchy in Cities and City Systems. In *Hierarchy in the Natural and Social Sciences*, ed. D. Pumain, 143–168. Dordrecht, Netherlands: Springer.

Batty, M. 2007. Visualizing Creative Destruction. CASA Working Paper 112. Centre for Advanced Spatial Analysis. London: University College. http://www.bartlett.ucl.ac.uk/casa/pdf/paper112.pdf. Accessed 16/10/2012.

Batty, M. 2008. The Size, Scale, and Shape of Cities. *Science* 319:769–771.

Batty, M. 2009a. Cities as Complex Systems: Scaling, Interactions, Networks, Dynamics and Urban Morphologies. In *Encyclopedia of Complexity and Systems Science*. vol. 1. ed. R. Meyers, 1041–1071. Berlin, DE: Springer.

Batty, M. 2009b. Urban Modeling. In *International Encyclopedia of Human Geography*. vol. 12. ed. R. Kitchin and N. Thrift, 51–58. Oxford: Elsevier.

Batty, M. 2010. Space, Scale and Scaling in Entropy-Maximizing. *Geographical Analysis* 42:395–421.

Batty, M. 2011. Commentary: When All the World's a City. *Environment and Planning A* 43:765–772.

Batty, M. 2012. A Generic Framework for Computational Spatial Modelling. In *Agent-Based Models of Geographical Systems*, ed. A. J. Heppenstall, A. T. Crooks, L. M. See, and M. Batty, 19–50. Berlin, New York: Springer.

Batty, M., P. G. Hall, and D. N. M. Starkie. 1974. The Impact of Fares-Free Public Transport upon Urban Land Use and Activity Patterns. *Journal of the Transportation Research Forum* 15:347–353.

Batty, M., and P. A. Longley. 1994. *Fractal Cities: A Geometry of Form and Function*. San Diego, CA: Academic Press.

Batty, M., and L. March. 1976. The Method of Residues in Urban Modeling. *Environment and Planning A* 8:189–214.

Batty, M., and S. Rana. 2004. The Automatic Definition and Generation of Axial Lines and Axial Maps. *Environment and Planning B* 31:615–640.

Batty, M., and N. Shiode. 2003. Population Growth Dynamics in Cities, Countries and Communication Systems. In *Advanced Spatial Analysis*, ed. P. A. Longley and M. Batty, 327–343. Redlands, CA: ESRI Press.

Batty, M., D. Smith, J. Reades, A. Johansson, J. Serras, and C. Vargas-Ruiz. 2011. Visually-Intelligible Land Use Transportation Models for the Rapid Assessment of Urban Futures. CASA Working Paper 163. Centre for Advanced Spatial Analysis. London: University College. http://www.bartlett.ucl.ac.uk/casa/pdf/paper163.pdf. Accessed 17/9/2012.

Batty, M., and K. J. Tinkler. 1979. Symmetric Structure in Spatial and Social Processes. *Environment and Planning B* 6:3–27.

Batty, M., and Y. Xie. 2005. Urban Growth using Cellular Automata Models. In *GIS, Spatial Analysis, and Modeling*, ed. D. J., Maguire, M. Batty, and M. F. Goodchild, 151–172. Redlands, CA: ESRI Press.

Baudrillard, J. 1983. *Simulations*. Cambridge, MA: MIT Press.

Bavaud, F. 2002. The Quasi-Symmetric Side of Gravity Modeling. *Environment and Planning A* 34:61–79.

Bazjanac, V. 1974. Architectural Design Theory: Models of the Design Process. In *Basic Questions of Design Theory*, ed. W. R. Spillers, 3–19. Amsterdam: North Holland.

Beckmann, M. J. 1958. City Hierarchies and the Distribution of City Size. *Economic Development and Cultural Change* 6:243–248.

Beckmann, M. J. 1968. *Location Theory*. New York: Random House.

Bellman, R. 1957a. A Markovian Decision Process. *Journal of Mathematics and Mechanics* 6:679–684.

Bellman, R. 1957b. *Dynamic Programming*. Princeton, NJ: Princeton University Press.

Ben-Akiva, M., and S. Lerman. 1985. *Discrete Choice Analysis: Theory and Application to Travel Demand*. Cambridge, MA: MIT Press.

Benguigui, L., E. Blumenfeld-Lieberthal, and M. Batty. 2009. Macro and Micro Dynamics of the City-Size Distribution: The Case of Israel. In *Complexity and Spatial Networks: In Search of Simplicity*, ed. A. Reggiani and P. Nijkamp, 33–49. Heidelberg: Springer-Verlag.

Bera, R., and C. Claramunt. 2002. Topology-Based Proximities in Spatial Systems. *Journal of Geographical Systems* 6:1–27.

Berger, J., and J. L. Snell. 1957. On the Concept of Equal Exchange. *Behavioral Science* 2:111–118.

Berry, B. J. L. 1964. Cities as Systems within Systems of Cities. *Papers of the Regional Science Association* 13:147–164.

Berry, B. J. L. 1967. *Geography of Market Centers and Retail Distribution*. Englewood Cliffs, NJ: Prentice-Hall.

Berry, B. J. L., and A. Okulicz-Kozaryn. 2012. The City Size Debate: Resolution for U.S. Urban Regions and Megalopolitan Areas. *Cities* 29 (Suppl. 1):S17–S23.

Bettencourt, L. M. A., J. Lobo, D. Helbing, C. Kuchnert, and G. B. West. 2007. Growth, Innovation, Scaling, and the Pace of Life in Cities. *Proceedings of the National Academy of Sciences of the United States of America* 104:7301–7306.

Bhat, U. N. 1972. *Elements of Applied Stochastic Processes*. New York: John Wiley.

Blank, A., and S. Solomon. 2000. Power Laws in Cities Population, Financial Markets and Internet Sites: Scaling and Systems with a Variable Number of Components. *Physica A* 287:279–288.

Blau, P. M. 1964. *Exchange and Power in Social Life*. New York: John Wiley.

Blondel, V. D., J. M. Hendrickx, A. Olshevsky, and J. N. Tsitsiklis. 2005. Convergence in Multiagent Coordination, Consensus, and Flocking. In *Proceedings of the Joint 44th IEEE Conference on Decision and Control*, European Control Conference. Seville, Spain, December 12–15, 2005. http://stuff.mit.edu/people/jnt/Papers/C-05-ol-flock-fin.pdf. Accessed 21/9/2012.

Boaden, N. 1971. *Urban Policy-Making*. Cambridge, UK: Cambridge University Press.

Bolan, R. S., and R. L. Nuttall. 1975. *Urban Planning and Politics*. Lexington, MA: Heath and Company.

Bonner, J. T. 2006. *Why Size Matters: From Bacteria To Blue Whales*. Princeton, NJ: Princeton University Press.

Borgatti, S. P., and M. G. Everett. 1997. Network Analysis of 2-Mode Data. *Social Networks* 19:243–269.

Bosker, M., S. Brakman, H. Garretsen, H. De Jong, and M. Schramm. 2007. The Development of Cities in Italy, 1300–1861. CESifo Working Paper No. 1893. Munich, Germany: CESifo. http://www.cesifo-group.de/ifoHome/publications.html. Accessed 16/10/2012.

Boulding, K. E. 1975. Reflections on Planning: The Value of Uncertainty. *Strategy and Leadership* 3:11–12.

Brail, R. K., ed. 2008. *Planning Support Systems for Cities and Regions*. Cambridge, MA: Lincoln Institute of Land Policy.

Brand, S. 2010. *Whole Earth Discipline: Why Dense Cities, Nuclear Power, Transgenic Crops, Restored Wildlands, Radical Science, and Geoengineering are Necessary*. New York: Atlantic Books.

Brewer, G. D. 1973. *Politicians, Bureaucrats and the Consultant: A Critique of Urban Problem Solving*. New York: Basic Books.

Buldyrev, S. V., R. Parshani, G. Paul, H. E. Stanley, and S. Havlin. 2010. Catastrophic Cascade of Failures in Interdependent Networks. *Nature* 464:1025–1028.

Bussiere, R. 1970. *The Spatial Distribution of Urban Populations*. Paris: Centre de Recherche d'Urbanisme.

Cairncross, F. 1997. *The Death of Distance: How the Communications Revolution Will Change Our Lives*. Cambridge, MA: Harvard Business School Press.

Carr, M. H., and P. D. Zwick. 2007. *Smart Land Use Analysis: The LUCIS Model*. Redlands, CA: ESRI Press.

Carvalho, R., and M. Batty. 2004. Automatic Extraction of Hierarchical Urban Networks: A Micro-Spatial Approach. In Computational Science: ICCS 2004, 4th International Conference, Part III, 1109–1116. Berlin: Springer.

Castells, M. 1989. *The Informational City: Information Technology, Economic Restructuring, and the Urban Regional Process*. New York: Wiley-Blackwell.

Cerda, I. 1999. *The Five Bases of the General Theory of Urbanization*. Ed. A. Soria y Puig. Madrid: Electa. Original work published 1859.

Chadwick, G. F. 1971. *A Systems View of Planning: Towards a Theory of the Urban and Regional Planning Process*. Oxford, UK: Pergamon Press.

Chandler, T. 1987. *Four Thousand Years of Urban Growth: An Historical Census*. Lampeter, UK: Edward Mellon.

Chatterjee, S., and E. Seneta. 1977. Toward Consensus: Some Convergence Theorems on Repeated Averaging. *Journal of Applied Probability* 14:89–97.

Choukroun, J.-M. 1984. The Validation of Models of Complex Systems. *Environment and Planning B* 11:263–277.

Christaller, W. 1966. *Central Places in Southern Germany*. Englewood Cliffs, NJ: Prentice-Hall. Original work published 1933 as *Die Zentralen Orte in Suddeutschland*, Jena, Germany: Gustav Fischer.

Chung, F., and L. Lu. 2002. The Average Distance in Random Graphs with Given Expected Degrees. *Proceedings of the National Academy of Sciences of the United States of America* 99:15879–15882.

Churchman, C. W. 1971. *The Design of Inquiring Systems*. New York: Basic Books.

Churchman, C. W., R. Ackoff, and E. L. Arnoff. 1957. *Introduction to Operations Research*. New York: John Wiley.

CIS. 1973. *The Recurrent Crisis of London*. Anti-Report on the Property Developers. London: Counter Information Services.

Clark, C. 1951. Urban Population Densities. *Journal of the Royal Statistical Society. Series A (General)* 114:490–496.

Clauset, A., C. R. Shalizi, and M. E. J. Newman. 2009. Power-Law Distributions in Empirical Data. *SIAM Review* 51 (4):661–703.

CNN. 2009. Fortune 500. *CNN Money*. http://money.cnn.com/magazines/fortune/fortune500 _archive/full/1955/index.html. Accessed 21/3/2009.

Coleman, J. S. 1964. *Mathematical Sociology*. Glencoe, NY: Free Press.

Coleman, J. S. 1966. Foundations for a Theory of Collective Decisions. *American Journal of Sociology* 71:615–627.

Coleman, J. S. 1972. Systems of Social Exchange. *Journal of Mathematical Sociology* 2:145–163.

Coleman, J. S. 1973. *The Mathematics of Collective Action*. London: Heinemann Educational Books.

Coleman, J. S. 1976. Social Structure and a Theory of Action. In *Approaches to the Study of Social Structure*, ed. P. M. Bleu, 76–93. New York: The Free Press.

Coleman, J. S. 1977. Social Action Systems. In *Problems of Formalization in the Social Sciences*, ed. K. Szaniawski, 11–50. UNESCO Division of International Development of Social Sciences and The Polish Academy for Science Division of Social Sciences, Warsaw: Ossolineum Publishers.

Coleman, J. S. 1998. *Foundations of Social Theory*. Cambridge, MA: Harvard University Press.

Creighton, R. L., Carroll Jr., D., and Finney, G. S. 1959. Data Processing for City Planning. *Journal of the American Institute of Planners* 25:96–103.

Cruz, P. M. 2012. Lisbon's Blood Vessels. http://vimeo.com/19116437 and http://pmcruz.com/ information-visualization/lisbons-blood-vessels. Accessed 1/9/2012.

Davis, A., B. B. Gardner, and M. R. Gardner. 1941. *Deep South: A Social Anthropological Study of Caste and Class*. Chicago, IL: University of Chicago Press.

Dawkins, R. 1986. *The Blind Watchmaker: Why the Evidence of Evolution Reveals a Universe without Design*. London: Longmans.

Dawson, R., J. Hal, S. Barr, M. Batty, A. Bristow, S. Carney, A. Dagoumas, et al. 2009. A Blueprint for the Integrated Assessment of Climate Change in Cities. In *Green Citynomics:*

The Urban War Against Climate Change, ed. K. Tang, 32–51. Chippenham, UK: Greenleaf Publishing.

De Groot, M. H. 1974. Reaching a Consensus. *Journal of the American Statistical Association* 69:118–121.

De Solla Price, D. 1965. Networks of Scientific Papers. *Science* 149 (3683):510–515.

De Vries, J. J., P. Nijkamp, and P. Rietveld. 2000. Alonso's General Theory of Movement. Discussion Paper TI 2000–062/3. Amsterdam: Tinbergen Institute.

Dearlove, J. 1973. *The Politics of Policy in Local Government*. Cambridge, UK: Cambridge University Press.

DeMarco, P., D. Vayanos, and J. Zwiebel. 2003. Persuasion Bias, Social Influence, and Unidimensional Opinions. *Quarterly Journal of Economics* 118:909–965.

Dennett, D. 1995. *Darwin's Dangerous Idea: Evolution and the Meanings of Life*. New York: Simon and Schuster.

Doig, A. 1963. The Minimum Number of Basic Feasible Solutions to a Transport Problem. *Journal of the Operational Research Society* 14:387–391.

Dorigo, G., and W. R. Tobler. 1983. Push-Pull Migration Laws. *Annals of the Association of American Geographers* 73:1–17.

Dorling, D. 2012. *The Visualization of Spatial Social Structure*. London: Wiley-Blackwell.

Dorogovtsev, S. N., and J. F. F. Mendes. 2003. *Evolution of Networks: From Biological Nets to the Internet and WWW*. Oxford, UK: Oxford University Press.

Dragulescu, A., and V. M. Yakovenko. 2000. Statistical Mechanics of Money. *European Physical Journal B* 17:723–729.

Dunbar, R. I. M. 1992. Neocortex Size as a Constraint on Group Size in Primates. *Journal of Human Evolution* 22:469–493.

Echenique, M. H. 2004. Econometric Models of Land Use and Transportation. In *Handbook of Transport Geography and Spatial Systems*, ed. D. A. Hensher, K. J. Button, K. E. Haynes, and P. R. Stopher, 185–202. Amsterdam: Pergamon.

Emporis. 2009. Emporis Building Information Global Database. http://www.emporis.com/statistics/most-skyscrapers. Accessed 23/5/2009.

Epstein, J. M. 2008. Why Model? *Journal of Artificial Societies and Social Simulation* 11(4):12. http://jasss.soc.surrey.ac.uk/11/4/12.html. Accessed 17/9/2012.

Feles, D., K. Gergely, A. Bujdosó, G. Hajdu, and L. Kiss. 2012. SubMap: Visualizing Locative and Time Based Data on Distorted Maps. http://submap.kibu.hu. Accessed 22/10/2012.

Feller, W. 1957. *An Introduction to Probability Theory and Its Applications*. vol. 1. New York: John Wiley.

Feynman, R. P. 1992. *The Character of the Physical Law*. London: Penguin Books. Original work published 1965.

Figueiredo, L., and L. Amorim. 2005. Continuity Lines in the Axial System. 5th International Space Syntax Symposium, TU-Delft. http://www.spacesyntax.tudelft.nl/index.html. Accessed 20/10/2012.

Fischer, E. 2012. Paths Through Cities: London. http://www.flickr.com/photos/walkingsf/6755911359. Accessed 1/9/2012.

Flament, C. 1963. *Applications of Graph Theory to Group Structure*. Englewood Cliffs, NJ: Prentice-Hall.

Forester, J. 1989. *Planning in the Face of Power*. Berkeley, CA: University of California Press.

Forester, J. 2009. *Dealing with Differences: Dramas of Mediating Public Disputes*. New York: Oxford University Press.

Forrester, J. W. 1969. *Urban Dynamics*. Cambridge, MA: MIT Press.

Frankhauser, P. 1994. *La Fractalité des Structures Urbaines, Collection Villes*. Paris, France: Anthropos.

Freeman, L. C. 1979. Centrality in Social Networks. *Social Networks* 1:215–239.

French, J. R. P. 1956. A Formal Theory of Social Power. *Psychological Review* 63: 181–194.

Friedmann, J. 1973. *Retracking America: A Theory of Transactive Planning*. New York: Doubleday Anchor Press.

Fujita, M., P. Krugman, and A. J. Venables. 1999. *The Spatial Economy: Cities, Regions, and International Trade*. Cambridge, MA: MIT Press.

Gabaix, X. 1999. Zipf's Law for Cities: An Explanation. *Quarterly Journal of Economics* 114:739–767.

Gabbour, I., and T. J. Cartwright. 1974. *Les Relations entre Organismes Facteurs Strategiques in Planification Urbaine*. Universite de Montreal: Centre de Recherches et d'Innovation Urbaines.

Gale, D. 1960. *The Theory of Linear Economic Models*. New York: McGraw-Hill.

Geddes, P. 1949. *Cities in Evolution*. London: Williams and Norgate. Original work published 1915.

Geddes, P. 2010. *Civics: as Applied Sociology*. Boston, MA: Qontro Classic Books. Original work published 1905.

Gibrat, R. 1931. *Les Inegalites Economiques, Librarie du Recueil*. Paris: Sirey.

Gibson, C. 1998. Population of the 100 Largest Cities and Other Urban Places in the United States: 1790 to 1990. Population Division Paper 27. Washington, DC: US Bureau of the Census.

Giddens, A. 1976. *New Rules of Sociological Method: A Positive Critique of Interpretative Sociologies*. London: Hutchinson.

Glaeser, E. L. 1996. Why Economists Still Like Cities. *City Journal* 6(2). http://www.city-journal.org/html/6_2_why_economists.html. Accessed 18/10/2012.

Glaeser, E. L. 2008. *Cities, Agglomeration and Spatial Equilibrium*. New York: Oxford University Press.

Glaeser, E. L. 2011. *Triumph of the City: How Our Greatest Invention Makes Us Richer, Smarter, Greener, Healthier, and Happier*. New York: Penguin.

Goethe, J. W. 2009. *The Metamorphosis of Plants*. Cambridge, MA: MIT Press. Original work published 1790.

Goldstein, H. A. 1984. Planning as Argumentation. *Environment and Planning B* 11: 297–312.

Golub, B., and M. O. Jackson. 2010. Naive Learning in Social Networks and the Wisdom of Crowds. *American Economic Journal: Microeconomics* 2:112–149.

Grant, D. P., and B. Thompson. 1971. Simulating Conflicts of Interest over the Location of Public Housing with the Aid of a Computer-Aided Space Allocation Technique. Proceedings of the 6th Urban Symposium. New York: Association of Computing Machinery.

Gruen, V. 1964. *The Heart of Our Cities: The Urban Crisis, Diagnosis and Cure*. New York: Simon and Schuster.

Guerin-Pace, F. 1995. Rank-Size Distribution and the Process of Urban Growth. *Urban Studies* 32:551–562.

Gurney, K. 1997. *An Introduction to Neural Networks*. London: UCL Press.

Haggett, P., and R. J. Chorley. 1969. *Network Analysis in Geography*. London: Edward Arnold.

Hall, J. W., R. J. Dawson, C. L. Walsh, T. Barker, S. L. Barr, M. Batty, A. L. Bristow, et al. 2009. Engineering Cities: How Can Cities Grow Whilst Reducing Emissions and Vulnerability? Department of Civil and Geomatic Engineering. Newcastle-upon-Tyne, UK: Newcastle University.

Hall, P. 1975. Tolmers Square. *New Society* 32:653.

Hamdi, N. 2004. *Small Change: About the Art of Practice and the Limits of Planning in Cities*. London: Earthscan.

Hansell, S. 2008. Zuckerberg's Law of Information Sharing. Bits [blog]. *New York Times*. http://bits.blogs.nytimes.com/2008/11/06/zuckerbergs-law-of-information-sharing/. Accessed 1/9/2012.

Hansen, W. G. 1959. How Accessibility Shapes Land Use. *Journal of the American Institute of Planners* 25:73–76.

Harary, F. 1959. A Criterion for Unanimity in French's Theory of Social Power. In *Studies in Social Power*, ed. D. Cartwright, 168–182. Ann Arbor, MI: Institute for Social Research.

Harary, F., and B. Lipstein. 1962. The Dynamics of Brand Loyalty: A Markovian Approach. *Operations Research* 10:19–40.

Harary, F., R. Z. Norman, and D. Cartwright. 1965. *Structural Models: An Introduction to the Theory of Directed Graphs*. New York: John Wiley.

Harel, D., and Y. Koren. 2002. Graph Drawing by High-Dimensional Embedding. Proceedings of the 9th International Symposium on Graph Drawing. Springer Lecture Notes in Computer Science 2265:207–219. http://www.graphdrawing.org/symposium/. Accessed 19/9/2012.

Harris, B. 1970. Change and Equilibrium in the Urban System. Highway Research Record No 309., 24–33. Washington, DC: Highway Research Board.

Harris, B. 1978. A Paradigm for Planning. Unpublished manuscript, Department of City and Regional Planning. Philadelphia, PA: University of Pennsylvania.

Harris, B. 1983. Positive and Normative Aspects of Modeling Large-Scale Social Systems. In *Systems Analysis in Urban Policy-making and Planning*, ed. M. Batty and B. Hutchinson, 475–490. New York: Plenum Press.

Harris, B. 1989. Beyond Geographic Information Systems: Computers and the Planning Professional. *Journal of the American Planning Association* 55:85–92.

Harris, B., and M. Batty. 1993. Locational Models, Geographic Information and Planning Support Systems. *Journal of Planning Education and Research* 12:184–198.

Harsanyi, J. C. 1976. *Essays on Ethics, Social Behavior, and Scientific Explanation*. Dordrecht, Netherlands: Reidel.

Healey, P. 2006. *Collaborative Planning: Shaping Places in Fragmented Societies*. London: Palgrave Macmillan.

Healey, P. 2007. *Urban Complexity and Spatial Strategies: Towards a Relational Planning for Our Times*. London: Routledge.

Hegselmann, R., and U. Krause. 2002. Opinion Dynamics and Bounded Confidence: Models, Analysis and Simulation. *Journal of Artificial Societies and Social Simulation* 5(3). http://jasss.soc.surrey.ac.uk/5/3/2.html. Accessed 20/10/2012.

Heppenstall, A. J., A. T. Crooks, L. M. See, and M. Batty, eds. 2012. *Agent-Based Models of Geographical Systems*. Berlin, New York: Springer.

Hillier, B. 1996. *Space is the Machine: A Configurational Theory of Architecture*. Cambridge, UK: Cambridge University Press.

Hillier, B., and J. Hanson. 1984. *The Social Logic of Space*. Cambridge, UK: Cambridge University Press.

Hillier, B., A. Leaman, P. Stansall, and M. Bedford. 1976. Space Syntax. *Environment and Planning B* 3:147–185.

Hillier, B., A. Penn, J. Hanson, T. Grajewski, and J. Xu. 1993. Natural Movement: Or Configuration and Attraction in Urban Pedestrian Movement. *Environment and Planning B* 20:29–66.

Holroyd, E. M. 1966. Average Journey Lengths in Circular Towns with Various Routeing Systems. RRL Report LR43. London: Road Research Laboratory.

Homans, G. C. 1974. *Social Behavior: Its Elementary Forms*. New York: Harcourt, Brace and Jovanovich.

Hopkins, L. D. 1977. Methods for Generating Land Suitability Maps: A Comparative Evaluation. *Journal of the American Institute of Planners* 43:386–400.

Howard, R. A. 1960. *Dynamic Programming and Markov Processes*. Cambridge, MA: MIT Press.

Howard, R. A. 1971. *Dynamic Probabilistic Systems, Vol. 1: Markov Models*. New York: John Wiley.

Hua, C.-I. 2002. Alonso's Systemic Model: A Review and Representation. *International Regional Science Review* 24:360–385.

Hunt, J. D., E. J. Miller, and D. S. Kriger. 2005. Current Operational Urban Land-Use-Transport Modeling Frameworks: A Review. *Transport Reviews* 25:329–376.

Iacono, I., D. Levinson, and A. El-Geneidy. 2008. Models of Transportation and Land Use Change: A Guide to the Territory. *Journal of Planning Literature* 22:323–340.

Isard, W. 1956. *Location And Space Economy: A General Theory Relating to Industrial Location, Market Areas, Land Use, Trade, and Urban Structure*. Cambridge, MA: MIT Press.

Ispolatov, S., P. L. Krapivsky, and S. Redner. 1998. Wealth Distribution in Asset Exchange Models. *European Physical Journal B* 2:267–276.

Jackson, M. O. 2010. *Social and Economic Networks*. Princeton, NJ: Princeton University Press.

Jackson, M. O. 2011. An Overview of Social Networks and Economic Applications. In *The Handbook of Social Economics*. vol. 1. ed. J. Benhabib, A. Bisin, and M. O. Jackson, 511–585. Amsterdam: Elsevier.

Jacobs, J. 1961. *The Death and Life of Great American Cities*. New York: Random House.

Jacobs, J. 1969. *The Economy of Cities*. New York: Random House.

Jiang, B., and C. Claramunt. 2000. Integration of Space Syntax into GIS: New Perspectives for Urban Morphology. *Transactions in GIS* 6:295–307.

Kansky, K. J. 1963. Structure of Transportation Networks: Relationships Between Network Geometry and Regional Characteristics. Research Paper 84. Department of Geography. Chicago: University of Chicago.

Keeble, L. 1952. *Principles and Practice of Town and Country Planning*. London: Estates Gazette.

Kelly, F. P. 1979. *Reversibility and Stochastic Networks*. New York: John Wiley.

Kelly, F. P. 1981. How a Group Reaches an Agreement: A Stochastic Model. *Mathematical Social Sciences* 2:1–8.

Kemeny, J. G., H. Mirkil, J. L. Snell, and G. L. Thompson. 1959. *Finite Mathematical Structures*. Englewood Cliffs, NJ: Prentice-Hall.

Kemeny, J. G., and J. L. Snell. 1960. *Finite Markov Chains*. Princeton, NJ: Van Nostrand.

Kemeny, J. G., and J. L. Snell. 1962. *Mathematical Models in the Social Sciences*. New York: Ginn and Company.

Kohl, J. G. 1841. *Der Verkehr und die Ansiedelung der Menschen in ihrer Abhangigkeit uon der Gestaltung der Eudoberflache*. Leipzig, Germany: Arnoldische Buchhandlung.

Kreweras, G. 1968. A Model for Opinion Change During Repeated Balloting. In *Readings in Mathematical Social Science*, ed. P. F. Lazarsfeld and N. W. Henry, 174–191. Cambridge, MA: MIT Press.

Kruger, M. 1989. On Node and Axial Maps: Distance Measures and Related Topics. Unpublished paper presented to the European Conference on the Representation and Management of Urban Change. http://www.ces.uc.pt/investigadores/cv/mario_kruger.php. Accessed 16/9/2012.

Krugman, P. 1996. Confronting the Mystery of Urban Hierarchy. *Journal of the Japanese and International Economies* 10:399–418.

Kruskal, J. B. 1956. On the Shortest Spanning Subtree of a Graph and the Traveling Salesman Problem. *Proceedings of the American Mathematical Society* 7:48–50.

Kullback, S. 1959. *Information Theory and Statistics*. New York: John Wiley.

Lambiotte, R., R. Sinatra, J. C. Delvenne, T. S. Evans, M. Barahona, and V. Latora. 2011. Flow Graphs: İnterweaving Dynamics and Structure. *Physical Review E: Statistical, Nonlinear, and Soft Matter Physics* 84:017102.

Lee, D. B. 1973. Requiem for Large-scale Models. *Journal of the American Institute of Planners* 39:163–178.

Lipp, M., D. Scherzer, P. Wonka, and M. Wimmer. 2011. Interactive Modeling of City Layouts using Layers of Procedural Content. *Computer Graphics Forum* 30:345–354.

Liu, Y. 2008. *Modelling Urban Development with Geographical Information Systems and Cellular Automata*. Boca Raton, FL: CRC Press.

Los, M. 1981. Some Reflections on Epistemology, Design and Planning Theory. In *Urbanization and Urban Planning in Capitalist Society*, ed. M. Dear and A. J. Scott, 63–88. London: Methuen.

Losch, A. 1954. *The Economics of Location*. 2nd edition. Trans. W. H. Woglom. New Haven, CT: Yale University Press. Original work published 1940.

Lowry, I. S. 1964. A Model of Metropolis. Memorandum RM-4035-RC. Santa Monica, CA: The Rand Corporation.

Lowry, I. S. 1967. Seven Models of Urban Development: A Structural Comparison. In *Urban Development Models, Special Report 97*, ed. G. C. Hemmens, 121–163. Washington, DC: Highway Research Board.

Lynch, K. 1960. *The Image of the City*. Cambridge, MA: MIT Press.

Malanima, P. 1998. Italian Cities 1300–1800: A Quantitative Approach. *Rivista di Storia Economica* 14:91–126.

Mandelbrot, B. B. 1967. How Long Is the Coast of Britain? Statistical Self-Similarity and Fractional Dimension. *Science* 156:636–638.

Mandelbrot, B. B. 1982. *The Fractal Geometry of Nature*. San Francisco, CA: W. H. Freeman.

Manrubia, S. C., and D. H. Zanette. 1998. Intermittency Model for Urban Development. *Physical Review E: Statistical Physics, Plasmas, Fluids, and Related Interdisciplinary Topics* 58:295–302.

March, L., and J. P. Steadman. 1971. *The Geometry of Environment*. London: RIBA Publications.

Marshall, A. 1890. *Principles of Economics*. London: Macmillan.

May, R. M. 1976. Simple Mathematical Models with Very Complicated Dynamics. *Nature* 261:459–467.

McHarg, I. L. 1969. *Design with Nature*. Garden City, New York: Doubleday.

McMeekin, G. C. 1974. An Empirical Test of Alternative Estimation Techniques for the Consensus and Coalition Approaches to the Committee Decision Problem. PhD thesis, Wayne State University, Detroit. Ann Arbor, MI: Xerox University Microfilms 75–13366.

Meier, R. L. 1962. *A Communications Theory of Urban Growth*. Cambridge, MA: MIT Press.

Michener, H. A., E. D. Cohen, and A. B. Sorensen. 1977. Social Exchange: Predicting Transactional Outcomes in Five-Event, Four-Person Systems. *American Sociological Review* 42:522–536.

Milgram, S. 1967. The Small World Problem. *Psychology Today* 2:60–67.

Miller, J. H., and S. E. Page. 2007. *Complex Adaptive Systems: An Introduction to Computational Models of Social Life*. Princeton, NJ: Princeton University Press.

Mitchell, R. B., and C. Rapkin. 1954. *Urban Traffic: A Function of Land Use*. New York: Columbia University Press.

Morishima, M. 1977. *Walras's Economics: A Pure Theory of Capital and Money*. Cambridge, UK: Cambridge University Press.

Morphet, R. 2010. Thermodynamic Potentials and Phase Change for Transport Systems. CASA Working Paper 156. Centre for Advanced Spatial Analysis. London: University College. http://www.bartlett.ucl.ac.uk/casa/pdf/paper156.pdf. Accessed 17/9/2012.

Morris, I. 2010. *Why the West Rules—for Now: The Patterns of History, and What They Reveal About the Future*. London: Profile Books.

Mumford, L. 1928. The Arts. In *Whither Mankind: a Panorama of Modern Civilization*, ed. C. A. Beard, 287–312. New York: Longmans, Green and Company.

Negroponte, N. 1995. *Being Digital*. New York: Vintage Books.

Newman, M. E. J. 2003. The Structure and Function of Complex Networks. *SIAM Review* 45 (2):167–256.

Newman, M. E. J. 2005. Power Laws, Pareto Distributions and Zipf's Law. *Contemporary Physics* 46:323–351.

Newman, M. E. J. 2010. *Networks: An Introduction*. New York: Oxford University Press.

Nystuen, J. D., and M. F. Dacey. 1961. A Graph Theory Interpretation of Nodal Regions. *Papers of the Regional Science Association* 7:29–42.

O'Brien, O. 2012. Rank Clock Visualiser. http://casa.oobrien.com/rankclocks. Accessed 13/9/12.

Olson, M. 1965. *The Logic of Collective Action: Public Goods and the Theory of Groups*. Cambridge, MA: Harvard University Press.

ONS. 2010. Census Dissemination Unit. London: Office of National Statistics. http://cdu.mimas.ac.uk/. Accessed 6/1/2010.

Ortuzar, J. de Dios and Willumsen L. G. 2011. *Modelling Transport*. New York: John Wiley.

O'Sullivan, A. 2011. *Urban Economics*. Irwin, New York: McGraw-Hill.

Pareto, V. 1967. La Courbe de la Repartition de la Richesse. In *Oevres Completes de Vilfredo Pareto*, ed. G. Busino. Geneva: Librairie Droz. Original work published 1896.

Peponis, J., J. Wineman, M. Rashid, S. H. Kim, and S. Bafna. 1997. On the Description of Shape and Spatial Configuration inside Buildings: Convex Partitions and their Local Properties. *Environment and Planning B* 24:761–781.

Peucker, T. K. 1968. Johann Georg Kohl, A Theoretical Geographer of the 19th Century. *Professional Geographer* 20:247–250.

Piaget, J. 1971. *Structuralism*, London: Routledge and Kegan Paul.

Popper, K. R. 1959. *The Logic of Scientific Discovery*. London: Hutchinson.

Popper, K. R. 1992. *The Open Universe: An Argument for Indeterminism*. London: Routledge.

Porta, S., P. Crucitti, and V. Latora. 2006a. The Network Analysis of Urban Streets: A Primal Approach. *Environment and Planning B* 33:705–725.

Porta, S., P. Crucitti, and V. Latora. 2006b. The Network Analysis of Urban Streets: A Dual Approach. *Physica A* 369:853–866.

Portugali, J. 2000. *Self-Organization and the City*. New York: Springer.

Prell, C. 2012. *Social Network Analysis: History, Theory, Methodology*. Los Angeles: Sage Publications.

Pumain, D. 2000. Settlement Systems in the Evolution. *Geografiska Annaler* 82B:73–87.

Pumain, D. 2006a. Alternative Explanations of Hierarchical Differentiation in Urban Systems. In *Hierarchy in Natural and Social Sciences*, ed. D. Pumain, 169–222. Dordrecht, The Netherlands: Kluwer Academic Publishers.

Pumain, D., ed. 2006b. *Hierarchy in the Natural and Social Sciences*. Dordrecht, The Netherlands: Springer.

Rae, A. 2009. From Spatial Interaction Data to Spatial Interaction Information? Geovisualisation and Spatial Structures of Migration from the 2001 UK Census. *Computers, Environment and Urban Systems* 33:161–178.

Rapoport, A. 1949. Outline of a Probabilistic Approach to Animal Sociology. *Bulletin of Mathematical Biophysics* 11:273–281.

Ravenstein, E. G. 1885. The Laws of Migration 1. *Journal of the Statistical Society of London* 48:167–235.

Ravenstein, E. G. 1889. The Laws of Migration 2. *Journal of the Statistical Society of London* 52:241–305.

Reades, J. E. 2012. Pulse of the City. https://vimeo.com/41760845. Accessed 1/9/2012.

Rittel, H. W. J., and M. M. Webber. 1973. Dilemmas in a General Theory of Planning. *Policy Sciences* 4:155–169.

Robinson, A. H. 1955. The 1837 Maps of Henry Drury Harness. *Geographical Journal* 121:440–450.

Roskamp, K. W., and G. C. McMeekin. 1970. The Symmetry Approach to Committee Decisions: An Empirical Study of a Local Government Budget Committee. *Zeitschrift fur die Gesamte Staatswissenschaft* 126:75–96.

Roth, C., S. M. Kang, M. Batty, and M. Barthelemy. 2011. Structure of Urban Movements: Polycentric Activity and Entangled Hierarchical Flows. *PLoS ONE* 6 (1):e15923.

Roth, C., S. M. Kang, M. Batty, and M. Barthelemy. 2012. A Long-Time Limit for World Subway Networks. *Journal of the Royal Society, Interface* 9 (75):2540–2550.

Rozenfeld, H. D., D. Rybski, J. S. Andrade, M. Batty, H. E. Stanley, and H. A. Makse. 2008. Laws of Population Growth. *Proceedings of the National Academy of Sciences of the United States of America* 105:18702–18707.

Saaty, T. L. 1980. *The Analytic Hierarchy Process: Planning, Priority Setting, Resources Allocation*. New York: McGraw-Hill.

Saaty, T. L. 2005. *Theory and Applications of the Analytic Network Process: Decision Making with Benefits, Opportunities, Costs, and Risks*. New York: RWS Publications.

Saichev, A., Y. Malevergne, and D. Sornette. 2010. *Theory of Zipf's Law and Beyond, Lecture Notes in Economics and Mathematical Systems 632*. Heidelberg, DE: Springer.

Salat, S. 2011. *Cities and Forms*. Paris, France: Editions Hermann.

Salingaros, N. A. 2005. *Principles of Urban Structure*. Amsterdam, Netherlands: Techne Press.

Samaniego, H., and M. Moses. 2008. Cities as Organisms: Allometric Scaling of Urban Road Networks. *Journal of Transport and Land Use* 1 (1):21–39.

Schelling, T. C. 1969. Models of Segregation. *American Economic Review* 58:488–493.

Schlager, K. J. 1965. A Land Use Plan Design Model. *Journal of the American Institute of Planners* 31:103–111.

Sevtsuk, A. 2010. Path and Place: A Study of Urban Geometry and Retail Activity in Cambridge and Somerville, MA. PhD Dissertation, Urban Design and Planning. Cambridge, MA: MIT. http://www.cityform.net/publications. Accessed 17/10/2012.

Shannon, C. E. 1948. A Mathematical Theory of Communication. *Bell System Technical Journal* 27:379–423 and 623–656.

Shannon, C. E., and W. Weaver. 1949. *The Mathematical Theory of Communication*. Urbana, IL: University of Illinois Press.

Simon, H. A. 1955. On a Class of Skew Distribution Functions. *Biometrika* 42:425–440.

Simon, H. A. 1956. Rational Choice and the Structure of the Environment. *Psychological Review* 63:129–138.

Simon, H. A. 1960. *The New Science of Management Decision*. New York: Harper and Row.

Simon, H. A. 1962. The Architecture of Complexity. *Proceedings of the American Philosophical Society* 106:467–482.

Simon, H. A. 1969. *The Sciences of the Artificial*. Cambridge, MA: MIT Press.

Simon, H. A. 1977. *Models of Discovery and Other Topics in the Methods of Science*. Dordrecht, Netherlands: D. Reidel.

Simon, H. A. 1983. *Reason in Human Affairs*. Oxford, UK: Basil Blackwell.

Smeed, R. J. 1961. *The Traffic Problem in Towns*. Manchester, UK: Manchester Statistical Society.

Smith, T. E., and S.-H. Hsieh. 1997. Gravity-Type Interactive Markov Models—Part I: A Programming Formulation of Steady States. *Journal of Regional Science* 37:653–682.

Sornette, D. 2004. *Critical Phenomena in Natural Sciences, Chaos, Fractals, Self-organization and Disorder: Concepts and Tools*. Heidelberg, DE: Springer.

Sornette, D., and R. Cont. 1997. Convergent Multiplicative Processes Repelled from Zero: Power Laws and Truncated Power Laws. *Journal de Physique I* 7:431–444.

Stanley, H. E., and N. Ostrowsky, eds. 1985. *On Growth and Form: Fractal and Non-Fractal Patterns in Physics*. New York: Springer.

Steadman, J. P. 1983. *Architectural Morphology: Introduction to the Geometry of Building Plans*. London: Pion Press.

Steadman, P. 2008. *The Evolution of Designs: Biological Analogy in Architecture and the Applied Arts*. London: Routledge.

Steinitz, C. 2012. *A Framework for Geodesign: Changing Geography by Design*. Redlands, CA: ESRI Press.

Steinitz, C., P. Parker, and L. Jordan. 1976. Hand-Drawn Overlays: Their History and Prospective Uses. *Landscape Architecture* 66:444–455.

Stewart, J. Q. 1941. An Inverse Distance Variation for Certain Social Influences. *Science* 93:89–90.

Stewart, J. Q. 1947. Suggested Principles of Social Physics. *Science* 106:179–180.

Stiny, G. 2006. *Shape: Talking about Seeing and Doing*. Cambridge, MA: The MIT Press.

Sugiyama, K., S. Tagawa, and M. Toda. 1981. Methods for Visual Understanding of Hierarchical System Structures. *IEEE Transactions on Systems, Man, and Cybernetics* 11: 109–125.

Taylor, D. G., and J. S. Coleman. 1979. Equilibrating Processes in Social Networks: A Model for Conceptualization and Analysis. In *Perspectives on Social Network Research*, ed. P. W. Holland and S. Leinhardt, 257–300. New York: Academic Press.

Teklenberg, J. A. F., H. T. P. Timmermans, and A. T. van Wagenberg. 1993. Space Syntax: Some Standard Integration Measures and Some Simulations. *Environment and Planning B* 20:347–357.

Theil, H. 1963. On the Symmetry Approach to the Committee Decision Problem. *Management Science* 9:380–393.

Theil, H. 1964. *Optimal Decision Rules for Government and Industry*. Amsterdam: North-Holland.

Theil, H. 1967. *Economics and Information Theory*. Amsterdam: North Holland.

Theil, H. 1972. *Statistical Decomposition Analysis*. Amsterdam, Netherlands: North Holland.

Theil, H. 1976. Disutility as a Probability: A Reconsideration of the Symmetry Approach to the Committee Decision Problem. *Management Science* 23:109–116.

Thompson, D. W. 1917. *On Growth and Form*. Cambridge, UK: Cambridge University Press.

Thurstain-Goodwin, M., and M. Batty. 2002. The Sustainable Town Centre. In *Planning for a Sustainable Future*, ed. A. Layard, S. Davoudi, and S. Batty, 253–268. London: Routledge/Spon Press.

Timmermans, H. 2006. The Saga of Integrated Land Use and Transport Modeling: How Many More Dreams Before We Wake Up? In *Moving Through Nets: The Physical and Social Dimensions of Travel*, ed. K. W. Axhausen, 219–248. Oxford, UK: Elsevier.

Tobler, W. 1970. A Computer Movie Simulating Urban Growth in the Detroit Region. *Economic Geography* 46:234–240.

Tobler, W. R. 1976. Spatial Interaction Patterns. *Journal of Environmental Systems* 6 (4):271–301.

Tobler, W. R. 1981. A Model of Geographical Movement. *Geographical Analysis* 13:1–20.

Tobler, W. R. 1983. An Alternative Formulation for Spatial-Interaction Modeling. *Environment and Planning A* 15:693–703.

Tobler, W. R. 1987. Experiments in Migration Mapping by Computer. *American Cartographer* 14:155–163.

Tobler, W. R. 1995. Migration: Ravenstein, Thornthwaite, and Beyond. *Urban Geography* 16:327–343.

Tufte, E. 1983. *The Visual Display of Quantitative Information*. Cheshire, CT: Graphics Press.

Turner, A., M. Doxa, D. O'Sullivan, and A. Penn. 2001. From Isovists to Visibility Graphs: A Methodology for the Analysis of Architectural Space. *Environment and Planning B* 28:103–121.

Van den Bogaard, P. J. M., and J. Verslius. 1962. The Design of Optimal Committee Decisions. *Statistica Neerlandica* 16:271–289.

Vicsek, T., A. Czirok, E. Ben-Jacob, I. Cohen, and O. Sochet. 1995. Novel Type of Phase Transition in a System of Self-Driven Particles. *Physical Review Letters* 75:1226–1229.

von Thunen, J. H. 1966. *Von Thunen's Isolated State*. Trans. C. M. Wartenberg, ed. P. Hall. Oxford, UK: Pergamon Press. Original work published 1826 as *Der Isolierte Staat in Beziehung auf Landwirtschaft und Nationaloekonomie*, Jena, Germany: Gustav Fischer.

Voogd, H. 1983. *Multicriteria Evaluation for Urban and Regional Planning*. London: Pion Press.

Wannop, U., ed. 1971. Coventry-Solihull-Warwickshire Subregional Planning Study: Supplementary Report 3: Alternatives. Coventry, UK: Coventry Corporation.

Ward, J. H. 1963. Hierarchical Grouping to Optimize an Objective Function. *Journal of the American Statistical Association* 58:236–244.

Wates, N. 1973. Tolmers Square: How a Property Developer Didn't Make £20m. *Community Action* 11:7–9.

Wates, N. 1975. The Tolmers Village Squatters. *New Society* 33:364–366.

Wates, N. 1976. *The Battle for Tolmers Square*. London: Routledge and Kegan Paul.

Watts, D. J. 1999. *Small Worlds: The Dynamics of Networks between Order and Randomness*. Princeton, NJ: Princeton University Press.

Watts, D. J. 2002. *Six Degrees: The Science of a Connected Age*. New York: W. W. Norton.

Watts, D. J., and S. H. Strogatz. 1998. Collective Dynamics of 'Small-World' Networks. *Nature* 393 (6684):440–442.

Weber, A. F. 1899. *The Growth of Cities in the Nineteenth Century*. New York: Macmillan.

Wegener, M. 2008. Multi-Scale Spatial Models: Linking Macro to Micro. Paper presented at the Centre for Advanced Spatial Analysis, University College London. http://www.spiekermann-wegener.de/pro/pdf/MW_CASA_090108.pdf. Accessed 17/9/2012.

Wegener, M., F. Gnad, and M. Vannahme. 1986. The Time Scale of Urban Change. In *Advances in Urban Systems Modeling*, ed. B. Hutchinson and M. Batty, 175–197. Amsterdam: North Holland.

West, G. B., J. H. Brown, and B. J. Enquist. 1997. A General Model for the Origin of Allometric Scaling Laws in Biology. *Science* 276:122–126.

Wikipedia. 2012. Squares of Savannah, Georgia. http://en.wikipedia.org/wiki/Squares_of_Savannah,_Georgia. Accessed 18/10/2012.

Wilson, A. G. 1970. *Entropy in Urban and Regional Modelling*. London: Pion Press.

Wilson, A. G. 1981. *Catastrophe Theory and Bifurcation; Applications to Urban and Regional Systems*. Berkeley, CA: University of California Press.

Wilson, A. G. 2008. Boltzmann, Lotka, and Volterra and Spatial Structural Evolution: An Integrated Methodology for Some Dynamical Systems. *Journal of the Royal Society, Interface* 5:865–871.

Wilson, A. G. 2009. The "Thermodynamics" of the City: Evolution and Complexity Science in Urban Modeling. In *Complexity and Spatial Networks*, ed. A. Reggiani and P. Nijkamp, 11–31. Berlin: Springer.

Wilson, A. G., Coelho, J., Macgill, S., and Williams, H. C. W. L. 1981. *Optimization in Land Use and Transport*. Chichester, UK: John Wiley and Sons Ltd.

Wolfe, P., and G. B. Dantzig. 1962. Linear Programming in a Markov Chain. *Operations Research* 10:702–710.

Wolfram, S. 2002. *A New Kind of Science*. Champaign, IL: Wolfram Media.

Wood, J., J. Dykes, and A. Slingsby. 2010. Visualization of Origins, Destinations and Flows with OD Maps. *Cartographic Journal* 47:117–129.

Wood, J., A. Slingsby, and J. Dykes. 2011. Visualizing the Dynamics of London's Bicycle Hire Scheme. *Cartographica* 46:239–251.

Xie, F., and D. M. Levinson. 2011. *Evolving Transportation Networks*. New York: Springer.

Yule, G. U. 1925. A Mathematical Theory of Evolution Based on the Conclusions of Dr. J.C. Willis, F.R.S. *Philosophical Transactions of the Royal Society of London B* 213:21–87.

Zipf, G. K. 1949. *Human Behavior and the Principle of Least Effort*. Cambridge, MA: Addison-Wesley.

Zollman, K. R. 2012. Social Network Structure and the Achievement of Consensus. *Politics, Philosophy and Economics* 11:26–44.

Author Index

Subject Index

absolute counts, 286
abstraction, 271
accessibility(ies), 20, 107, 116, 180,
    189–190, 197, 217, 236, 257
  changes, 296
  line, 188, 236
  log sum, 280
  point, 237
  potential, 265
  surfaces, 230, 297
  weighted, 218
action,
  chains of collective, 368
  collective, 4, 303, 365, 411, 434
action-at-a-distance, 265
actor(s), 381, 387, 398, 408, 418,
    424
  interactions, 420
  power, 379
actual ownership (control), 416
additive exchange, 126, 129
adjacency, 238
  matrix, 319–320
agent-based models (ABM), xix, 3, 11, 263,
    272, 412, 458
agglomeration, 13, 16–17
  economies, 29, 36, 164, 253
aggregate distances, 135
aggregate flows, 78
aggregation, 326
allometric scaling, 39
allometry, 41, 107
  negative, 22, 41
  positive, 41, 110

  spatial, 110
  urban, 8
Alonso's general theory of movement,
    73
Alonso-Wilson law, 39
analytic hierarchy process, 317
argumentation, 365
assignment, 50
asymmetry, 49, 75, 153
average distance in a graph, 94
average trip costs (of a journey to work),
    292
averaging, 98, 101, 196, 227, 233, 332,
    345, 419, 447
axial line(s), 184, 214, 231
axial map(s), 184, 193–214, 220

backward process, 97–98, 346
Baltimore, MD, 258–259
baseline model of symmetry in flow
    systems, 70
Battle for Tolmers Square, The, 413, 420,
    427
Beckmann's model, 172
Bellman's principle of optimality, 352
benefits as measure of solution variables,
    307
Bettencourt-West law, 39
betweenness centrality, 107–108, 169
between-set entropy, 327
bifurcation, 28
binary bipartite system, 88
binary graphs, 81, 88, 181–2
bipartite flow graph, 89, 97